T0289786

Densification Impact on Raw, Chemically and Thermally Pretreated Biomass

Physical Properties and Biofuels Production

Sustainable Chemistry Series

ISSN: 2514-3042

The concept of Green Chemistry was first introduced in 1998 with the publication of Anastas and Warner's "12 Principles of Green Chemistry". Today, these principles are becoming adopted as general practice in the chemical industries in order to reduce or eliminate the use and generation of hazardous materials, reduce waste, and make use of sustainable resources. New, safer materials and products are being released all the time. Alternative technologies are being developed to improve the efficiency of the chemical industry, while reducing its environmental impact. Sustainable resources are being investigated to replace our reliance on fossil fuels – not only as source of energy but also a source of chemicals — be they feedstock, bulk, or fine. Consideration is now given to the whole life cycle of a product or chemical — from design to disposal. And, as more of the Earth's resources become scarce so new alternatives must be found.

As the world works towards meeting the needs of the present generation without compromising the needs of the future, this series presents comprehensive books from leaders in the field of green and sustainable chemistry. The volumes will offer an excellent source of information for professional researchers in academia and industry, and postgraduate students across the multiple disciplines involved.

Published

Vol. 6 *Densification Impact on Raw, Chemically and Thermally Pretreated Biomass: Physical Properties and Biofuels Production*
edited by Jaya Shankar Tumuluru

Vol. 5 *Solution Combustion Synthesis of Nanostructured Solid Catalysts for Sustainable Chemistry*
edited by Sergio L. González-Cortés

Vol. 4 *Silica-Based Organic–Inorganic Hybrid Nanomaterials: Synthesis, Functionalization and Applications in the Field of Catalysis*
by Rakesh Kumar Sharma

Vol. 3 *Functional Materials from Lignin: Methods and Advances*
edited by Xian Jun Loh, Dan Kai and Zibiao Li

For the complete list of volumes in this series, please visit
www.worldscientific.com/series/suschem

Sustainable Chemistry Series

Volume 6

Densification Impact on Raw, Chemically and Thermally Pretreated Biomass

Physical Properties and Biofuels Production

Series Editor

Nicholas Gathergood
University of Lincoln, UK

Editor

Jaya Shankar Tumuluru
Southwestern Cotton Ginning Research Laboratory, USA

NEW JERSEY • LONDON • SINGAPORE • BEIJING • SHANGHAI • HONG KONG • TAIPEI • CHENNAI • TOKYO

Published by

World Scientific Publishing Europe Ltd.
57 Shelton Street, Covent Garden, London WC2H 9HE
Head office: 5 Toh Tuck Link, Singapore 596224
USA office: 27 Warren Street, Suite 401-402, Hackensack, NJ 07601

Library of Congress Cataloging-in-Publication Data
Names: Tumuluru, Jaya Shankar, editor.
Title: Densification impact on raw, chemically and thermally pretreated biomass : physical properties and biofuels production / editor, Jaya Shankar Tumuluru, Southwestern Cotton Ginning Research Laboratory, USA.
Description: New Jersey : World Scientific, 2023. | Series: Sustainable chemistry series, ISSN 2514-3042 ; volume 6 | Includes bibliographical references and index.
Identifiers: LCCN 2022058158| ISBN 9781800613782 (hardcover) | ISBN 9781800613799 (ebook) | ISBN 9781800613805 (ebook other)
Subjects: LCSH: Biomass chemicals. | Biomass conversion. | Biomass energy. | Plant biomass. | Compacting. | Pelletizing.
Classification: LCC TP248.B55 D465 2023 | DDC 333.95/39 23/eng/20230--dc11
LC record available at https://lccn.loc.gov/2022058158

British Library Cataloguing-in-Publication Data
A catalogue record for this book is available from the British Library.

For any available supplementary material, please visit
https://www.worldscientific.com/worldscibooks/10.1142/Q0406#t=suppl

Desk Editors: Logeshwaran Arumugam/Adam Binnie/Shi Ying Koe

Typeset by Stallion Press
Email: enquiries@stallionpress.com

I dedicate this book to my wife, Naga Sri Valli Gali, two daughters, Priya lasya Tumuluru and Siva Lalasa Tumuluru, my parents Tumuluru Sivakama Sundaram and Vasundhara Devi, and my in-laws Gali Ramasubrahmanya Sarma and Jaya Lakshmi for their extraordinary support all these years.

https://doi.org/10.1142/9781800613799_fmatter

Preface

First-generation ethanol plants were designed based on a dense, stable, storable, shippable commodity-type product with multiple applications. With these properties, corn was used as a feedstock for large-scale biorefineries without any challenges and has grown exponentially (Dale, 2017). In the second-generation biofuels, the feedstocks used are low-cost carbon resources such as crop and forest residues and municipal solid waste. These materials are not dense; they have irregular size and shape, have variable moisture, and are not readily storable and shippable. Robust preprocessing technologies were not developed for these feedstocks as they do not have existing markets (Dale, 2017). When the industry tested these feedstocks for biofuel production, they faced flowability, storage, transportation, and conversion issues. Also, these low-cost carbon resources are not widely used for large-scale production of biofuels or biopower, resulting in less emphasis on these materials by the farmers or wood processing industries.

The physical properties of low-cost carbon resources limit the second-generation biorefineries from operating at the designed capacities. These physical properties, including variable moisture, irregular particle size and shape, and low bulk density, create storage, handling, and transportation challenges and can cause biomass particle plugging and bridging issues at commercial biorefineries, resulting in lower fuel production. For example, research has documented the difficulties of moving the variable moisture fibrous biomass through the conveyor systems and feeding it evenly into the bioreactors. Furthermore, biomass storage has emerged as a crucial upstream problem. For example, cellulosic ethanol plants have experienced fires in bale storage facilities due to spontaneous combustion issues. It is difficult to imagine how a large cellulosic biofuel industry will grow unless the variable moisture, particle size and shape, low bulk density, proximate, ultimate, and biochemical compositional challenges are solved.

Mechanical and thermal densification can overcome some challenges associated with the variable physical and chemical properties of low-cost carbon resources. Studies have indicated that mechanical densification is critical to converting low-density materials to products with the desired physical properties and aerobic stability (as densified products such as pellet and briquettes have a final moisture content of <10%, w.b.). The other advantages of feedstock densification are as follows: (i) improved handling and conveyance efficiencies throughout the supply system and biorefinery infeed, (ii) controlled particle size distribution for improved feedstock uniformity and density, (iii) fractionated structural components for improved compositional quality, and (iv) conformance to predetermined conversion technology and supply system specifications. Common biomass densification systems used are as follows: (i) pellet mill, (ii) cuber, (iii) briquette press, and (iv) screw extruder. Among these, the pellet mill is the most commonly used for bioenergy production. The quality of densified biomass produced using these systems is evaluated with the international standards established for physical properties such as size, shape, moisture, and density and chemical composition such as ash content.

The book focuses on understanding the challenges in biomass feedstock properties, woody and herbaceous biomass densification, and their conversion performance for biofuels production. Experts in biomass preprocessing, pretreatment, and conversion have contributed chapters on the challenges associated with the variable properties of various biomass feedstocks and how densification helps overcome the challenges. The chapters include research conducted by the authors and a review of the literature on biomass preprocessing and pretreatments for biofuels production.

Chapter 1 discusses biomasses such as agricultural residues, energy crops, forest residues, and other wastes, their physical properties and chemical composition and their challenges as biorefinery feedstocks, and how preprocessing helps overcome these challenges. Chapter 2 focuses on densification process models commonly used to understand the compression characteristics. Also, this chapter discusses the various densification systems widely used to convert biomass into an aerobically stable, high-density product. Chapter 3 discusses the densification behavior of different raw and pretreated herbaceous biomasses, their blends, and how the physical properties and chemical composition of different biomass impact the quality of the densified product. Chapter 4 discusses the densification of blends of woody and herbaceous biomasses. The chapter emphasizes why blending is important and how pelleting facilities in the future will need to accommodate variable feedstocks blended from biomass residues with different chemical compositions and physical properties. Chapter 5 discusses the flow behavior of various biomass types and how densification can improve the flow of the biomass feedstocks within conveyor and feed systems. Chapter 6 outlines how thermal pretreatment of biomass enhances the product's physical properties and chemical composition and discusses the densification characteristics of the torrefied biomass and the quality of the pellets produced.

Chapters 7 to 11 discuss how the densified products perform in the downstream conversion for fuels and biopower production. Chapter 7 focuses on biomass densification and how it improves the gasification process. Chapter 8 addresses the hydrothermal liquefaction of biomass and how densification produces higher yields and improves the bio-oil quality. Chapter 9 discusses the densification of microwave-torrefied oat hulls for a solid fuel and details how the quality properties of oat hull pellets were determined. Chapter 10 discusses the densification characteristics of ammonia fiber explosion pretreated biomass, the pretreatment applications, and how the approach can overcome issues of transporting and storing biomass sold as commodity products. Finally, Chapter 11 discusses biomass densification and how it impacts sugar production using enzymatic hydrolysis.

Readership

This book aims to advance the use of low-cost carbon resources such as agricultural and forest residues for bioenergy production by addressing the inherent physical properties, moisture content and chemical compositional challenges. Students, researchers in biomass for biofuels production, and chemical, agricultural, and biochemical engineers working in academic institutes, research institutes, industries, and governmental agencies can benefit from the book. In addition, biomass engineers and biorefinery managers can use this book to understand the low-cost carbon resources challenges and the technologies available to overcome them. The book can also help energy planners, policymakers, and others to understand the low-cost carbon resources limitations and the commonly used technologies to overcome these limitations and help biorefineries operate at their designed capacities.

Reference

Dale, B. (2017). A sober view of the difficulties in scaling cellulosic biofuels. *Biofuels Bioprod. Bioref.*, *11*, 5–7. https://doi.org/10.1002/bbb.1745.

https://doi.org/10.1142/9781800613799_fmatter

About the Editor

Jaya Shankar Tumuluru currently works as a Research Agricultural Engineer at the Southwestern Cotton Ginning Research Laboratory, the United States Department of Agriculture (USDA), Agricultural Research Service (ARS), Las Cruces, New Mexico, USA. His research focuses on improving the cotton ginning process and using cotton waste for fuels and bioproducts applications. Before joining the USDA, Dr. Tumuluru worked as a Distinguished Staff Engineer at Idaho National Laboratory (INL), one of 17 national labs in the US Department of Energy complex. He was the principal investigator for multiple US Department of Energy-funded projects on biomass processing for biofuels and bioproducts production. Dr. Tumuluru's research at INL was focused on mechanical preprocessing and thermal pretreatments of biomass. Dr. Tumuluru has published papers on biomass preprocessing and pretreatments, food processing, preservation and storage, and prediction and optimization of the bioprocess using artificial neural networks and evolutionary algorithms. Dr. Tumuluru has published more than 100 papers in journals and conference proceedings; presented more than 85 papers in national and international conferences; edited books on biomass preprocessing, valorization, and volume estimation; edited a monograph on biomass densification; and published more than 15 book chapters. He was named an R&D 100 Award Finalist in 2018 and 2020; received the INL Outstanding Engineering Achievement Award for 2019; the

Asian American Engineer of the Year Award for 2018; the INL Outstanding Achievement in Scientific and Technical Publication Award in 2014; the Outstanding Reviewer Award for the American Society of Agricultural and Biological Engineers Food and Process Engineering Division in 2013, 2014, and 2016; the Outstanding Reviewer for the *Biomass and Bioenergy Journal* in 2015; and the Outstanding Reviewer for the UK Institute of Chemical Engineers in 2011.

https://doi.org/10.1142/9781800613799_fmatter

Acknowledgments

The editor would like to thank Mr. Neal Yancey, Manager, Idaho National Laboratory, for supporting the book's completion. The editor also takes the opportunity to thank Dr. Derek Whitelock, Research Leader, Southwestern Cotton Ginning Research Laboratory, United States Department of Agriculture, Agricultural Research Service, Las Cruces, New Mexico for his support in completing the book. In addition, the editor would like to thank Mr. David Combs, INL Art Director and Branding Specialist, and Mr. Gordon Holt, INL Senior Editor, for their invaluable support. The editor would also like to thank all the authors who contributed to this book, without whom it would have been impossible to publish. Finally, the editor would like to express extreme gratitude to his family for their help and support.

https://doi.org/10.1142/9781800613799_fmatter

Disclaimer

US Department of Energy Disclaimer

The views and opinions of the authors expressed herein do not necessarily state or reflect those of the United States government or any agency thereof. Neither the US government nor any agency thereof, nor any of their employees, makes any warranty, express or implied, or assumes any legal liability or responsibility for the accuracy, completeness, or usefulness of any information, apparatus, product, or process disclosed, or represents that its use would not infringe privately owned rights. References herein to any specific commercial product, process, or service by trade name, trademark, manufacturer, or otherwise do not necessarily constitute or imply its endorsement, recommendation, or favoring by the US government or any agency thereof. The views and opinions of authors expressed herein do not necessarily state or reflect those of the US government or any agency thereof. Accordingly, the publisher, by accepting the article for publication, acknowledges that the US government retains a nonexclusive, paid-up, irrevocable, worldwide license to publish or reproduce the published form of this manuscript or allow others to do so for US government purposes.

US Department of Agriculture Disclaimer

The findings and conclusions in this [publication/presentation/blog/report] are those of the author(s) and should not be construed to represent any official USDA or US Government determination or policy. Mention of trade names or commercial products in this publication is solely for the purpose of providing specific information and does not imply recommendation or endorsement by the US Department of Agriculture. USDA is an equal opportunity provider and employer.

https://doi.org/10.1142/9781800613799_fmatter

Contents

https://doi.org/10.1142/9781800613799_fmatter

List of Abbreviations

AD	Anaerobic digestion
AFEX	Ammonia fiber expansion
COND	Condenser
COMPR	Compressor
EA	Extractive ammonia
GHG	Greenhouse gas
in.	Inch
LBPD	Local biomass processing depot
LCC	Lignin carbohydrate complex
PSV	Pressure safety valve
RD	Rotating disc gate valve
SG	Switchgrass
TRL	Technology readiness level

https://doi.org/10.1142/9781800613799_0001

Chapter 1

An Overview of Preprocessing and Pretreatment Technologies for Biomass

Jaya Shankar Tumuluru
Southwestern Cotton Ginning Laboratory
United States Department of Agriculture
Agriculture Research Service, Las Cruces, New Mexico, USA
jayashankar.tumuluru@usda.gov

Abstract

The biomasses such as agricultural residues, energy crops, forest residues, and other wastes usually have irregular shapes and a low-bulk density, resulting in loose harvest formats and a corresponding energy density lower than coal or other fossil fuels. They also have high moisture content, speeding up degradation during storage. The low-bulk density of the biomass creates problems, such as logistics and storage, handling and feeding, transportation costs, a large storage footprint, inefficiencies in processing in the biorefineries that can increase the downtime, and the overall size of the reactor and the conversion equipment. Besides the physical properties challenges, other chemical composition and energy content challenges also exist. These challenges limit the biorefineries from operating at their designed capacities. Various chemicals and thermal and mechanical pretreatment and preprocessing methods are used to improve the biomass quality and make it suitable for the biorefineries to operate at their designed capacities. This chapter discusses enhancing biomass quality in terms of physical properties and chemical composition using mechanical, thermal, and chemical preprocessing and pretreatment technologies.

Keywords: Herbaceous and woody biomass, biomass challenges, physical properties and chemical composition, mechanical preprocessing, thermal pretreatment, chemical pretreatment

1.1. Introduction

There is currently a great emphasis on climate change, which is considered a threat to global security. The United Kingdom hosted the 26th United Nations Climate Change Conference of the Parties (COP26) in Glasgow from 31 October to 13 November 2021. The COP26 summit brought these parties together to speed up the actions to reach the UN Framework Convention on Climate Change goals established in Paris Agreement. To meet the COP26 goals, a significant reduction in greenhouse gas emissions and the pursuit of renewable energy from various sources such as wind, water, solar, and biomass should be a major part of the energy portfolio. The key is to reduce greenhouse gas emissions, achieve carbon neutrality, and efficiently utilize natural resources for energy applications.

According to Paul and Dutta (2018) and Hu and Ragauskas (2012), lignocellulosic biomass materials, such as agricultural residues (crop residues), wood, and grass, are promising sources of alternative energy. About 181.5 billion tons of lignocellulosic biomass are available annually (Kumar *et al.*, 2008). In the United States, more than 1 billion tons of lignocellulosic biomass are available annually, and out of this, 30% is from forests (Limayem and Ricke, 2012). In Canada, 22.5 million tons of sustainably harvested agricultural residue were available in the Canadian province of Ontario in 2014 alone, out of which 12.78 million tons come exclusively from agriculture crop residue (Hewson, 2010). Canada's estimated annual agricultural crop residue is 69.25 million tons, where 18.44% is in Ontario (Melin, 2013).

Even though biomass can be one of the largest carbon sources for producing various fuels, bioproducts, and chemicals, after harvesting, woody and herbaceous biomasses are available in less convenient forms for storage, transportation, and handling. Tumuluru (2018a) pointed out that the major limitation of biomass is lower energy and mass density; the feedstock is not available at one definite location (it is more dispersed). Heterogeneous, complex, and rigid biomass structures require much energy to convert to a wide range of value-added products. Biomass has a relatively low bulk and energy density and requires more biomass feedstocks to produce equivalent energy as a traditional hydrocarbon fuel. High-oxygen contents of biomass materials can also negatively affect their conversion to various products, such as fuels. For example, biomass oxygen content must be reduced, and energy content needs to be increased to produce hydrocarbon fuels comparable to petroleum-based ones. The efficiency of conversion processes can also vary depending on the biomass types (e.g., hardwood, softwood, and grass).

Biomass provides many opportunities to produce bioenergy and bio-based products, which can effectively replace the products and energy produced using fossil-based energy. However, while the potential of biomass for energy applications is widely recognized and actively studied, generating energy from biomass is still rather expensive because of both technological limits and logistical constraints. This necessitates developing innovative and energy-efficient preprocessing technologies to improve feedstock quality and availability, reduce the cost to scale up, and expand biomass utilization.

1.2. Biomass Challenges

Biomass after harvesting is available at high and variable moisture content. Typically, woody biomass is available at about 50% moisture content, and herbaceous biomass is between 10 and 35% (w.b.). The high variability in the moisture content results in challenges such as (a) increased transportation costs, (b) feedstock handling and conveying issues, and (c) quality loss during storage, making it difficult to use for downstream conversions, including hydrolysis, gasification and pyrolysis. High moisture in the biomass can reduce conversion efficiency. In addition, the fuel that is produced will end up with a high moisture content, which is undesirable (Tumuluru, 2018a). High moisture content in the biomass creates uncertainty in its physical, chemical, and microbial properties. For example, the moisture in the biomass impacts how the material grinds and the final quality of the grind (e.g., significant changes in particle size distribution). High moisture in the biomass makes it fibrous and difficult to grind and exponentially increases grinding energy (Tumuluru, 2019; Tumuluru and Heikkila, 2019). The elastic and cohesive nature of the biomass at high moisture content creates feeding and handling issues, such as plugging the grinder screens, screw conveyors, drop chutes, and conversion reactors. In addition, the variability in moisture content creates irregular particle sizes after grinding, which will further react inconsistently in a reactor, thus reducing process efficiencies. These challenges impact the biochemical and thermochemical process efficiency and quality of products produced. The other challenge of biomass is low-bulk density. Biomass materials are heterogeneous and have a low-bulk density ranging from 60–80 kg/m^3 for grasses and agricultural straws and 200–250 kg/m^3 for wood chips (Tumuluru *et al.*, 2011b). This creates difficulties in handling large volumes of feedstocks and increases transportation and storage costs (Tumuluru, 2019). Size, shape, moisture content, particle density, and surface characteristics are the major factors affecting the bulk density of a material.

Raw biomass chemical composition is different compared to fossil fuels. It has a higher oxygen content but a lower carbon content. This can significantly impact thermochemical conversions, such as gasification and pyrolysis. According to Hernandez *et al.* (2010), the biomass's physical properties and chemical composition significantly impact the gasifier operating conditions and quality of the syngas produced. The issues related to biomass's physical properties and chemical composition are limiting factors in the successful commercialization of the gasification process (Hernandez *et al.*, 2010).

Moreover, different biomass feedstocks exhibit different physicochemical properties (e.g., particle size and shape, chemical composition, and heating value), resulting in an irregular material flow within the reactor. Table 1.1 indicates some of the woody and herbaceous biomass's physical properties and biochemical chemical composition (Tumuluru, 2018b). The proximate and ultimate composition and heating values of woody and herbaceous biomass are shown in Table 1.2. It is very clear from the tables that the composition changes based on the feedstock type. Also, the other factors impacting the feedstock's quality are the harvesting and

Table 1.1. Physical and biochemical composition of woody and herbaceous biomass (Tumuluru, 2018b).

Feedstock	Moisture content (%, w.b.)	Bulk density (kg/m^3) (ground in a hammer mill fitted with a 4.8 mm screen)	% Total proteins	% Extractives (% EtOH)	% Extractives (% H_2O)	% Lignin
Corn stover	11.0	157.9	2.54	2.64	6.52	45.29
Wheat straw	6.37	151.8	3.59	2.41	7.34	13.19
Sorghum	10.9	223.7	4.03	2.89	13.49	12.90
Lodgepole pine	7.54	246.2	0.21	4.43	4.06	26.78
Pinyon-juniper	16.1	257.47	1.83	5.40	7.65	32.86

Table 1.2. Carbon, nitrogen, and hydrogen, volatiles, and energy content of corn stover, switchgrass, and lodgepole pine (Tumuluru and Fillerup, 2020).

Composition	Carbon (%)	Hydrogen (%)	Nitrogen (%)	Volatiles (%)	Ash (%)	Fixed carbon (%)	Lower heating value (MJ/kg)	Higher heating value (MJ/kg)
Corn stover	38.64	4.77	0.66	66.93	22.60	10.47	13.36	15.90
Switchgrass	46.84	5.64	0.64	79.53	5.42	15.05	16.18	19.19
Lodgepole pine	50.48	5.88	0.12	85.46	0.8	13.73	17.10	20.24

storage methods. For example, corn stover harvested using multi-pass methods results in higher ash content and biomass feedstock storage in humid conditions, resulting in dry matter losses. Currently there is a big push for using blends of woody and herbaceous biomass and municipal solid waste for biofuels applications as it significantly impacts the feedstock cost for biofuels production. Modeling the feedstock cost, which includes grower payments, pre-processing, storage, harvesting, and transportation for the blends of woody, herbaceous, and municipal solid waste, reduced the feedstock cost from \$149.58/dry tonne for pure corn stover to \$82.86/dry tonne using blends (Tumuluru and Fillerup, 2020). According to

Tumuluru and Fillerup (2020), intercropping (or polyculture) is of great interest as it allows better utilization of resources (water and nutrients), provides efficient land management, and increases yields. The challenge is intercropping of various woody and herbaceous biomass results in blends of residues, which can further increase the physical properties and chemical composition variability challenges. The variability in physical properties and chemical and biochemical composition leads to inconsistent gasifier operation; hence, the syngas quality is negatively affected. Therefore, a biomass feedstock with consistent quality in size, shape, density, moisture content, and chemical composition is desirable to achieve a reliable and trouble-free gasifier operation.

Furthermore, different biomass types, such as grassy and woody materials, and their physiochemical properties pose difficulties in mechanizing continuous feeding and controlling the burning rate (Luo *et al.*, 2011). The difference in densities of biomass and coal also causes feeding problems for their mixture in the existing feed systems of the coal-based boiler and power plants. The challenges associated with lignocellulosic biomass compared to coal reduce the efficiency of thermochemical conversion processes. Variability in physical properties, such as moisture content, foreign matter (soil), particle size, and particle size distribution, result in reductions in grinding throughput, equipment wear, plugging during conveyance, and reactor feeding problems. These problems all contribute to a decrease in production capacity, an interruption of normal operations, an increase in preprocessing and conversion costs, and a major reduction in yield.

1.3. Biomass Preprocessing and Pretreatments

Biomass preprocessing and pretreatment technologies can help overcome some of the biomass challenges. Tumuluru (2018a) concluded that solving the biomass's inherent physical, chemical, and rheological challenges is critical to solving upstream logistics and downstream conversion issues. Mechanical biomass preprocessing using grinding and densification can help overcome the density challenges. The low density is due to the different sizes and shapes of the biomass. Biomass materials are mechanically compressed in this process to increase their density and convert them to uniform shapes and sizes. The densification helps to produce a commodity-type product from biomass feedstocks. Densification enables several advantages, including (1) improved handling and conveyance efficiencies throughout the supply system, and biorefinery infeed; (2) controlled particle size distribution for improved feedstock uniformity and density; (3) fractionated structural components for improved compositional quality and conversion; and (4) conformance to predetermined conversion technology and supply system specifications (Tumuluru *et al.*, 2011b). Another major advantage of densification is the blends of the various woody and herbaceous biomasses, which have different physical properties, can be converted into a product with uniform size, shape, and density, making the blend easier to feed, store, and transport. The other advantage of blending woody biomass with herbaceous biomass is that it can improve the chemical position and energy content. The studies conducted by Tumuluru (2019) and Tumuluru and Fillerup (2020) on pelleting and briquetting of blends of woody and herbaceous biomass have indicated that blending woody biomass, which has high lignin content, with herbaceous biomass (which has lower lignin content) improved the particle binding. These authors have concluded that a higher percentage of pine in the corn stover and switchgrass blend improved the chemical composition and the quality of the pellets and briquettes (density and durability) produced. The mechanical densification systems that are used for biomass and their blends are pellet mill, cuber, briquette press, screw extruder, tabletizer, and agglomerator. Various researchers has used these systems to densify the biomass and its blends (Li and Liu, 2000; Mani *et al.*, 2006; Ndiema *et al.*, 2002; Tabil and Sokhansanj, 1996; Tumuluru *et al.*, 2010; Tumuluru, 2014, 2019; Tumuluru and Fillerup, 2020; Tumuluru *et al.*, 2021; Tumuluru, 2018b, 2016).

Among these densification systems, the pellet mill, briquette press, and screw extruder are the most commonly used for producing a densified product from various biomass feedstocks. The quality of

Figure 1.1. (a) 25.4-mm and (b) 2-mm screen size ground corn stover (Tumuluru and Heikkila, 2019).

densified biomass produced using these systems is evaluated with the international standards developed for pellet mills and briquette press systems, but no specific standards are available for densified products produced using an agglomerator or tablet press. Figures 1.1 and 1.2 indicate some of the ground and densified biomass products produced using woody and herbaceous biomass. Figure 1.2 shows the pellets produced using woody and herbaceous biomass (Tumuluru, 2018b). It is very clear from the figures that densification using pelleting helps to convert biomass into a product with uniform size, shape, and high density. Typically, herbaceous biomass pellets have densities of >400 kg/m^3 based on the densification process conditions selected (Tumuluru, 2014, 2018b, 2019).

Thermal treatment methods such as drying and torrefaction make biomass aerobically stable. They also help improve the biomass's physical properties by making them less fibrous with improved chemical compositions (proximate and ultimate composition). Temperature ranges for drying and torrefaction are 100–200°C and 200–300°C (Tumuluru *et al.*, 2011a). Figure 1.3 shows the lodgepole pine deep dried and torrefied at various temperatures (ranging from 180–270°C and different residence times). Typically, biomass is dried in rotary driers to a lower moisture

Figure 1.2. Pellets from herbaceous and woody biomass (Tumuluru, 2018b).

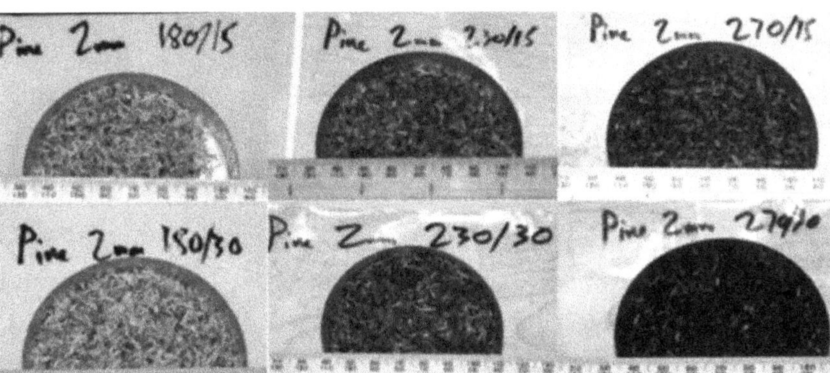

Figure 1.3. Pine torrefied at various temperatures and residence times (Tumuluru, 2016).

content of 10% (w.b.) (Yancey *et al.*, 2013; Tumuluru, 2016) to make it aerobically stable and reduce the dry matter losses during storage. High-temperature thermal methods, such as torrefaction, are successfully tested to improve biomass's physical, chemical, and energy properties. Typically, 70% of the mass is retained as a solid product during torrefaction, containing 90% of the initial energy content (Tumuluru *et al.*, 2011a). Torrefaction makes biomass brittle and decreases grinding energy by about 70%. This ground torrefied biomass has improved sphericity, particle surface area, and particle size distribution (Phanphanich and Mani, 2011). In addition, torrefaction increases the carbon content, and the calorific value of the biomass by 15–25% wt, while the moisture content decreases to <3% (w.b.) (Tumuluru *et al.*, 2011a). The major limitation of torrefaction is the loss of mass density, which can be overcome by densifying the torrefied biomass using a pellet mill or a briquette press. Figure 1.4 indicates the briquettes produced using torrefied biomass (Aamiri *et al.*, 2019). The Energy Research Center of the Netherlands developed a torrefaction and pelletization process (TOP), where the hot torrefied material is compressed and formed into a pellet (Bergman, 2005). This research also indicated that pelleting the torrefied biomass at 225°C increases the mill's throughput twice and reduces the specific energy consumption by two times. Another major advantage of torrefied material is its hydrophobic nature. Loss of the hydroxyl (OH) functional group during torrefaction makes the biomass hydrophobic (i.e., it loses the ability to attract water molecules), making it more stable against chemical oxidation and microbial degradation during storage.

The ash content is another major challenge in using agricultural biomass. Ash in biomass is inorganic and derived from soil physiological sources and, therefore, is not convertible to sugars. Also, high ash content results in equipment wear during preprocessing and acts as an inhibitor during the biochemical and thermochemical conversion. Chemical pretreatment using acid or alkali makes biomass undergo structural changes that can further help the mechanical preprocessing, such as grinding and densification, and reduce the equipment's wear. Various chemical preprocessing technologies are tested to reduce the ash content in the biomass. Washing and leaching of raw biomass can also help minimize ash content, reducing corrosion, slagging, fouling (ash deposition), sintering, and the agglomeration of the bed. Ash in the biomass during the biochemical conversion can reduce the effectiveness of dilute acid pretreatment,

Figure 1.4. Briquettes produced using lightly and heavily torrefied biomass blends (Aamiri *et al.*, 2019).

cell growth, and ethanol yields (Carpenter *et al.*, 2014; Zhang *et al.*, 2013; Palmqvist and Hahn-Hägerdal, 2000; Casey *et al.*, 2013). The presence of ash affects the decomposition mechanism and reduces the product yield during pyrolysis and gasification. In the case of biopower, it can result in slagging and fouling surfaces.

Integrating mechanical, chemical, and thermal preprocessing and pretreatment technologies helps overcome biomass's physical, chemical, and energy property limitations and makes it more suitable for biofuels production. Individually, these preprocessing methods can be expensive, but combining them creates process intensification opportunities, which can help produce biomass of high quality and lower the cost. For example, thermal pretreatment technologies, combined with mechanical methods such as densification, will help improve the process efficiencies of the preprocessing systems, save costs, and minimize product losses. Chemical pretreatments can also be integrated with other mechanical preprocessing methods such as grinding and densification to meet the feedstock specifications necessary for biochemical conversion. Some studies at Idaho National Laboratory on integrating chemical pretreatment with mechanical preprocessing, such as grinding and densification, have resulted in less preprocessing energy than raw biomass. Demonstrating preprocessing technologies (e.g., chemical, mechanical, and thermal) or integrated preprocessing systems will help produce a consistent feedstock with desirable physical, chemical, and biochemical composition and overcome the upstream handling operations issues and downstream conversion challenges. This integration can also help supply biorefineries with a consistent feedstock with the desirable attributes reducing the quality and supply risk.

1.4. Conclusions

Mechanical, chemical, and thermal preprocessing technologies will significantly impact biomass's physical properties and chemical composition. Mechanical densification systems increase the bulk density of biomass by four to six times as compared to raw biomass. Thermal processing techniques using torrefaction reduce the grinding energy by 70%. Also, the ground biomass has improved physical properties, such as sphericity, particle size, shape and surface area, and particle size distribution. The major limitation of torrefaction is the loss of bulk density, which can be overcome by mechanical densification. Another biomass specification that negatively impacts biochemical and thermochemical conversion is ash. The alkaline earth and alkali metals, such as calcium, magnesium, sodium, and potassium, reduce product yield during thermochemical and biochemical conversion. Chemical preprocessing methods, such as washing, leaching, acid, alkali, and ammonia fiber explosion pretreatments, reduce the recalcitrance in biomass and improve sugars or bio-oil yields. To meet the specifications of the biochemical, thermochemical, and cofiring applications, developing a way to integrate the mechanical, thermal, and chemical preprocessing and pretreatments steps is critical. Combining mechanical, thermal, and chemical preprocessing and pretreatments methods can help the biomass meet the carbon content ash and carbohydrate specifications and improve particle size and density.

References

Aamiri, O. B., Thilakaratne, R., Tumuluru, J. S., & Satyavolu, J. (2019). An "in-situ binding" approach to produce torrefied biomass briquettes. *Bioengineering*, *6*, 87. https://doi.org/10.3390/bioengineering6040087.

Bergman, P. C. A. (2005). Torrefaction in combination with pelletization — The TOP process, ECN Report, ECN-C-05-073, Petten.

Carpenter, D., Westover, T. L., Czernik, S., & Jablonski, W. (2014). Biomass feedstocks for renewable fuel production: A review of the impacts of feedstock and pretreatment 170 on the yield and product distribution of fast pyrolysis bio-oils and vapors. *Green Chem.*, *16*(2), 384–406.

Casey, E., Mosier, N. S., Adamec, J., Stockdale, Z., Ho, N., & Sedlak, M. (2013). Effect of salts on the co-fermentation of glucose and xylose by a genetically engineered strain of Saccharomyces cerevisiae. *Biotechnol. Biofuels*, *6*, 83.

Hernandez, J. J., Aranda-Almansa, G., & Bula, A. (2010). Gasification of biomass wastes in an entrained flow gasifier: Effect of the particle size and the residence time. *Fuel Process. Technol.*, *91*, 681–692.

Hewson, D. (2010). Assessment of agricultural residuals as a biomass fuel for Ontario Power Generation. Canada.

Hu, F. & Ragauskas, A. (2012). Pretreatment and lignocellulosic chemistry. *Bioenergy Res.*, *5*(4), 1043–1066.

Kumar, R., Singh, S., & Singh, O. V. (2008). Bioconversion of lignocellulosic biomass: Biochemical and molecular perspectives. *J. Ind. Microbiol. Biotechnol.*, *35*(5), 377–391.

Li, Y. & Liu, H. (2000). High-pressure densification of wood residues to form an upgraded fuel. *Biomass Bioenerg*, *19*, 177–186.

Limayem, A. & Ricke, S. C. (2012). Lignocellulosic biomass for bioethanol production: Current perspectives, potential issues, and future prospects. *Prog. Energy Comb. Sci.*, *38*(4), 449–467.

Luo, S. Y., Liu, C., Xiao, B., & Xiao, L. (2011). A novel biomass pulverization technology. *Renew. Energ.*, *36*, 578–582.

Mani, S., Tabil, L. G., & Sokhansanj, S. (2006). Specific energy requirement for compacting corn stover. *Bioresour. Technol.*, *97*, 1420–1426.

Melin, S. (2013). Consideration for Grading Agricultural Residue. Ontario Federation of Agriculture. Available at: https://ofa.on.ca/wp-content/uploads/2017/11/GradingAgriculturalResidues-FINAL1.pdf.

Ndiema, C. K. W., Manga, P. N., & Ruttoh, C. R. (2002). Influence of die pressure on relaxation characteristics of briquetted biomass. *Energy Convers. Manag.*, *43*, 2157–2161.

Palmqvist, E. & Hahn-Hägerdal, B. (2000). Fermentation of lignocellulosic hydrolysates. II: Inhibitors and mechanisms of inhibition. *Bioresour. Technol.*, *74*, 25–33.

Paul, S. & Dutta, A. (2018). Challenges and opportunities of lignocellulosic biomass for anaerobic digestion. *Resources, Conservation and Recycling*, *130*, 164–174.

Phanphanich, M. & Mani, S. (2011). Impact of torrefaction on the grindability and fuel characteristics of forest biomass. *Bioresour. Technol.*, *102*, 1246–1253.

Tabil, L. G. & Sokhansanj, S. (1996). Process conditions affecting the physical quality of alfalfa pellets. *Trans. ASAE*, *12*(3), 345–350.

Tumuluru, J. S. (2014). Effect of process variables on the density and durability of the pellets made from high moisture corn stover. *Biosyst. Eng.*, *119*, 44–57.

Tumuluru, J. S. (2016). Specific energy consumption and quality of wood pellets produced using high-moisture lodgepole pine grind in a flat die pellet mill. *Chem. Eng. Res. Des.*, *110*, 82–97.

Tumuluru, J. S. (2018a). Why biomass preprocessing and pretreatments? In J. S. Tumuluru (Ed.), *Biomass Preprocessing and Pretreatments for Production of Biofuels.* Boca Raton, FL: CRC Press, pp. 1–13.

Tumuluru, J. S. (2018b). Effect of pellet die diameter on the density and durability of pellets made from high moisture wood and herbaceous biomass. *Carbon Resour. Convers.*, *1*(1), 44–54.

Tumuluru, J. S. (2019). Pelleting of pine and switchgrass blends: Effect of process variables and blend ratio on the pellet quality and energy consumption. *Energies*, *12*, 1198.

Tumuluru, J. S. & Fillerup, E. (2020). Briquetting characteristics of wood and herbaceous biomass blends: Impact on the physical properties, chemical composition, and calorific value. *Biofuel. Bioprod. Biorefin.*, *14*, 1105–1124. doi: 10.1002/bbb.2121.

Tumuluru, J. S. & Heikkila, D. J. (2019). Biomass grinding process optimization using response surface methodology and a hybrid genetic algorithm. *Bioengineering*, *6*(1), 12. https://doi.org/10.3390/bioengineering6010012.

Tumuluru, J. S., Sokhansanj, S., Hess, J. R., Christopher T. Wright, C. T., & Richard D. Boardman, R. D. (2011a). A review on biomass torrefaction process and product properties for energy applications. *Ind. Biotechnol.*, *7*(5), 384–401.

Tumuluru, J. S., Wright, C. T., Hess, J. R., & Kenney, K. L. (2011b). A review of biomass densification systems to develop uniform feedstock commodities for bioenergy application. *Biofuel. Bioprod. Biorefin.*, *5*, 683–707.

Tumuluru, J. S., Wright, C. T., Kenney, K. L., & Hess, J. R. (2010). A technical review on biomass processing: Densification, preprocessing, modeling, and optimization. Paper No: 1009401. ASABE Annual International Meeting, Pittsburgh, PA, USA.

Yancey, N. A., Tumuluru, J. S., & Wright, C. T. (2013). Drying, grinding, and pelletization studies on raw and formulated biomass feedstocks for bioenergy applications. *J. Biobased Mater. Bioenergy*, *7*, 549–558.

Zhang, W., Yi, Z., Huang, J., Li, F., Hao, B., Li, M., Hong, S., Lv, Y., Sun, W., Ragauskas, A., Hu, F., Peng, J., & Peng, L. (2013). Three lignocellulose features that distinctively affect biomass enzymatic digestibility under NaOH and H2SO4 pretreatments in *Miscanthus. Bioresour. Technol.*, *130*, 30–37.

https://doi.org/10.1142/9781800613799_0002

Chapter 2

Densification Process Modeling and Commonly Used Systems

Jaya Shankar Tumuluru
Southwestern Cotton Ginning Laboratory
United States Department of Agriculture
Agriculture Research Service
Las Cruces, New Mexico, USA
jayashankar.tumuluru@usda.gov

Abstract

Various fundamental models were developed to understand the compaction characteristics of powders. Many authors have used these models to understand the compaction behavior of biomass particles. Biomass's compression, relaxation, and frictional (adhesion) properties depend on compressive force, moisture content, and particle size and shape. The various models commonly used for biomass are the Spencer and Heckel model, the Walker model, the Jones model, the Cooper–Eaton model, the Kawakita–Lüdde model, the Sonnergaard model, the Panelli–Filho model, and the asymptotic modulus model. Various researchers tested all these models for different biomass grinds. For example, the Cooper–Eaton model parameters for biomass grinds can help understand prominent compaction mechanisms, such as particle rearrangement and elastic and plastic deformation. However, the mechanism of mechanical interlocking and the ingredient melting phenomenon during biomass compression are also important to understanding the compaction mechanism. The Kawakita–Lüdde model helps relate the biomass grinds' initial porosity and the compacts' yield strength. Numerical software, such as Pro/Engineer, PFC3D, and others, are gaining importance in understanding biomass grinds' compaction behavior during the densification process. The common systems used for biomass densification are pellet mill, briquette press, and screw extruder. Among the densified products, the pellets produced using a pellet mill are widely used worldwide for biopower production.

Keywords: Biomass, compression characteristics, process conditions, models, densification systems, pellet mill, briquette press, screw extruder

2.1. Introduction

Densification is widely used in fields such as (a) coal pelleting for power generation, (b) extrusion cooking for food and feed processing, (c) agglomerates and granules in chemical industries, and (d) compact tablets in pharmaceutical industries. However, the low-bulk density of the biomass creates a problem during storage, handling, and transportation for further processing. Typically, the low-bulk densities of about 40–130 kg/m^3 (Tumuluru, 2014; Tumuluru, 2015; Tumuluru and Mwamufiya, 2021) were observed for ground herbaceous biomass and municipal solid waste, whereas the bulk density of woody chips is about 200–265 kg/m^3 (Tumuluru *et al.*, 2015). One way to address the density challenges is to densify the biomass using densification systems, such as briquette press, pellet mill, cuber, extruder, or others, which were developed for the food, feed, and pharmaceutical industries. These systems can increase the biomass bulk density by about 3–7 times. The increase in biomass bulk density depends on the process conditions, such as pressure, residence time, preheating temperature, and feedstock properties (e.g., particle size and shape and moisture content). Indirect burning of raw biomass is inefficient, primarily due to low-bulk density. Biomass is generally densified using a pellet mill or a briquette press before it is used for biopower. These systems increase the bulk density and the volumetric energy content and reduce the transportation cost, thus making the biomass aerobically stable (Tumuluru *et al.*, 2011; Tumuluru, 2014, 2016). During densification, the biomass is subjected to compression pressure where the particles come closer and form compacted biomass. The compaction of the particle increases the density. This increase in density makes biomass easier to handle and store than the original raw material. Densification of biomass is mostly called pelleting or briquetting is used to produce products for energy production. The fuel briquettes produced can be used for direct burning in combustion stoves, furnaces, and wood pellets in power generation plants. Also, the densified biomass pellets can be used as feed by the biorefineries for ethanol production. The major challenge of biomass densification is the cost.

During the densification process, the compressive forces form pellets and briquettes, as first described by Rumpf (1962). The mechanical forces in play during tumbling, kneading, agitation, extrusion, rolling, and compression result in contact with the particles, resulting in the particles' agglomeration. The mechanisms that result in possible agglomeration of the particles are as follows: (1) the forces between the attraction of solid particles; (2) interfacial forces and capillary pressure in the movable liquid surfaces; (3) adhesion and cohesion forces; (4) solid bridges; and (5) mechanical interlocking or form-closed bonds. During the first compression stage, particles rearrange themselves to form a closely packed mass. The particles keep most of their original properties, although energy is dissipated due to inter-particle and particle-to-wall friction. As the compaction pressures increase, particles are forced against each other while undergoing elastic and plastic deformations. This increases the inter-particle contact area. As a result, bonding forces like van der Waals forces become effective (Rumpf, 1962; Sastry and Fuerstenau, 1973; Pietsch, 1984). In addition, brittle particles may fracture under stress, leading to mechanical interlocking.

The compaction behavior of biomass particles depends on their mechanical properties. Therefore, a better understanding of the fundamental mechanism of the biomass compaction process is needed to design energy-efficient compaction equipment to mitigate the cost of production. In addition, the effect of various process variables impacts and enhances the quality of the compacted product. The compressive force is one of the major variables affecting the compacted product's quality. At higher compressive forces, the particles and internal pores within those particles rupture until the density of the compacted bulk approaches the true or solid densities of the component ingredients. If the melting point of the ingredients in the mix that form a eutectic mixture is favorable, the heat generated at the point of contact can lead to the local melting of materials. Once cooled, the molten material forms strong solid bridges (Ghebre-Sellassie, 1989).

Various biomass components include cellulose, hemicellulose, protein, lignin, crude fiber, wax and ash. Among these lignin, wax, and protein have a low melting point of less than 140°C and can take an active part during biomass compaction process. Figures 2.1(a)–2.1(c) show how moisture, pressure, and temperature impact the various biomass components (Tumuluru, 2018). According to this author, various biomass components, such as lignin, protein, and wax water-soluble carbohydrates in the woody and herbaceous biomasses, affect the quality of the pellets. Most of these biomass components below the glass transition temperature act brittle. Once these materials reach above the glass transition temperature, they become viscous and flow easily during the compression and extrusion in the pellet die. For example, lignin once it reaches the glass transition temperature, it moves to the surface and provides a shining appearance. Once the biomass components cool, they resolidify, bind with the other biomass particles, and form a compact mass. When densified products are formed below the glass transition temperature of the biomass components, they will not have enough strength and will crumble during storage, and handling.

2.2. Compression Models

When heated during densification, the biomass components melt and soften, thus exhibiting thermosetting properties (van Dam *et al.*, 2004). Tabil and Sokhansanj (1996) studied a similar compaction mechanism during the alfalfa pelleting process. Tabil (1996) reported that when alfalfa is compacted in a single pellet press die, the temperature of the alfalfa compact increases due to preheating and heat generated due to friction between the die and the materials. Alderborn and Wikberg (1996) found a series of compression

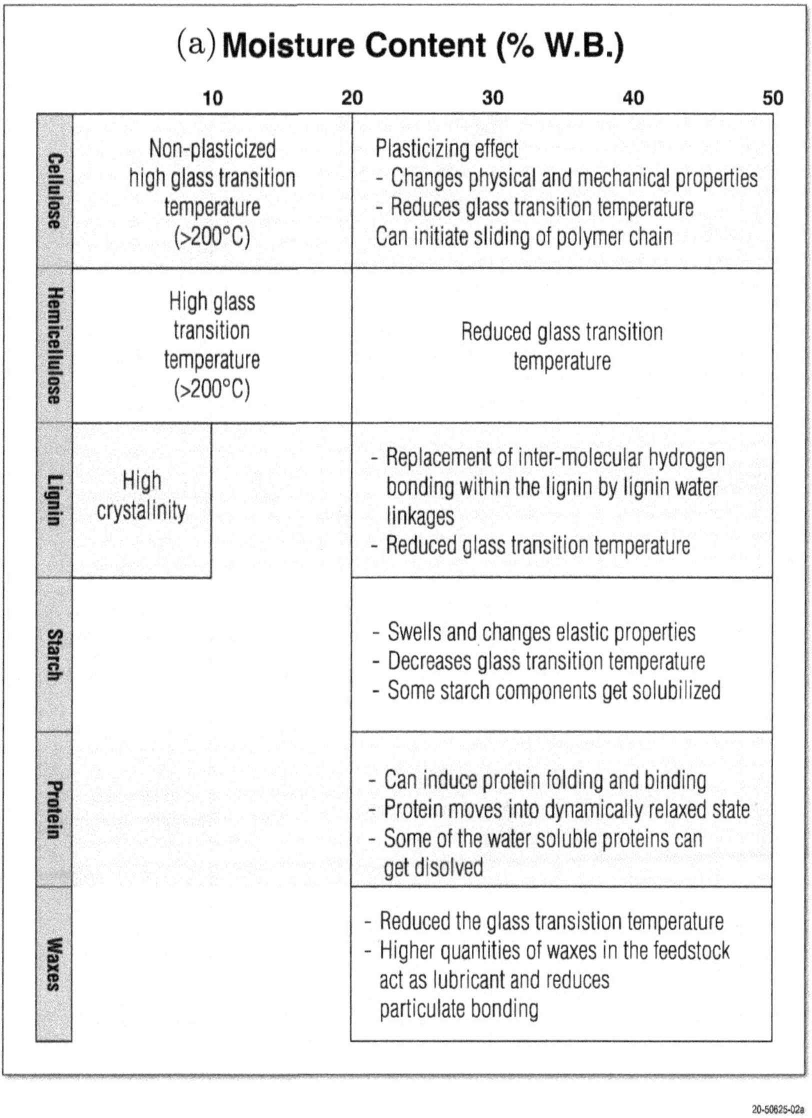

(a) **Moisture Content (% W.B.)**

	10	20 30 40 50
Cellulose	Non-plasticized high glass transition temperature (>200°C)	Plasticizing effect - Changes physical and mechanical properties - Reduces glass transition temperature Can initiate sliding of polymer chain
Hemicellulose	High glass transition temperature (>200°C)	Reduced glass transition temperature
Lignin	High crystalinity	- Replacement of inter-molecular hydrogen bonding within the lignin by lignin water linkages - Reduced glass transition temperature
Starch		- Swells and changes elastic properties - Decreases glass transition temperature - Some starch components get solubilized
Protein		- Can induce protein folding and binding - Protein moves into dynamically relaxed state - Some of the water soluble proteins can get disolved
Waxes		- Reduced the glass transistion temperature - Higher quantities of waxes in the feedstock act as lubricant and reduces particulate bonding

20-50625-02a

Figure 2.1. Temperature, pressure, and moisture impact on the biomass components (a) impact of moisture; (b) impact of pressure; (c) impact of temperature (Tumuluru, 2018).

(b) **Pressure (MPa)**

50 (70) 100 150 200 250 300

	100–200	200–300
Cellulose	High pressure can decrease cellulose crystallization	
Hemicellulose	Can break down the hemicellulose structures	Can initiate decomposition reactions
Lignin	- Can lower the glass transition temperature	Can result in charing
Starch	- Breakdown of starch structures) - Does not result in gelatinization	Can reduce enzyme activity and can make it less reactive
Protein	- Protein ligand binding - Reduced protein-protein interactions	Can denature the protein, resulting in high compressibility
Waxes	Reduces the glass transition temperature, due to break down into smaller molecules	Waxes precipitate

20-60625-02b

Figure 2.1. (*Continued*)

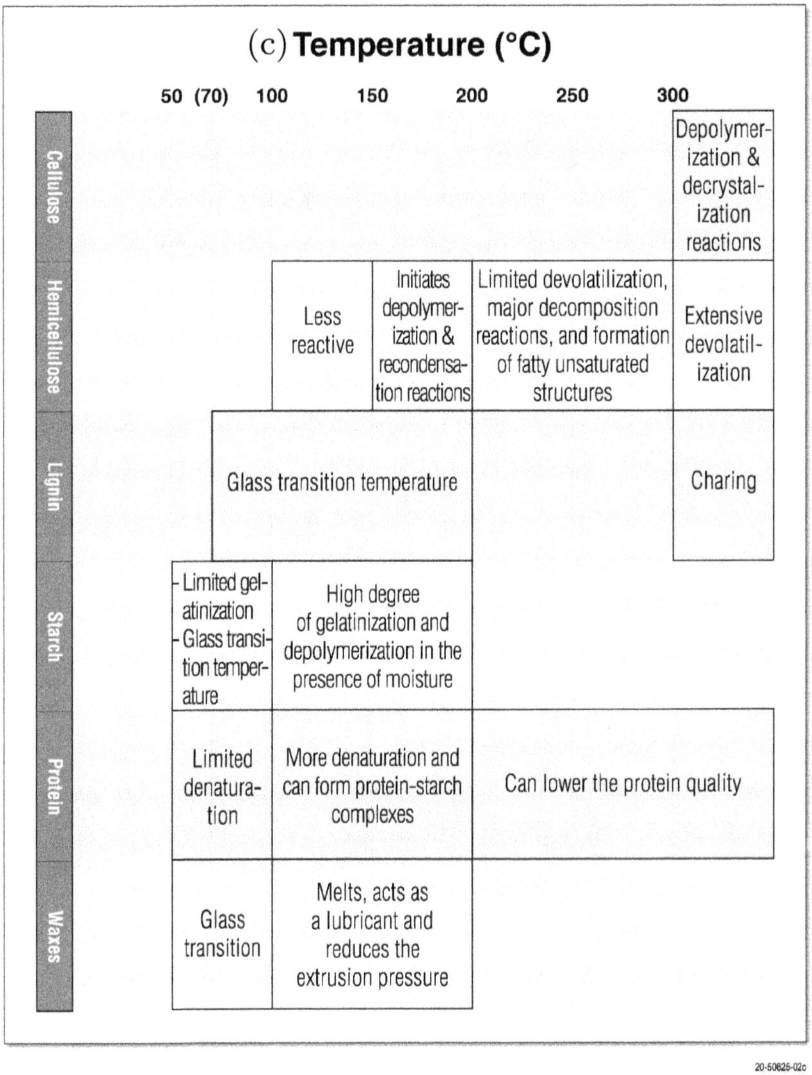

(c) Temperature (°C)

	50 (70)	100	150	200	250	300
Cellulose						Depolymerization & decrystalization reactions
Hemicellulose		Less reactive	Initiates depolymerization & recondensation reactions	Limited devolatilization, major decomposition reactions, and formation of fatty unsaturated structures		Extensive devolatilization
Lignin	Glass transition temperature					Charing
Starch	- Limited gelatinization - Glass transition temperature	High degree of gelatinization and depolymerization in the presence of moisture				
Protein	Limited denaturation	More denaturation and can form protein-starch complexes		Can lower the protein quality		
Waxes	Glass transition	Melts, acts as a lubricant and reduces the extrusion pressure				

20-50625-02c

Figure 2.1. (*Continued*)

mechanisms during the compression of pharmaceutical powders, which resulted in particle rearrangement, deformation, densification, fragmentation, and attrition.

Adapa *et al.* (2005) conducted compression tests on fractionated sun-cured and dehydrated alfalfa chops into cubes and developed models for pressure and density. Johansson *et al.* (1995) and Johansson and Alderborn (1996) studied the compression behavior of microcrystalline cellulose. These authors studied the permanent deformation (e.g., change in the shape of the individual particles) and densification (e.g., contraction or porosity reduction of the individual compacts) mechanisms. Various researchers have proposed several models to understand compaction behavior (Walker, 1923; Heckel, 1961; Cooper and Eaton, 1962; Kawakita and Lüdde, 1971). These models were successfully applied to understand the compaction behavior of pharmaceutical and other biobased powders. For example, Denny (2002), Mani *et al.* (2003b), and Tumuluru (2020) used these models to understand the compaction behavior of the biomass grinds and process scale-up.

Among the various proposed models, the Heckel (1961) and Cooper–Eaton (1962) models are still used to study the compaction mechanism of pharmaceutical and cellulosic materials. The Kawakita–Lüdde (1971) model is used for soft and fluffy materials. Tabil and Sokhansanj (1996) studied the applicability of such models for alfalfa pellets. They concluded that the Heckel and Cooper–Eaton models fit well with the alfalfa compression data.

2.2.1. *Spencer and Heckel Model*

The Spencer and Heckel (Heckel, 1961) model equation expresses the packing density of the fractions as a function of applied pressure:

$$\ln \frac{1}{1-\rho_f} = m\rho + b \tag{2.1}$$

where

$$\rho_f = \frac{\rho}{\rho_1 x_1 + \rho_2 x_2} \tag{2.2}$$

Equations (2.1) and (2.2) describe the powder materials' compression behavior. The constants m and b represent the two stages of compression: (1) pre-occupation and (2) particle rearrangement due to densification, x_1 and x_2 are mass fraction of components of the mixture, ρ_1 and ρ_2 are the particle densities of the component of the mixture (kg/m^3), ρ is the bulk density of the compacted power mixture (kg/m^3), P is the pressure (MPa), and ρf is packing fraction or relative density of the material after particle rearrangement.

Shivanand and Sprockel (1992) indicated that constant b is related to the relative density at particle rearrangement (ρ_f) by the following equation:

$$b = \ln \frac{1}{1 - \rho_f} \tag{2.3}$$

Equation (2.3) indicates that higher ρ_f will result in greater volume reduction due to more particle rearrangement, while m is the reciprocal of the mean yield pressure required to induce elastic deformation. Higher m values indicate the onset of plastic deformation due to low-yield pressures, showing the material is more compressible.

2.2.2. *Walker Model*

Walker (1923) developed a model based on the experimental data on the compressibility of powders and expressed the volume ratio (V_R) as a function of applied pressure (P).

$$V_R = m \ln P + b \tag{2.4}$$

$$V_R = \frac{V}{V_s} \tag{2.5}$$

where P is the applied pressure (MPa), V_R is the volume ratio, V is the volume of the compact at pressure P (m^3), and V_s is the void-free solid material (m^3).

2.2.3. *Jones Model*

Jones (1960) developed a model for the density and pressure data of the compacted metal powder.

$$\mathrm{Ln}\rho = m \ln P + b \tag{2.6}$$

$$b = \ln\left(\frac{1}{1-\rho_0}\right) \text{ and } \rho f = \frac{\rho}{\rho_1 x_1 + \rho_2 x_2} \tag{2.7}$$

where ρf is the relative density of the fractions after particle rearrangement, ρ_0 is the relative density of the powder mixture (kg/m^3), ρ_1 and ρ_2 are the particle densities of the component of the mixture (kg/m^3), and x_1 and x_2 are the mass fractions of the component of the mixture.

Constants b and m are determined based on the intercept and slope of the extrapolated linear region of the plot $\ln\left(\frac{1}{1-\rho f}\right)$ vs. P.

Higher values of ρf will indicate increased volume reduction of the samples due to particle rearrangement.

Constant m is reciprocal to the mean yield pressure required to induce the elastic deformation (York and Pilpel, 1973). A large m value with yield pressure is low, and plastic deformation onsets at relatively low pressures, showing that the material is more compressible.

2.2.4. *Cooper–Eaton Model*

The Cooper–Eaton (1962) model assumes compression as nearly two independent probabilistic processes: (1) filling of voids of the same size as the particles and (2) filling of voids smaller than the particles.

$$\frac{V_0 - V}{V_0 - V_s} = a_1 e^{-\frac{k_1}{P}} + a_2 e^{-\frac{k_2}{P}} \tag{2.8}$$

where V_0 is the volume of the compact at zero pressure (m^3), a_1, a_2, and k_1, k_2 are Cooper–Eaton model constants, V is volume of compact at pressure P (m^3), and V_s is void-free solid material volume (m^3).

The practical difficulty in applying the Cooper–Eaton (1962) model is understanding the physical significance of the constants in the equation (Comoglu, 2007). Therefore, it is more suitable for single-component systems (Comoglu, 2007).

2.2.5. *Kawakita–Lüdde Model*

The Kawakita–Lüdde (1971) model includes the pressure and volume factors.

$$\frac{P}{C} = \frac{1}{ab} + \frac{P}{a} \tag{2.9}$$

$$C = \frac{V_0 - V}{V_0} \tag{2.10}$$

where C is the degree of volume reduction or engineering strain and a and b are the Kawakita–Lüdde (1971) model constants related to powder characteristics, V is net volume of the powder (m^3), V_0 is volume of compact at zero pressure (m^3) and P is the pressure (MPa).

The linear relationship between $\frac{P}{C}$ and P allows the constants to be evaluated graphically. This compression equation holds true for soft and fluffy powders (Denny, 2002; Kawakita and Lüdde, 1971). Any deviations from this expression are sometimes due to fluctuations in the measured value of V_0. Mani *et al.* (2004) indicated that constant a is equal to the initial porosity of the sample, while constant $\frac{1}{b}$ is related to the failure stress in the case of piston compression.

2.2.6. *Sonnergaard Model (Log–Exp Equation)*

The Sonnergaard (2001) model is a log–exp equation that considers two processes simultaneously: (1) a logarithmic decrease in volume by fragmentation and (2) an exponential decay representing plastic deformation of powders.

$$V = V_1 - w \log P + V_0 \exp(-P/P_m) \tag{2.11}$$

where V_1 is the volume at pressure 1 (MPa), P_m is the mean pressure (MPa), and w is the constant.

Sonnergaard (2001) suggests that his model provides better regression values than either the Cooper–Eaton (1962) or the

Kawakita and Lüdde (1971) models. Therefore, this model is suitable for medium-pressure applications (approximately 50 MPa).

2.2.7. *Panelli–Filho Model*

The following expression gives the Panelli–Filho (2001) model:

$$\ln \frac{1}{1-\rho_r} = A\sqrt{P} + B \tag{2.12}$$

where ρ_r is the relative density of the compact, P is the pressure (MPa), A is the parameter related to densification of the compact based on particle deformation, and B is the parameter related to powder density at the beginning of compression.

2.2.8. *Peleg and Moreyra Model*

According to Peleg and Moreyra (1979), stress relaxation is a function of the physical changes under constant strain. This phenomenon can be interpreted as internal flow and rearrangement of liquid bridges or plasticizing of the particle's texture. Peleg (1979) presented a method for normalizing relaxation data from solid foods, which can be applied to powder compaction. Moreyra and Peleg (1980) further explained the normalization equation Peleg and Moreyra (1979) proposed, to which the force–time relaxation data were fitted.

$$\frac{F_0 t}{F_0 - F(t)} = k_1 + k_2 t \tag{2.13}$$

Moreyra and Peleg (1980) explained that the slope of Eq. (2.13) (k_2) could be considered an index of how "solid" the compacted specimen is on a short timescale. Liquids would have a slope of unity ($k_2 = 1$), indicating that the stresses will eventually relax to zero. Therefore, the value of k_2 for any solid must be greater than 1. It was further noted that any larger slope value indicates that there are stresses that will eventually remain unrelaxed (e.g., a solid-state property).

The constant k_2 was used to calculate an asymptotic modulus (EA) in Eq. (2.14), proposed by Scoville and Peleg (1981) and used by Moreyra and Peleg (1981) on food powders.

$$E_A = \frac{F_0}{A\varepsilon}\left(1 - \frac{1}{k_2}\right) \tag{2.14}$$

2.2.9. *Testing the Compression Models on Various Biomass Grinds*

Shaw and Tabil (2007) studied the compression, relaxation, and adhesion properties of peat moss, wheat straw, oat hulls, and flax shives. These authors considered the five loads (e.g., 1000 N, 2000 N, 3000 N, 4000 N, and 4400 N). These authors studied the compressibility, asymptotic modulus, coefficient of external friction, and adhesion coefficient using Instron. The results from this study indicated that compression force increases gradually and spikes just before achieving the pre-set maximum load for each test, where the maximum load achieved was always slightly higher than the pre-set load value. Peat moss required more time to reach the pre-set load than the other three grinds tested for the four different biomasses. These authors concluded that lower bulk density and higher standard deviation of the geometric mean particle size might have contributed to this effect. Peat moss was the only feedstock that would produce cohesive pellets at the 500 N pre-set load and was subsequently the only material for which this load was used in compression modeling. These authors used the Jones and Walker models to understand the compressibility (e.g., m and m', respectively) as given in Table 2.1. Based on the R^2 values, both models fit well with the experimental data; however, the Jones model had a slightly better fit. The data have shown that the compressibility is consistent between the two models, with flax shives having the highest, followed by peat moss, wheat straw, and oat hulls.

Table 2.2 presents the effect of pre-set load on the diametrical and longitudinal expansions of the single pellets. The experimental data indicated that diametrical measurement of the pellets (e.g., immediately after removal from the die) was higher, which suggests that the samples expanded after compressions and extrusion from the die (e.g., pellets diameter exceeded 6.35 mm). Pellet length depended

Table 2.1. Empirically established constant values for compression models.

Biomass grind	Compressibility (slope m or m')	Intercept (b or c)	R^2	Standard error
Jones: $\ln p = m \ln P + b$				
Peat moss	0.3	5.91	0.97	0.007
Wheat straw	0.28	6.02	0.96	0.008
Oat hulls	0.23	6.24	0.98	0.005
Flax shives	0.35	5.62	0.93	0.013
Walker: $\text{VR} = m' \ln P + c$				
Peat moss	−0.34	2.48	0.93	0.012
Wheat straw	−0.27	2.15	0.94	0.010
Oat hulls	−0.24	2.06	0.97	0.006
Flax shives	−0.41	2.86	0.89	0.021

Table 2.2. Effect of pre-set compression load and biomass feedstock on the diametrical and longitudinal expansions (after 14 days) of single pellets produced by compression.

Pre-set load (N)	Diametrical expansion (%)	Longitudinal expansion (%)
1000	0.66	3.67
2000	0.39	3.10
3000	0.34	1.38
4000	0.32	1.55
4400	0.37	1.34
Feedstock	**Diametrical expansion (%)**	**Longitudinal expansion (%)**
Peat moss	0.25	0.95
Wheat straw	0.73	4.27
Oat hulls	0.27	1.99
Flax shives	0.42	1.63

on the applied load and mass of the material fed to the die. This study indicated that the interaction between biomass feedstock and pre-set load did not significantly affect diametrical and longitudinal expansions (Shaw and Tabil, 2007).

Increasing the load reduced pellet expansion both diametrically and longitudinally. Table 2.2 also highlights the effect of biomass feedstock on the diametrical and longitudinal expansion of the single pellets produced in compression testing. Wheat straw showed the highest diametrical and longitudinal expansion values, whereas peat moss expanded the least. The change in the expansion properties of the various biomasses tested can be due to changes in the chemical composition of these feedstocks. Peat moss and flax shives had a higher lignin content than the other feedstocks. In general, lignin is the major component that helps the biomass grinds to bind (Anglès *et al.*, 2001). According to Granada *et al.* (2002) and Tumuluru (2018), the adhesive properties of thermally softened lignin and other biomass components, such as protein, and extractives contribute to the strength of the densified product produced. In addition, these authors found that peat moss had a higher protein content than the other three feedstocks. Tabil (1996) explained that if a sufficient natural protein is available, it will plasticize under heat, thereby improving the quality of the pellets.

2.2.10. *Asymptotic Modulus*

The asymptotic modulus values resulting from the relaxation data are shown in Table 2.3. As Scoville and Peleg (1981) reported, the asymptotic modulus is indicative of compact solidity and the ability of the compacts to sustain the unrelaxed stresses. Peat moss and oat hulls had significantly higher asymptotic modulus (EA) values than the other two feedstocks at pre-set loads of 4000 N and 4400 N. Peat moss alone had the highest EA value at 3000 N. There was no significant difference in asymptotic modulus values at the lower two loads (e.g., 1000 N and 2000 N). Higher asymptotic modulus values also can be seen to correlate to lower-dimensional expansions. The chemical composition did not appear to have any bearing on the relaxation characteristics of the four biomass grinds. Relaxation data for the pre-set compression test peak loads were fitted to a power-law model, to show the effect of initial stress (i.e., pre-set compressive load) on EA, as demonstrated by Tabil and Sokhansanj (1997).

$$E_A = a\sigma_0^d \tag{2.15}$$

Table 2.3. Mean asymptotic modulus values for peat moss, wheat straw, oat hulls, and flax shives.

Pre-set load		Asymptotic modulus, EA (MPa)		
	Peat moss	Wheat straw	Oat hulls	Flax shives
1000	24.7	25.48	24.0	26.2
2000	54.9	53.52	53.4	54.7
3000	93.9	84.86	86.2	86.1
4000	125.1	117.06	122.6	118.2
4400	139.1	129.83	138.0	132.1
Power law				
a	0.4732	0.5709	0.4121	0.6340
d	1.1304	1.0904	1.1656	1.0728
R^2	0.99	0.99	0.99	0.99

The asymptotic modulus values increased with increasing initial stress, and the power-law model was an excellent fit for the experimental data ($R^2 = 0.99$).

2.2.11. *Adhesion*

Figure 2.2 demonstrates the relationship between the normal and shear stresses and shows the values of the coefficient of external friction and the adhesion coefficient for the biomass grinds. The literature examining the frictional properties of biomass grinds is limited. Mani *et al.* (2004) reported significant increases in the coefficient of friction of corn stover with particle size and moisture. The values of the adhesion coefficient for wheat straw and flax shives are higher than those found for corn stover (Mani *et al.*, 2004). Usrey *et al.* (1992) found that the friction coefficient of rice straw was 0.489 on polished steel. The frictional coefficient varies based on the material used in the die.

Although the Cooper–Eaton model had the best fit for most cases, the regression values (R^2 values) were low, especially for sun-cured alfalfa, as observed in Table 2.4 (Adapa *et al.*, 2005).

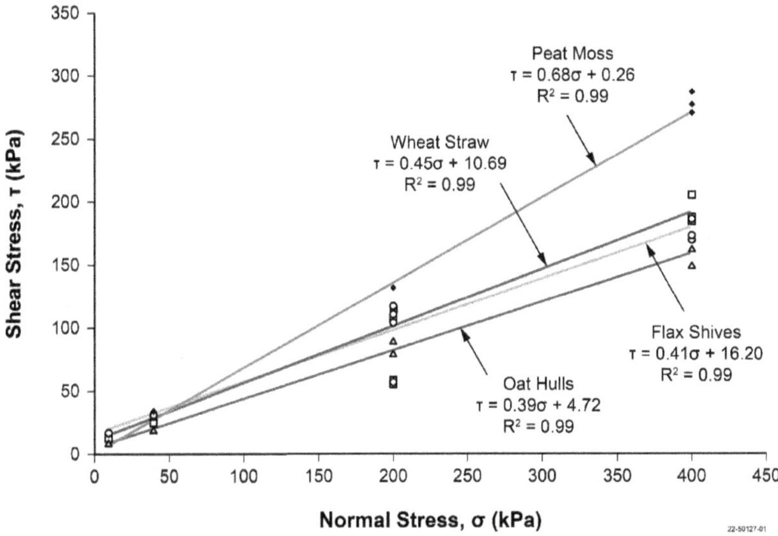

Figure 2.2. Shear stress resulting from applied normal stress for frictional analysis (e.g., slope represents the coefficient of external friction and intercept represents the coefficient of adhesion in kPa).

Therefore, using multiple regression analysis, these authors have also developed new statistical models to predict the compression characteristics of sun-cured and dehydrated alfalfa. For multiple linear regression analysis, density (D) was the dependent variable while pressure (P), holding time (t), and leaf content (x) were independent variables. These authors conducted separate analyses for the sun-cured and dehydrated chops to determine individual equations for each material, and multiple linear regression models were developed.

The regression model tested was

$$\rho = \beta_0 + \beta_1 t + \beta_2 P + \beta_3 x \tag{2.16}$$

where $\rho is the bulk density of compact power mixture$ $\left(\frac{\text{kg}}{\text{m}^3}\right)$, t is the holding time (s), P is pressure (MPa), x is the leaf content (%), and β_0, β_1, β_2, β_3 are the constants.

A linear relationship was observed between dependent and independent variables. No collinearity was detected using the variance

Table 2.4. Constants' values for the Cooper–Eaton model (Eq. (2.8)) for dehydrated and sun-cured alfalfa (Adapa *et al.*, 2005).

Holding time (s)	Leaf content (%)	Constants				R^2 values	SSE*
		a_1	a_2	k_1	k_2		
10 (dehydrated alfalfa)	0	0.5857	0.5000	5.1340	−2.1321	0.8406	0.0008
	25	0.5881	0.5000	6.0749	−2.5867	0.6418	0.0012
	50	0.5183	0.5002	0.6682	0.6682	0.7771	0.0015
	75	0.5172	0.5000	2.0940	−0.9177	0.7772	0.0006
	100	0.5249	0.5000	3.7221	−1.6808	0.8996	0.0002
30 (dehydrated alfalfa)	0	0.5339	0.5011	0.6112	0.6116	0.7953	0.0011
	25	0.5473	0.5001	0.7852	0.7854	0.8603	0.0012
	50	0.6236	0.5000	7.3853	−2.7185	0.7523	0.0015
	75	0.5099	0.5000	2.3291	−1.3324	0.5269	0.0005
	100	0.4940	0.5000	0.3658	0.3656	0.7413	0.0005
10 (sun-cured alfalfa)	0	0.5211	0.4983	0.5579	0.5775	0.5266	0.0035
	25	0.5769	0.5000	7.3733	−2.9516	0.2383	0.0024
	50	0.4652	0.5037	0.4340	0.4339	0.3839	0.0033
	75	0.4528	0.5080	0.2380	0.2376	0.1356	0.0039
	100	0.4401	0.5132	0.2963	0.2961	0.1011	0.0084
30 (sun-cured alfalfa)	0	0.5336	0.5001	0.5755	0.5754	0.7985	0.0010
	25	0.5266	0.5000	3.1938	−1.5485	0.8158	0.0004
	50	0.9570	0.5000	17.4401	−4.3325	0.2774	0.0048
	75	0.4923	0.5000	0.4180	0.4179	0.3335	0.0039
	100	0.4636	0.5002	0.2279	0.2278	0.0608	0.0089

Note: *Sum of Square of Errors.

inflation factor and condition index (Adapa *et al.*, 2005). Regression analysis on the full model indicated that the model with all three variables was significant for both sun-cured and dehydrated experiments ($P < 0.0001$). In addition, all the independent variables were significantly different than zero ($P < 0.0001$). Based on the R^2 value, the full model with all three independent variables was accepted as the best predictor model for both experiments. The regression equations obtained were

(a) Dehydrated model ($R^2 = 0.85$):

$$\rho = 609.27 + 0.91t + 19.29P + 1.84x \tag{2.17}$$

(b) Sun-cured model ($R^2 = 0.59$):

$$\rho = 622.32 + 3.23t + 11.65P + 1.34x \qquad (2.18)$$

Mani *et al.* (2004) evaluated barley straws, corn stover, wheat, and switchgrass compaction equations. The compression tests were conducted at different applied forces, moisture contents, and particle sizes using the single pellet press (Instron tester). Among the four biomass grinds studied, the corn stover grind reached its maximum density at low pressure, whereas the other biomass grinds required high pressure to achieve maximum density. The experimental data were used to fit the three compaction models to explain the mechanisms. Among the three models, the Kawakita–Lüdde and Cooper–Eaton models described well for all biomass grind samples. Based on the Cooper–Eaton model parameters, the dominant compaction mechanisms for biomass grinds were a rearrangement of particles followed by elastic and plastic deformation, whereas the mechanical interlocking was found to be negligible.

The Kawakita–Lüdde model indicated that the switchgrass grind compact has a higher yield strength than compacts made from different biomass grinds (Mani *et al.*, 2004). Among the four grinds tested, the compacts made from corn stover had lower yield strength based on the Kawakita–Lüdde model. The Cooper–Eaton model was used to plot the biomass grinds with a 3.2-mm screen size at a moisture content of 12% (w.b.), as given in Fig. 2.3. The Cooper–Eaton model's two intercepts, a_1 and a_2, indicated the relative density after particle rearrangement and relative density after deformation. The compaction of biomass grinds occurs partly by particle rearrangement and partly by particle deformation. For all the cases, the biomass grinds exhibit slightly lower a_1 values than a_2 values, which shows that the elastic and plastic deformations are the major reason for creating a densified compact. These authors found that the theoretical density for all biomass grinds from the 3.2-mm screen size at 12% moisture content corroborated well with the earlier published literature by Shivanand and Sprockel (1992) for cellulose acetate and cellulose acetate propionate and by Tabil and Sokhansanj (1996) for alfalfa grind.

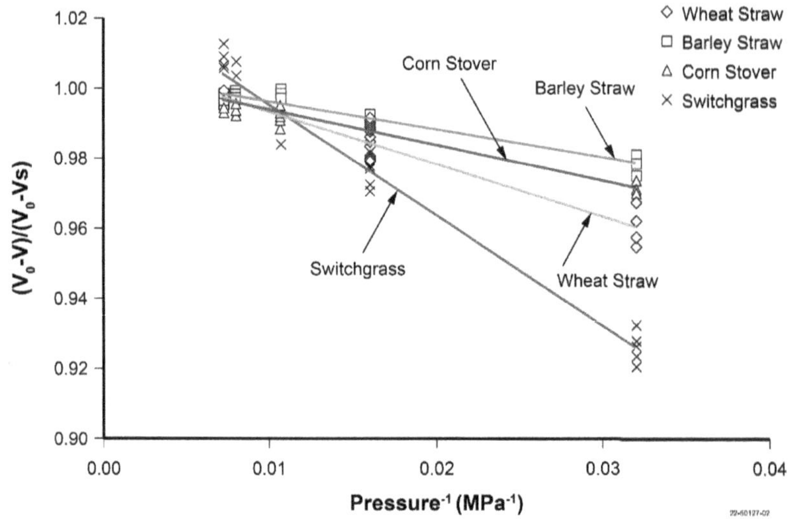

Figure 2.3. Cooper–Eaton plot for biomass grinds milled with a 3.2-mm screen size at a moisture content of 12% (w.b.) (Mani *et al.*, 2004).

Based on Shivanand and Sprockel (1992), k_1 in the Cooper–Eaton model represents the pressure required to initiate the densification by particle rearrangement (P_r). The k_2 represents the pressure required to induce densification through deformation (P_d). The P_r values were slightly lower than the P_d values for wheat and barley straw grinds, indicating that the straw grinds required marginally less pressure for particle rearrangement than particle deformation (Table 2.5). Overall, both wheat and barley straw had similar densification characteristics. Observations from the P_r and P_d parameters for corn stover grind indicated that all particles required equal pressure for particle rearrangement and deformation (Table 2.5). In the case of switchgrass grind densification, the high P_r and low P values indicate that higher pressure is needed for particle rearrangement than particle deformation (Table 2.5). Mani *et al.* (2004) concluded that switchgrass grind was more difficult to densify by particle rearrangement than by particle deformation. The reason can be due to fibrous nature of switchgrass than other biomass grinds (Mani *et al.*, 2004). When comparing the P_r and P_d values for the biomass grinds, higher values were observed for the switchgrass grind, while

Table 2.5. Model parameters for Cooper–Eaton and Kawakita–Lüdde models of all biomass grinds (Mani *et al.*, 2004).

Model parameters	Hammer mill screen size (mm)	Wheat straw		Barley straw		Corn stover		Switchgrass	
		12	15	12	15	12	15	12	15
P_r (MPa)	3.2	2.0	1.7	0.80	1.0	1.2	0.70	3.3	4.8
	1.6	0.8	1.7	1.3	0.6	0.89	0.5	2.3	4.8
	0.8	0.4	0.6	3.1	1.9	0.90	0.6	5.5	5.1
P_d (MPa)	3.2	1.1	1.0	0.8	1.0	1.2	0.7	3.3	2.0
	1.6	2.0	1.7	1.8	2.0	0.8	0.5	2.0	2.0
	0.8	2.0	0.6	2.0	1.9	0.9	0.6	2.0	2.0
$a_1 + a_2$	3.2	1.01	1.00	1.00	0.99	1.01	0.99	1.03	1.03
	1.6	0.99	0.98	0.99	0.97	0.99	0.98	0.99	0.98
	0.8	0.97	0.95	0.99	0.98	0.98	0.97	1.0	0.97
a	3.2	0.91	0.91	0.91	0.90	0.89	0.88	0.91	0.91
	1.6	0.91	0.89	0.91	0.89	0.87	0.86	0.85	0.85
	0.8	0.89	0.86	0.91	0.89	0.87	0.85	0.85	0.82
a^*	3.2	0.91	—	0.91	—	0.89	—	0.88	—
	1.6	0.92	—	0.91	—	0.88	—	0.86	—
	0.8	0.91	—	0.91	—	0.89	—	0.85	—
$1/b$ (MPa)	3.2	1.64	1.60	0.71	1.07	1.09	0.63	3.65	3.92
	1.6	1.71	1.15	1.78	1.68	0.59	0.44	2.04	4.03
	0.8	1.29	1.32	3.05	1.70	0.75	0.60	3.97	4.03

Note: a^* indicates the initial theoretical porosity of the biomass grinds.

lower values were observed for the corn stover grind. This study has indicated that the switchgrass grind was more difficult to densify, whereas the corn stover was the easiest (Mani *et al.*, 2004). In addition, the Kawakita–Lüdde parameter can be related to the biomass grinds' initial porosity (a^*). Table 2.5 shows that initial porosity (e.g., parameter a) was almost similar for all biomass grinds (Mani *et al.*, 2004). Parameter $1/b$ represents the yield strength or failure stress of the compact. A higher $1/b$ value directly relates to higher yield strength. Table 2.5 shows that switchgrass compacts have higher yield strength, while corn stover has the lowest. Low $1/b$ values were observed for compacts from corn stover grinds. These authors from these studies have concluded that the compact from the corn stover grind may have less failure stress, whereas the compacts made from switchgrass will be harder to break.

Timothy hay compression and relaxation characteristics were investigated with respect to moisture content, applied load, and hay quality by Talebi *et al.* (2011). A Baldwin hydraulic universal testing machine model (e.g., the 60 HVL-1254) was used to complete the experiments. The applied loads ranged from 90 kN to 240 kN to 30 kN. Two qualities of timothy hay were used in the tests (a) high quality with moisture content values of 7.44%, 10.17%, 12.97%, and 16.42% wet basis (w.b.) and (b) low quality with moisture content values of 6.38%, 8.67%, 16.24%, and 18.94% (w.b.).

The outcome of the work indicated that the compact density of hay samples increased with increasing moisture content and applied pressure. The lower applied pressure was required to achieve the same compact density at higher moisture content values. The use of the Faborode–O'Callaghan model for bale densities less than 500 kg/m^3 and the simple power-law model for bale densities greater than 500 kg/m^3 were found to be the most relevant models to relate the density and pressure during the compression of timothy hay. These authors concluded that the relaxation of the hay samples was affected by the initial maximum applied load or pressure, and the moisture content and higher moisture content resulted in higher relaxation rates. The percent relaxation values ranged from 27–54% for the two qualities of hay tested. These authors also concluded that the asymptotic values (EA) (Eq. 2.15) are influenced by the maximum applied pressure and hay moisture content.

Mani *et al.* (2003a) modeled the densification of biomass grinds using the discrete element method (DEM) by Particle Flow Code in 3 Dimensions (PFC3D). A numerical model was developed using DEM, to understand the dynamic behavior of particle–particle and particle–wall interactions and the force–deformation relationship during compression. These authors used a cylindrical geometry (boundary) with biomass particles inside the geometry to develop a compaction model. The model was generated using the PFC3D software package. The specific properties of biomass particles, such as particle size distribution, particle density, particle stiffness, particle–particle friction, and particle–wall friction, were incorporated into the model. These authors have concluded that PFC3D software had the flexibility to develop the structure's geometry and generate particles inside the geometry to simulate the compaction process. The geometry of a single compaction die was generated using the FISH programming language embedded in the PFC3D software. A robust code was developed to fill the particles inside the enclosures. In this analysis the authors has considered the particles as spherical. Preliminary computation results did not match the experimental results. These authors concluded that further research should focus on particle shapes, such as adding clusters to the existing particle assembly to represent physically realistic particle assembly.

2.3. Densification Systems

Among the various densification methods used by the industry, pelleting, briquetting and extrusion are the most commonly used for biomass. The pelleting and briquetting are also called binderless technologies as they bind the particles using high pressures at elevated die temperatures and in the presence of moisture.

2.3.1. *Pellet Mills*

The pellet mill has perforated hard steel die with two or three rollers. Rotating die or rollers force the feedstock to form densified pellets through the perforations. In pellet mills, the raw biomass is compressed and extruded to produce smaller densified products called pellets. Two types of pellet mills exist — the flat die pellet mill and the ring die pellet mill. The principle of operation in both is the same: the raw biomass is compressed and extruded. Still, the quality of the pellets, die dimensions, and how the raw material is fed into the die are different.

Biomass is initially fed into screw conditioners where the biomass is pretreated with steam. The addition of steam helps soften some biomass components, such as mobilizing the lignin and gelatinizing the starch, which results in better binding of the particles in the pellet die. The steam-conditioned biomass is further discharged into

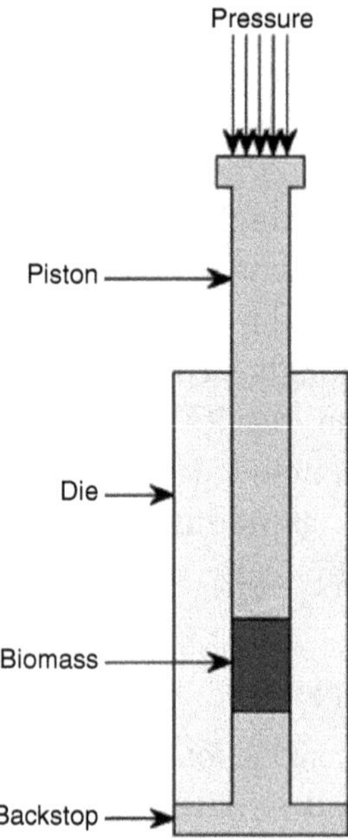

Figure 2.4. Single pellet press commonly used to make individual pellets (Tumuluru *et al.*, 2019).

the feed spout, which leads to the pellet die. The flights provided in the die cover help feed the biomass evenly in each roller, and the feed distributors spread the feed evenly on the face of the die. Some pellet mills are provided with force-feed augurs, which can help move the material into the die more uniformly and consistently. In addition, force-feed augurs help feed biomass material that is low and variable in density. Finally, pellet mills are provided with knives to cut the extruded pellets to their desired lengths. Figure 2.4 indicates the single pellet press, commonly used for conducting compression tests on biomass grinds to precisely understand the impact of pelleting process conditions and biomass properties on the unit density of the pellets.

2.3.1.1. *Flat die pellet mill*

Flat die pellet mills are of two types: rotating rollers and rotating dies. In the rotating roller, the roller moves where the die is

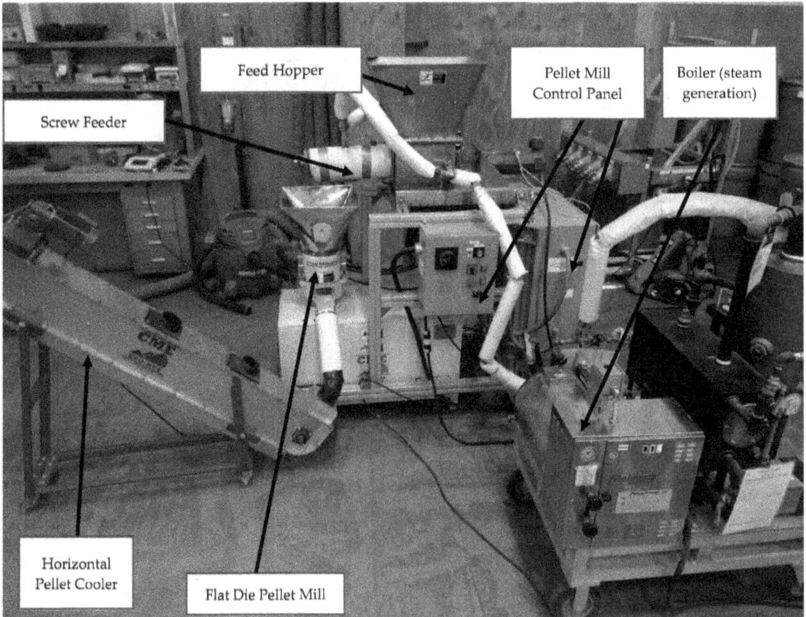

Figure 2.5. Flat die pellet mill used for pellet production (Tumuluru, 2019).

stationary, whereas, in the rotating die, the pellet die moves where the rollers sit on the die. Two or more rollers are typically used in flat die mills, where the biomass is fed into the pellet mill chamber by gravity, where it gets compressed between the rollers and the die and extrudes pellets through the die hole. Figure 2.5 indicates the flat die pellet mill. The major advantages of the flat die pellet mill are its small size and lightweight, compact design, relatively easy cleaning and maintenance. The major challenge with flat die mills is adjusting the gap between the rollers. In addition, the die can be tricky and needs to be adjusted based on the type of raw materials being fed into it.

2.3.1.2. *Ring die pellet mill*

The basic principle of pelleting biomass in a ring die pellet mill is its distribution over the inner surface of the rotating perforated die ahead of the rollers, which is further compressed into the die holes

Figure 2.6. Ring die pellet mill at Idaho National Laboratory used for large-scale production of pellets.

Notes: A: Ribbon blender (for blending, moisture addition and conditioning); B: Pellet mill feed bin; C: Pellet mill steam conditioner; D: Ring die pellet mill; E: Pellet mill outfeed conveyor; F: Weighing balance; G: Pellet mill data logger.

to form the pellets. Figure 2.6 indicates the pelleting process in a ring die pellet mill. In general, ring die pellet mills generate less wear and tear as the roller travels the same distance. Generally, a ring die pellet mill is used to produce fuel and feed pellets. The major drawback of a ring die pellet mill is the slip of the biomass during squeezing into the pellet die. However, proper die design can help overcome this issue.

2.3.2. *Briquette Press*

There are two types of briquette presses (a) hydraulic and (b) mechanical briquette presses. Figure 2.7 shows the hydraulic briquette press used for biomass briquette production. A mechanical press is used for larger-scale production, generates more pressure, and produces higher-density briquettes. In the briquetting process, the plunger connected to the continuously rotating center mass forces the biomass through a restricted opening. However, the challenge with hydraulic motors is the oil heats up when operated for long

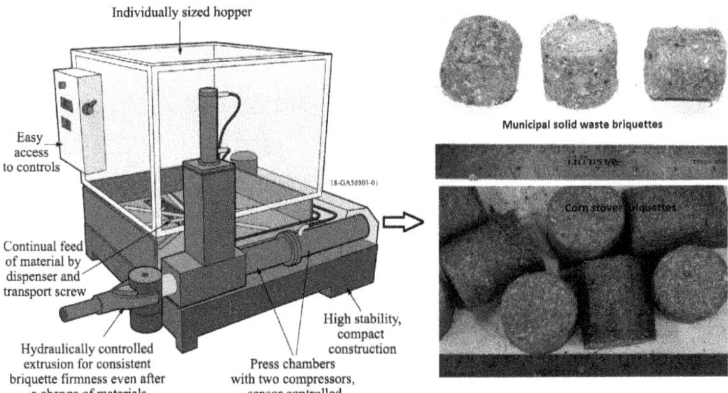

Figure 2.7. Hydraulic briquette press used to produce biomass briquettes (Tumuluru, 2019).

hours. Therefore, some hydraulic presses are provided with cooling fans to reduce the heat in the hydraulic oil. During the briquetting process, the biomass is preprocessed in the augur connected to a hopper which is further processed using a load pusher. In the third stage, applying pressure using a hydraulic ram briquette is produced. Finally, the formed briquette is discharged from the press through the die opening. For the briquetting process, bigger particles are desirable as the particle binding is more due to mechanical interlocking. The biomass briquettes are typically used for heating furnaces because they have higher heating values and less particulate emissions. The other advantage of briquettes is that they can be stored and transported more efficiently due to their uniform size and shape.

2.3.3. *Screw Extruder*

In a screw extruder, a helical screw rod is used to extrude biomass material where the extruder barrel is heated from outside to maintain a constant temperature during compression and extrusion. Figure 2.8 shows the extruder used for biomass densification. During extrusion, the biomass undergoes mixing, shearing, and compression, where the lignin and starch content in the biomass is softened, which helps form a densified biomass. When the biomass material is passed through

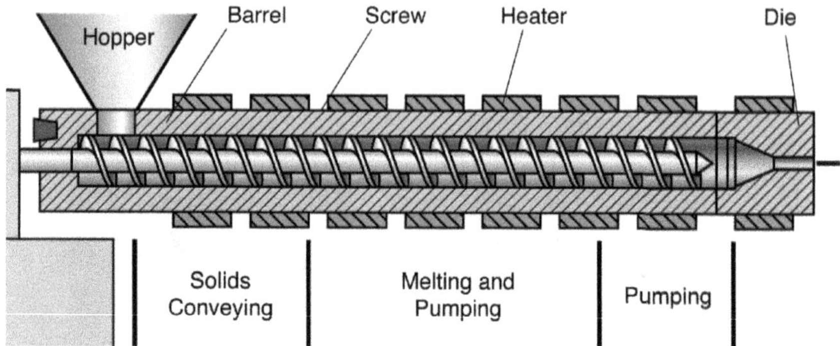

Figure 2.8. Extruder used for producing densified biomass products (Tumuluru *et al.*, 2010).

the extruder the material undergoes shearing and mixing. This increases the internal temperature in the extruder. If the generated temperature is insufficient, external heaters are provided to maintain the extruder barrel at the desired temperature. The heated, sheared, and mixed biomass is further extruded through a constricted die hole to form densified biomass; as such, the design of the screw has a great impact on the resulting quality of the extruded pellets produced. Table 2.6 compares the various densification systems discussed regarding the processing conditions and suitability of the product for downstream conversion.

Recent technological advances in the pelleting process have made pelleting more economical compared to the briquetting process. Tumuluru (2014) and Tumuluru *et al.* (2022) developed the high moisture pelleting process, which helps reduce the cost of pellet production by more than 50% compared to the method followed by the industry (Lamers *et al.*, 2015). Figure 2.9 shows the conventional and high moisture pelleting process (Lamer *et al.*, 2015). For densification of biomass using conventional methods followed by the industry, the biomass is dried to about 10% moisture content on a wet basis (w.b.) (Fig. 2.9(b)). Woody biomass after harvesting has a moisture content between 30 and 50% (w.b.), and most herbaceous biomasses are between 10 and 35% (w.b.). Hence, drying the high moisture biomass to <10% (w.b.) moisture content using

Table 2.6. Comparison of different densification equipment (Tumuluru *et al.*, 2011; Tumuluru, 2014, 2016; Tumuluru and Fillerup, 2020).

	Screw press	Briquette press	Pellet mill
The optimum moisture content of the raw material	8–9%	10–21%	10–30%
Hammer mill screen size (inch)	Smaller (<1/4 inch)	Larger (1/2–3/4 inch)	Larger (1/4–7/16 inch)
Wear contact parts	High	Low	High
Output from machine	Continuous	In strokes	Continuous
Specific energy consumption (kWh/t)	36.8–150	37.4–100	50–150
Throughputs (t/h)	0.5	2.5	5–10
Unit density	1–1.4 g/cm^3	1–1.2 g/cm^3	1.2–1.3 g/cm^3
Maintenance	Low	High	Low
Combustion performance of briquettes	Very good	Moderate	Very good
Carbonization of charcoal	Makes good charcoal	Not possible	Not possible
Suitability in gasifiers	Suitable	Suitable	Suitable
Suitability for cofiring	Suitable	Suitable	Suitable
Suitability for biochemical conversion	Not suitable	Suitable	Suitable
Homogeneity of densified biomass	Homogenous	Not homogenous	Homogeneous

conventional rotary dryers adds to the energy and capital costs (Tumuluru, 2016) (Fig. 2.9(b)). Another major challenge of high-temperature drying is volatile organic emissions which form into photo-oxidants that are hazardous to human health if inhaled. The

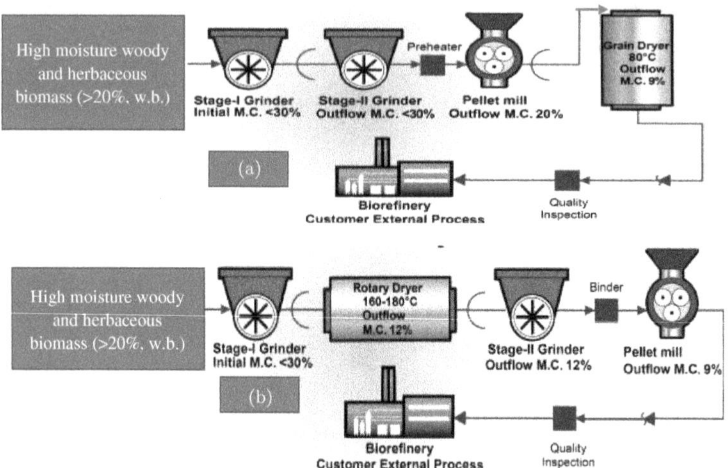

Figure 2.9. (a) High moisture pelleting process and (b) conventional pelleting processes.

high-moisture pelleting process eliminates the rotary drying step and reduces energy and capital costs (Fig. 2.9(a)). In this process, the biomass is pelleted at higher moisture levels >20% (w.b.). This process has been tested on feedstocks, such as corn stover, lodgepole pine, municipal solid waste, and blends (Tumuluru, 2014, 2016, 2019; Tumuluru and Mwamufiya, 2021). The resulting pellets had good quality in terms of density and durability. Recent study on municipal solid waste (MSW) also showed that high moisture pelleting could save about 40–46% of the pelleting cost and lower 46% of greenhouse gas emissions (Tumuluru and Mwamufiya, 2021). The resulting MSW pellets exhibited durability between 90% and 98% and a bulk density of about 450–550 kg/m^3 (Tumuluru and Mwamufiya, 2021). The research has indicated that high moisture pelleting is a good alternative *in place* of conventional pelleting for the densification of woody, herbaceous, municipal solid waste, and their blends.

2.4. Conclusions

The compression force and residence time influence the compression and relaxation characteristics of biomass grinds. The compact density generally increases with increasing the applied pressure. The research also indicated that the lower applied pressure was required to achieve the same compact density with increasing moisture content. The Faborode–O'Callaghan model and a simple power-law model have appropriately expressed the relationship between pressure and density during the compression of timothy hay. During cubing, the biomass density increased by more than 5% when the holding time increased from 10 to 30 s. The compression behavior of wheat and barley straws, corn stover, and switchgrass grinds indicated that Heckel, Cooper–Eaton, and Kawakita–Lüdde's Cooper–Eaton model had the best fit for the experimental data. The Cooper–Eaton model parameters for the biomass grinds showed that the prominent compaction mechanisms are particle rearrangement and elastic and plastic deformation. The mechanical interlocking mechanism and the biomass components' melting phenomenon during biomass compression must be studied more extensively to understand the particle binding mechanism. The Kawakita–Lüdde model helps relate the initial porosity of biomass grinds and the yield strength of compacts. Numerical software, such as Pro/Engineer, PFC3D, and others, can be used to develop a numerical model for the compression and extrusion process during biomass densification. Various densification systems, such as pellet mill, briquette press, and screw extruder, are commonly used for making densified biomass products. The pellet mill is widely used, and pellets produced have higher bulk density than briquettes or logs produced using a screw extruder. The biomass pellets produced using pellet mill are the most commonly used densified products for biopower generation.

References

Adapa, P., Schoenau, G., Tabil, L., Sokhansanj, S., & Singh, A. (2005). Compression of fractionated sun-cured and dehydrated alfalfa chops into cubes: Pressure and density models. *Can. Biosyst. Eng.*, *47*(3), 33–39.

Alderborn, G. & Wikberg, M. (1996). Granule properties. In G. Alderborn & Nystrom, C. (Eds.), *Pharmaceutical Powder Compaction Technology* (pp. 323–373). New York, NY, USA: Marcel Dekker, Inc.

Anglès, M. N., Ferrando, F., Farriol, X., & Salvadó, J. (2001). Suitability of steam exploded residual softwood for the production of binderless panels: Effect of the pre-treatment severity and lignin addition. *Biomass Bioenerg.*, *21*, 211–224.

Comoglu, T. (2007). An overview of compaction equations. *J. Fac. Pharm. Ankara, 36*(2), 123–133.

Cooper, A. R. & Eaton, L. E. (1962). Compaction behavior of several ceramic powders. *J. Am. Ceram. Soc.*, *45*(3), 97–101.

Denny, P. J. (2002). Compaction equations: A comparison of the Heckel and Kawakita equations. *Powder Technol.*, *127*, 162–172.

Ghebre-Sellassie, I. (1989). *Pharmaceutical Pelletization Technology. Drugs and the Pharmaceutical Sciences Series.* New York, NY, USA: M. Dekker.

Granada, E., López González, L. M., Míguez, J. L., & Moran, J. (2002). Fuel lignocellulosic briquettes, die design and products study. *Renew. Energy*, *27*, 561–573.

Heckel, R. W. (1961). An analysis of powder compaction phenomena. *Trans. Met. Soc. AIME*, *221*, 1001–1008.

Johansson, B. & Alderborn, G. (1996). Degree of pellet deformation during compaction and its relationship to the tensile strength of tablets formed of microcrystalline cellulose. *Int. J. Pharm.*, *132*, 207–220.

Johansson, B., Wikberg, M., Ek, R., & Alderborn, G. (1995). Compression behavior and compactability of microcrystalline cellulose compacts in relationship to their pore structure and mechanical properties. *Int. J. Pharm.*, *117*, 57–73.

Jones, W. D. (1960). *Fundamental Principles of Powder Metallurgy.* London, UK: Edward Arnold Publishers Ltd.

Kawakita, K. & Lüdde, K.-H. (1971). Some considerations on powder compression equations. *Powder Technol.*, *4*, 61–68.

Mani, S., Roberge, M., Tabil, L. G., & Sokhansanj, S. (2003a). Modeling of densification of biomass grinds using discrete element method by PFC3D. *CSAE/SCGR 2003 Meeting*, 6–9 July 2003, Montréal, Québec, Canada.

Mani, S., Tabil, L. G., & Sokhansanj, S. (2003b). An overview of compaction of biomass grinds: Powder handling and processing. *Int. J. Storing Handling Process. Powder*, *15*(3), 160–168.

Mani, S., Tabil, L. G., & Sokhansanj, S. (2004). Evaluation of compaction equations applied to four biomass species. *Can. Biosyst. Eng.*, *46*(3), 55–61.

Moreyra, R. & Peleg, M. (1980). Compressive deformation patterns of selected food powders. *J. Food Sci.*, *45*, 864–868.

Moreyra, R. & Peleg, M. (1981). Effect of equilibrium water activity on the bulk properties of selected food powders. *J. Food Sci.*, *46*, 1918–1922.

Panelli, R. & Filho, F. A. (2001). A study of a new phenomenological compacting equation. *Powder Technol.*, *114*, 255–261.

Peleg, M., & Moreyra, R. (1979). Effect of moisture on the stress relaxation of compacted powders. *Powder Technol.*, *23*(2), 277–279.

Pietsch, W. B. (1984). Size enlargement methods and equipment, Part 2: Agglomerate bonding and strength. In M. E. Fayed & L. O. Van Nostrand (Eds.), *Handbook of Powder Science and Technology* (pp. 231–252). New York, NY, USA: Reinhold Co.

Rumpf, H. (1962). The strength of granules and agglomerates. In W. A. Knepper (Ed.), *Agglomeration* (pp. 379–418), New York, NY, USA: Interscience Publishers.

Sastry, K. V. S. & Fuerstenau, D. W. (1973). Mechanisms of agglomerate growth in green pelletization. *Powder Technol.*, *7*, 97–105.

Scoville, E. & Peleg, M. (1981). Evaluation of the effects of liquid bridges on the bulk properties of model powders. *J. Food Sci.*, *46*(1), 174–177.

Shaw, M. D. & Tabil, L. G. (2007). Compression and relaxation characteristics of selected biomass grinds. *2007 American Society of Agricultural and Biological Engineers (ASABE) Annual International Meeting*, Minneapolis, MN, Paper Number 076183, St. Joseph, MI, USA.

Shivanand, P. & Sprockel, O. L. (1992). Compaction behavior of cellulose polymers. *Powder Technol.*, *69*, 177–184.

Sonnergaard, J. (2001). Investigation of a new mathematical model for compression of pharmaceutical powders. *Eur. J. Pharm. Sci.*, *14*(2), 149–157.

Tabil, L. G. & Sokhansanj, S. (1996). Compression and compaction behavior of alfalfa grinds. Part 1: Compression behavior. *Powder Handling Process.*, *8*(1), 7–23.

Tabil, L. G. & Sokhansanj, S. (1997). Bulk properties of alfalfa grind in relation to its compaction characteristics. *Appl. Eng. Agric.*, *13*(4), 499–505.

Tabil, L. G. (1996). Binding and pelleting characteristics of alfalfa. PhD dissertation, Department of Agricultural and Bioresource Engineering, University of Saskatchewan, Saskatoon, Saskatchewan, Canada.

Talebi, S., Tabil, L., Opoku, A., & Shaw, M. (2011). Compression and relaxation properties of timothy hay. *Int. J. Agric. Biol. Eng.*, *4*(3), 69–78.

Tumuluru, J. S. (2014). Effect of process variables on the density and durability of the pellets made from high moisture corn stover. *Biosyst. Eng.*, *119*, 44–57.

Tumuluru, J. S. (2015). High moisture corn stover pelleting in a flat die pellet mill fitted with a 6 mm die: Physical properties and specific energy consumption. *Energy Sci. Eng.*, *3*(4), 327–341.

Tumuluru, J. S. (2016). Specific energy consumption and quality of wood pellets produced using high-moisture lodgepole pine grind in a flat die pellet mill. *Chem. Eng. Res. Des.*, *110*, 82–97. doi: 10.1002/bbb.2121.

Tumuluru, J. S. (2018). Effect of pellet die diameter on density and durability of pellets made from high moisture woody and herbaceous biomass. *Carbon Resour. Convers.* *1*, 44–54. doi: 10.1016/j.crcon.2018.06.002.

Tumuluru, J. S. (2019). Effect of moisture content and hammer mill screen size on the briquetting characteristics of Woody and Herbaceous biomass. *KONA Powder Part. J.*, *36*, 241–251.

Tumuluru, J. S. (2020). Densification process models and optimization. In *Biomass Densification.* Cham: Springer. https://doi.org/10.1007/978-3-030-62888-8_3.

Tumuluru, J. S. & Fillerup, E. (2020). Briquetting characteristics of Woody and Herbaceous biomass blends: Impact on physical properties, chemical composition, and calorific value. *Biofuels, Bioprod. Bioref.*, *14*, 1105–1124. doi: 10.1002/bbb.2121.

Tumuluru, J. S. & Mwamufiya, M. (2021). FCIC DFO — Moisture management and optimization in municipal solid waste feedstock through mechanical processing. Retrieved from https://www.energy.gov/sites/default/files/2021-04/beto-12-peer-review-2021-fcic-tumuluru.pdf (Accessed August 15, 2021).

Tumuluru, J. S., Wright, C. T., Kenney, K. L., & Hess, J. R. (2010). A review on biomass densification technologies for energy applications. [Online]. Tech. Report INL/EXT-10-18420, Idaho National Laboratory, Idaho Falls, Idaho, USA. Available at: https://inldigitallibrary.inl.gov/sites/sti/sti/4559449.pdf (Accessed November 26, 2022).

Tumuluru, J. S., Wright, C. T., Hess, J. R., & Kenney, K. (2011). A review of biomass densification systems to develop uniform feedstock commodities for bioenergy application. *Biofuel. Bioprod. Biorefin. 5*(6), 683–707.

Tumuluru, J. S., Lim, C. J., Bi, X. T., Kuang, X., Melin, S., Yazdanpanah, F., & Sokhansanj, S. (2015). Analysis on Storage Off-Gas Emissions from Woody, Herbaceous, and Torrefied Biomass. *Energies*, *8*, 1745–1759. https://doi.org/10.3390/en8031745.

Tumuluru, J. S., Fillerup, E., Kane, J., & Murray, D. J. (2019). 1.2.1.2 Biomass engineering: Size reduction, drying and densification of high moisture biomass, technology area session: Feedstock supply & logistics, U.S. Department of Energy (DOE) Bioenergy Technologies Office (BETO) 2019 Project Peer Review, March 6th, 2019. Retrieved from https://www.energy.gov/sites/default/files/2019/04/f61/Size%20Reduction%2C%20Drying%20and%20Densification%20of%20High%20Moisture%20Biomass_NL0026654.pdf (Accessed July 13, 2022).

Tumuluru, J. S., Yaney, N. A., & Joshua, J. J. (2021). Pilot scale grinding and briquetting studies on variable moisture content municipal solid waste bales-imapct on physical properties, chemical composition and calorific value. *Waste Manage.*, *125*, 316–327.

Tumuluru, J. S., Rajan, K., Hamilton, C., Pope, C., Rials, T. G., McCord, J., Labbé, N., & André, N. O. (2022). Pilot-Scale pelleting tests on high-moisture pine, switchgrass, and their blends: Impact on pellet physical properties, chemical composition, and heating values. *Front. Energy Res.*, *9*, 788284. doi: 10.3389/fenrg.2021.788284.

Usrey, L. J., Walker, J. T., & Loewer, O. J. (1992). Physical characteristics of rice straw for harvesting simulation. *Trans. ASAE*, *35*(3), 923–930.

van Dam, J. E. G., van den Oever, M. J. A., Teunissen, W., Keijsers, E. R. P., & Peralta, A. G. (2004). Process for production of high density/high performance binderless boards from whole coconut husk. Part 1: Lignin as intrinsic thermosetting binder resin. *Ind. Crops Prod.*, *19*(3), 207–216.

Walker, E. E. (1923). The properties of powders. Part VI: The compressibility of powders. *T. Faraday Soc.*, *19*(1), 73–82.

York, P. & Pilpel, N. (1973). The tensile strength and compression behavior of lactose: Four fatty acids and their mixture in relation to tableting. *J. Pharm. Pharmacol.*, *25*, 1–11.

https://doi.org/10.1142/9781800613799_0003

Chapter 3

Densification Characteristics of Raw and Pretreated Herbaceous Biomasses and Their Blends

Stefan Frodeson
Environmental and Energy Systems
Department of Engineering and Chemical Science
Karlstad University, Karlstad, Sweden
stefan.frodeson@kau.se

Abstract

Different biomaterials have different chemical compositions and since the densification properties are strongly related to feedstock, it is important to increase knowledge about biomass relationship to densification properties. The purpose of this chapter is to give a brief introduction to the development of the plant kingdom, the chemical composition of biomasses, and how different components can affect the densification characteristics. A study where 11 different pure substances (cellulose, hemicelluloses, lignin, etc.), added to pine and beech, are pelletized in a single pellet press and results are presented showing that polysaccharides can play an important role when biomasses are densified as single sources or blends solutions in a densification process.

Keywords: Raw biomass, pretreated biomass, pure substance pelleting, cellulose, hemicellulose, lignin

3.1. Background

It has become clear that it is necessary to significantly reduce the use of fossil resources and increase the use of renewable resources. Thus, it is also necessary to increase the utilization of all kinds of biomasses, including waste-based biomasses (EC, 2012) which for biofuel pellets should lead to increased intake of different types of woods and herbs as well as increased intake of waste products. However, to increase the use of a wider range of biomasses for pelletizing, it is important to understand how the chemical compositions of different biomasses relate to pelletizing properties. This chapter offers a brief introduction to the chemical composition of biomasses and how these substances can affect the pelletizing characteristics of wood and herbaceous materials. There is a paucity of knowledge on the ways in which chemical composition affects pelletizing properties (Rumpf, 1962; Mani *et al.*, 2003; Holm *et al.*, 2006; Mani *et al.*, 2006; Kaliyan and Morey, 2009; Kaliyan and Morey, 2010; Stelte *et al.*, 2012b; Poddar *et al.*, 2014; Ramírez-Gómez, 2016), and it is hoped that this chapter leads to further research in this field so that a greater variety of biomass feedstocks can be used in pellet production.

This paucity of knowledge about how chemical composition affects pelletizing properties means that pellet producers strive for feedstocks with chemical compositions that are as uniform as possible, which works well in countries with a wealth of raw materials such as spruce or pine. However, in areas or countries where there is a wider range of raw materials, such as tropical areas, this leads to a greater risk of production problems during pellet production. These are countries where increased use of fuel pellets has potential to reduce both health problems and environmental effects. For example, in Zambia, the energy sources used for cooking are mainly firewood and charcoal (Kamanula *et al.*, 2010; Janssen and Rutz, 2012; Matakala *et al.*, 2015; World Energy, 2016), which has resulted in Zambia being in the top 10 list of countries with the high deforestation rate (Matakala *et al.*, 2015). Energy utilization is very low (Pennise, 2001; Kisiangani and Masters, 2011; Shane *et al.*, 2016; World Energy, 2016), and poisonous gases associated with conventional charcoal cooking cause millions of premature deaths every year (WHO, 2016). A solution could be the increased use of pellets since it is possible to cook food and generate heat from pellet stoves specially designed for cooking (Bhattacharya and Abdulsalam, 2002; Peša, 2017; Souza *et al.*, 2017). However, pellets must be produced, and since these countries have a wider range of raw materials as feedstock solutions, increased knowledge about how different biomasses and their chemical compositions affect pelleting characteristics is of importance. The goal in these environments is pellet production units where pellets can be based on local crops, harvested throughout the year, residues from small sawmills as well as wastes from forest resources and various invasive species. An increased understanding of how chemical components in the biomass affect pelletizing properties would also lead to less production problems related to variations in the raw material as well as wider operating properties for pellet production in general.

3.2. Chemical Composition in Wood and Herbaceous Materials

The most important phyla in the plant kingdom are shown in Fig. 3.1, which is presented as a simplified evolutionary tree. The first plants were algae, which evolved in the sea. Mosses, hornworts, and liverworts, which do not have root systems and absorb water through their entire surface, evolved from algae during the colonization of land 470 million years ago. The next level of plant life, ferns and horsetails, is a more advanced group of vascular plants that contain the lignin which allowed for the creation of a newer cell type called the tracheid. The tracheid is a long cell type that allows both water transport and increased plant strength. The tracheid's root system allows for larger plants and growth under drier conditions. There are two main types of vascular plants. The first type is herbs. Herbs are either non-woody plants in soil that generally die after one or a few seasons or woody plants, including bushes and trees, which permanently grow in soil and have stiff stems and branches. Herbs are not necessarily small. In tropical areas, there are herbs that can be over 10 m in height, which is larger than many trees. However, herbs and woody plants are not evolutionary groups. Notable is that almost all families of vascular plants contain both herbs and woody plants, and it can also be difficult to identify a plant as either a herb or woody plant. The second and largest group of vascular plants are

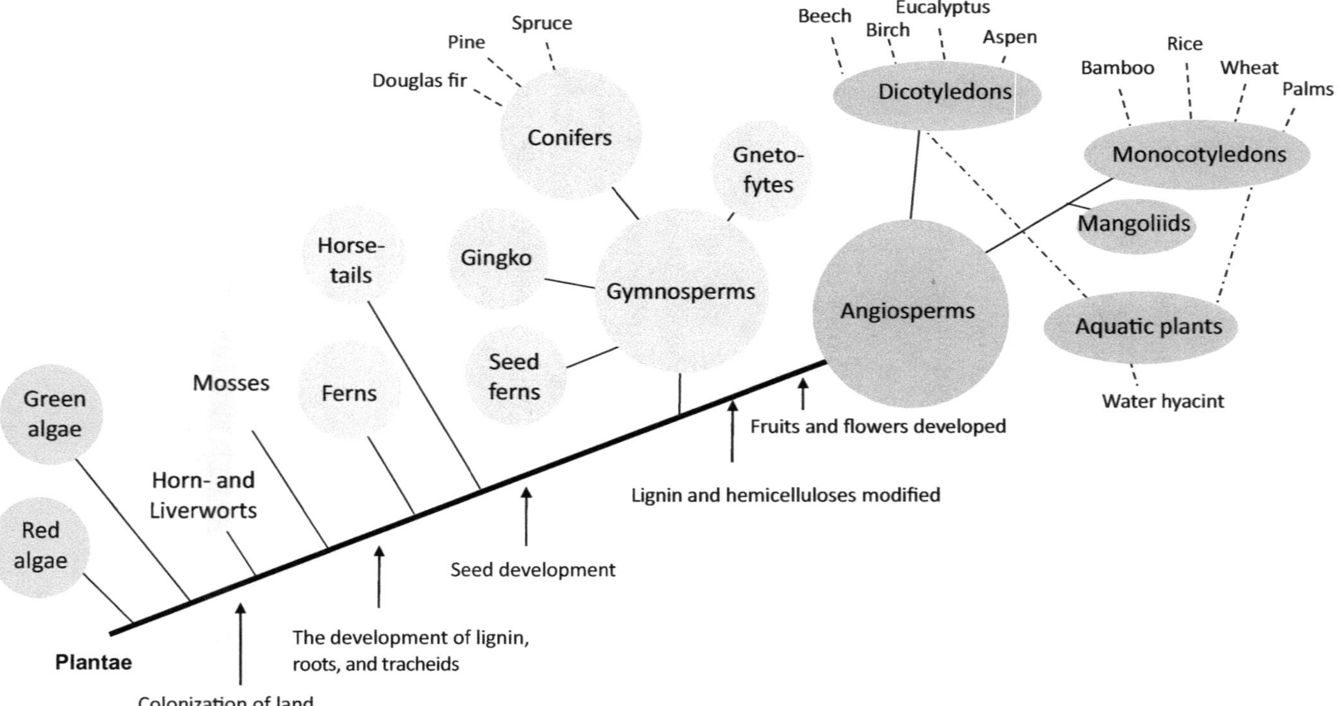

Figure 3.1. Simplified evolution tree of the plant kingdom, showing the diversity of various biomass species (reproduced from Ek *et al.*, 2009; Rabemanolontsoa and Saka, 2013).

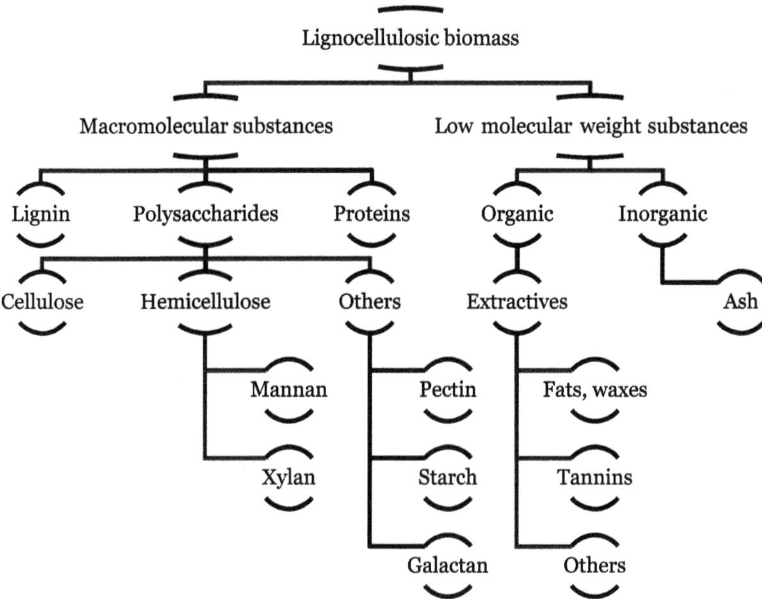

Figure 3.2. General structure scheme for lignocellulosic biomass substances (Fengel and Wegener, 1989; Frodeson *et al.*, 2018).

the angiosperms, which represent approximately 80% of all living green plants, including all food plants.

Biomasses, therefore, have evolved over a long period of time and, thus, have developed a variety of heterogeneous biological materials. One way to describe this heterogeneity is by looking at chemical composition. In general, lignocellulosic biomasses can be divided into two broad categories: macromolecular and low molecular weight substances (see Fig. 3.2). Macromolecular substances include lignin, polysaccharides,[1] and proteins. In addition, polysaccharides can be categorized as cellulose, hemicellulose — mainly glucomannan and xylan — and other polysaccharides, such as pectin, starch, and galactan. Low molecular weight substances include organic

[1]Polysaccharides are polymeric carbohydrate molecules composed of long chains of monosaccharide units bound together.

substances or extractives such as fat, waxes, and tannin substances, as well as inorganic substances such as ashes (see Fig. 3.2).

The largest group of substances within the biomasses are polysaccharides. Polysaccharides can be categorized as those that are more or less branched, with the former being more flexible and the latter being stiffer. Polysaccharides with more side chains bound to long chains of monosaccharides can be more branched or flexible, whereas those with fewer side chains are stiffer, which correlates with them having fewer possibilities to rotate (Fengel and Wegener, 1989).

The most common polysaccharide substances in nature are cellulose. Cellulose has a simple structure, which consists of repeating -(1, 4) linked glucose units with two glucose units being repeated. Together, these units form a long, linear, unbranched chain. The degree of polymerization (number of glucose units) is very high, and values of 15,000 are found (Ek *et al.*, 2009).

The second largest group of polysaccharides are the hemicelluloses. The hemicelluloses represent about 20–35% of lignocellulosic biomass, whereby xylan is likely the most abundant hemicellulose (Saha, 2003). Often, hemicelluloses are discussed as one substance; however, they are in fact a group of substances mostly based on glucomannan and xylan. Galactan is also sometimes included within the hemicelluloses (Barsett *et al.*, 2005), but within this chapter, it is separated from the most common ones and viewed as a substance related to other polysaccharides. However, although the hemicelluloses are found in the matrix between cellulose fibrils in the cell wall and are often associated with cellulose, they have different compositions. While cellulose only contains glucose units, hemicelluloses have a range of different sugar monomers. For instance, besides glucose, sugar monomers found in hemicelluloses can include the pentoses (the five-carbon sugars xylose and arabinose) and hexoses (the six-carbon sugars mannose and galactose, and the six-carbon deoxy sugar rhamnose) (Ek *et al.*, 2009). This means that different types of hemicelluloses exist depending on plant material and type of tissue, such as xylan, glucuronoxylan, arabinoxylan, glucomannan, and xyloglucan, and the differences between these components. For example, xylan has more flexible chains bound to its main chain compared to glucomannan (Berglund *et al.*, 2016). It is likely that the hemicelluloses contribute to the mechanical properties of the cell wall (Ek *et al.*, 2009) and influence the moisture equilibrium of the living tree. The water storage properties of macromolecules in the cell can be grouped as pectin > hemicellulose > cellulose > lignin (Ek *et al.*, 2009).

The third group of polysaccharides is referred to as the "others" and includes substances such as pectin, starches, and galactan. Pectins are a diverse group of irregular and amorphous polysaccharides with a highly variable chemical composition. The total amount of pectin present in wood is normally low, but when it is found, it is present in both the central lamella and the primary layer of the fiber cell and contributes to the mechanical properties of the cell wall (Ek *et al.*, 2009). Pectin is a polysaccharide that is rather branched and flexible in character (Fengel and Wegener, 1989). Starch is a polymeric carbohydrate consisting of numerous glucose units joined by glycosidic bonds, produced by most green plants for its ability to be stored as energy. Starches consist of two types of molecules: the stiffer amylose and the more flexible amylopectin (Fengel and Wegener, 1989). Amylopectin has α-1,4-linked glucose chains joined by α-1,6-linked branch points, while amylose is composed of long α-1,4-linked linear chains with rare α-1,6-linked branch points (Seung, 2020). Depending on the plant, starches generally contain less than 35% amylose by weight (Seung, 2020).

The principal chain in galactan-containing polysaccharides consists of 1–3, 1–4, 1–6, and α- and β-bonded D-galactopyranoses. Galactans have been found in at least 30 plant families, where arabinogalactans are common galactan-containing polysaccharides. Arabinogalactans have been categorized into two types: arabinogalactan I and II (Arifkhodzhaev, 2000). Arabinogalactan occurs in large amounts in larch wood and heartwood and can include 10–25% by dry weight, while the amount is generally less than 1% in other softwoods (Ek *et al.*, 2009). Larch arabinogalactan is water soluble, has a high degree of branching, and, thus, is flexible in character (Fengel and Wegener, 1989) and has a very low viscosity in water (Ek *et al.*, 2009). In Fig. 3.3, four polysaccharides are presented

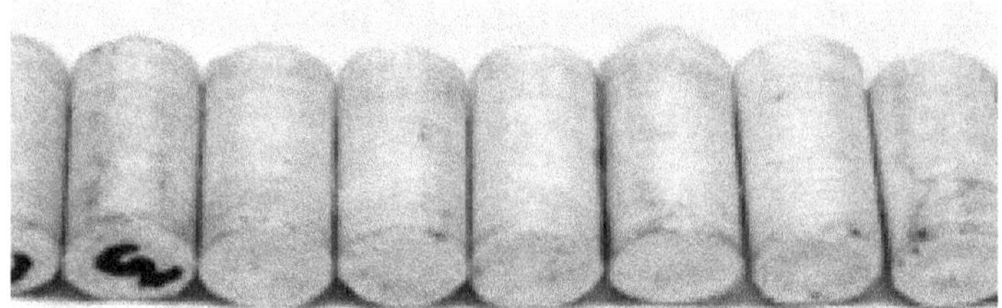

Cellulose (Avicel)

Mannan (Galactomannan)

Xylan (Rice xylan)

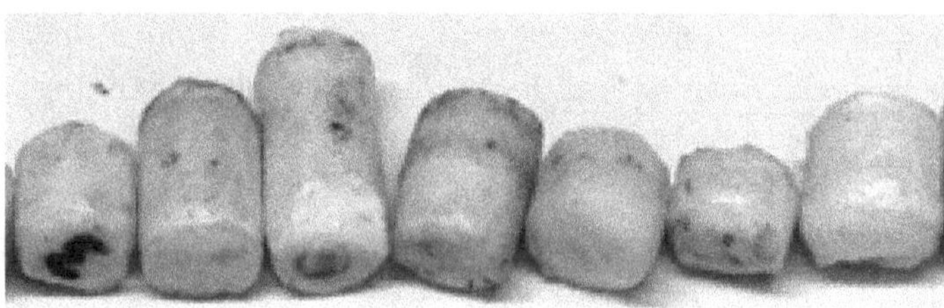

Galactan (Larch arbinogalactan)

Figure 3.3. Examples of four different polysaccharides pellets, cellulose, mannan, xylan, and galactan, pelletized in a single pellet press (Frodeson *et al.*, 2018).

as examples of pellets produced from cellulose, mannan, xylan, and galactan.

Probably the most discussed substances within pellet production are lignins, which are the substances with the most complex structures. Lignin has at least four important roles in plants: it (i) gives stiffness to the cell walls, (ii) makes the cell wall hydrophobic, (iii) glues cells together, and (iv) gives protection against the microbial degradation of wood (Ek *et al.*, 2009). Lignin is a heterogeneous polymer derived from cross-link lignols in diverse ways. These cross-link lignols have been categorized into three main types: coniferyl alcohol, sinapyl alcohol, and paracoumaryl alcohol, where hardwoods are rich in coniferyl and sinapyl units, softwoods are rich in coniferyl units, and grasses are rich in coniferyl and sinapyl units (Boerjan *et al.*, 2003). The amount of lignin in biomass is variable (from 15% to 40%) in wood, while herbaceous plants are less lignified (Fengel and Wegener, 1989).

The group of extractives covers many different low molecular weight substances and are commonly categorized as extractives being soluble in either neutral solvents, such as water, or liquids with low polarity. Water-soluble extractives include sugars, lignans, and other phenolic compounds that are extractable from wood with various neutral solvents (Ek *et al.*, 2009). Extractives that are soluble in liquids of low polarity (the standard method is to use acetone), often referred to as wood resins, can be divided into four classes: (i) fats and fatty acids, (ii) steryl esters and sterols, (iii) terpenoids, including terpenes and polyisoprenes, and (iv) waxes, that is, fatty alcohols and their esters with fatty acids (Ek *et al.*, 2009). The content of extractives differs significantly between different types of biomasses and concentrations can vary from 1–2% to 40% (Roffael, 2016).

The final group of substances are those that are inorganic, namely ashes. The ash content in wood plant fibers is low, at around only a few percent (Ek *et al.*, 2009); however, the variation can be wide with the exception of herbaceous biomasses (Bakker and Elbersen, 2005). The exact amount and composition may vary considerably, even within the same species, and are related to the soil in which the plant grows (Bakker and Elbersen, 2005; Ek *et al.*, 2009). This inorganic material consists mainly of different metal salts, as well as some nitrogen and phosphorus, where calcium ions are the most common inorganic compound (Ek *et al.*, 2009).

Examples of biomass species and their chemical composition are presented in Table 3.1, where it can also be seen how a specific species can vary between its different parts (e.g., stem and branches), as is the case with both cassava and corn.

3.3. Pelleting Characteristics Versus Chemical Composition

There are several parameters that affect pelletability in biomass and, in general, these parameters can be defined by the feedstock species and technical parameters. The actual densification process that occurs inside a pelletizer, independent of whether it is a flat- or ring-die pelletizer, begins when the feedstock is being compressed by the roller wheels. One way to describe this process is by dividing it into three stages (Nielsen, 2009): (i) The compression stage, where the roller compresses the feedstock into a thin layer; (ii) the flow stage, where the compressed layer under pressure flows into the die channels and is partly compressed further from the sides in the cone; and (iii) the friction stage, where the compressed feedstock is further transported through the die channel. As the pressure increases during the compression and flow stages, particles are forced against each other while undergoing elastic and plastic deformation; this increases the inter-particle contact area and, as a result, binding forces come into effect (Mani *et al.*, 2004). Therefore, most of the increase in density has occurred before the particles reach the friction stage of the channel. However, the die is drilled with holes, and each of these holes is composed of a conical entrance (mainly), an active part of the press channel, and an inactive part with a large diameter (relief) (Nielsen, 2009). The active part of the press channel, often refereed as the presslength together with the height of the conical entrance, gives the total pressway, which generates a backpressure that is sufficient for the rollers to create pressure. The friction generates the appropriate die temperature, and both the correct backpressure

Table 3.1. Chemical composition of selected biomass species (%) (Ek *et al.*, 2009; Vassilev *et al.*, 2010; Pooja and Padmaja, 2015).

	Polysaccharides								
		Hemicelluloses							
Species	Cellulose	Total	Glucomannan	Glucuroxylan	Others	Proteins	Lignin	Extractives	Residuals
Green algae (*Chladophora*)	5.0	34.9	—	—	7.3	11.1	0.0	0.2	41.4
Norway spruce (*Picea abies*)	41.7	24.9	16.3	8.6	3.4	—	27.4	1.7	0.9
Scotch pine (*Pinus sylvestris*)	40.0	24.9	16.0	8.9	3.6	—	27.7	3.5	0.3
Cassava stem	22.8	28.8	—	—	15.0	3.7	22.1	—	1.9
Cassava peels	14.2	23.4	—	—	29.8	5.3	10.9	—	3.7
Cassava leaves	17.3	27.6	—	—	2.4	1.2	20.1	—	2.5
Rice straw	34.5	21.8	—	—	0.9	4.7	20.2	4.5	13.3
Corn cob	34.3	32.8	—	—	1.9	5.8	18.0	2.8	3.5
Corn leaves	26.8	24.8	—	—	0.3	16.5	15.1	5.1	10.9
Bamboo	39.4	31.1	—	—	1.1	1.8	20.6	3.8	1.2
Birch (*Betula verrucose*)	41.0	29.8	2.3	27.5	2.6	—	22.0	3.2	1.4
Beech (*Fagus sylvatica*)	39.4	29.1	1.3	27.8	4.2	—	24.8	1.2	1.3
River red gum (*Eucalyptus calmaldulensis*)	45.0	17.2	3.1	14.1	2.0	—	31.3	2.8	1.7

and die temperature are necessary for the production of high-quality pellets (Nielsen *et al.*, 2009). Therefore, from a technical perspective, it is important to understand how different biomass species affect both compression energy and friction, especially if a more variated feedstock solution is to be used.

So, the exact knowledge regarding the actions that the chemical components generate within the die is limited, and the reason for this is that it is difficult to clarify because in their natural context, they are all combined and strongly connected to each other. However, if more knowledge is found about each component's role during pelletizing, the possibilities to predict the pelleting properties of different biomaterials based on its chemical composition would increase and thus its ability to mix and contribute to the pelletizing process. One way to study substances within the chemical composition is to pelletize these as pure substances, after being extracted from biomaterials through chemical processing. The author has conducted several studies (Frodeson *et al.*, 2018; Frodeson, 2019; Frodeson *et al.*, 2019a, 2019b; Frodeson *et al.*, 2021) with substances that originate from biomasses, with the purpose of studying the behavior of during densification, either as pure substances or mixed with wood. Some of the conclusions, made in this chapter, are thus based on studies on a total of twenty-one different pure substances, tested in a single pellet press to evaluate their influence on the densification process. It is of course different to pelletize extracted pure substances versus biomasses in its origin and it must be remembered that these substances may have been chemically affected during the chemical process of extracting them; however, they are used in other research fields to reflect their properties in a biomass (Abbaszadeh *et al.*, 2016; Bi *et al.*, 2016; Kulasinski *et al.*, 2016; Kudahettige-Nilsson *et al.*, 2018).

In a study, done by the author, 11 different substances at an amount of 10% (d.b.) were mixed with pine and beech and pelletized in a single pellet press at a moisture content of 10% (w.b.), the die temperature was set to 100°C (Frodeson *et al.*, 2021). These 11 different substances were as follows: seven polysaccharides (Avicel/cellulose, Locust bean gum mannan, Eucalypt xylan, larch

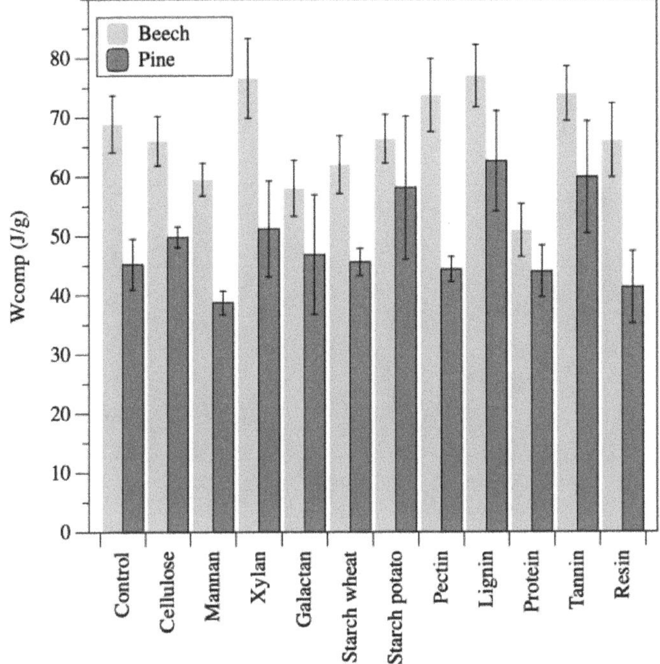

Figure 3.4. Compression work when 10% of 11 different substances are added to beech and pine and pelletized in a single pellet press at approximately 10% MC and at a die temperature of 100°C.

arbinogalactan, potato and wheat starch, and pectin), one lignin (Lignoboost), one protein (Soy), and two extractives (wood resin and tannin).

Within Fig. 3.4, the compression work is presented as J/g when the pellets are produced within a single pellet press. It is clear that adding mannan to pine and beech decreases the energy for compression while adding xylan shows the opposite. It is also clear that lignin and tannin increase the energy needed for compression. Compression is one way to evaluate the pelletizing properties; another is backpressure or F_{max}, which represents the highest force required to set a pellet in motion after the compression step.

Within Fig. 3.5, F_{max} together with hardness are presented, where hardness represents different substances' possibilities to affect the bondings within the pellet. These results, together with other

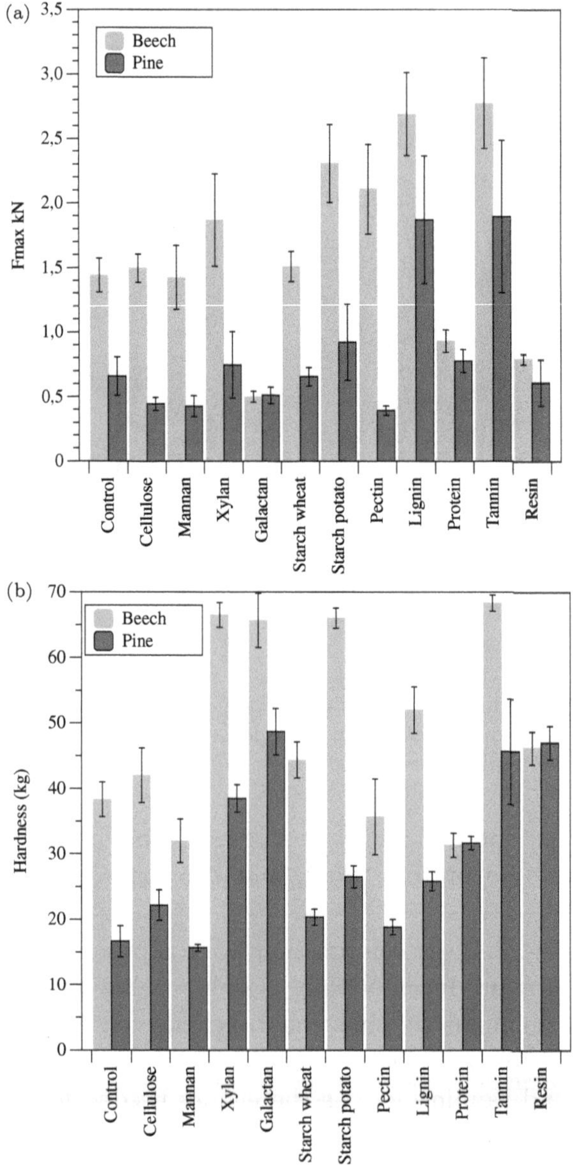

Figure 3.5. Results showing how F_{max} (a) and hardness (b) are affected when 10% of 11 different substances are added to beech and pine and pelletized in a single pellet press at approximately 10% MC and at a die temperature of 100°C. Results are published in (Frodeson *et al.*, 2021) and presented with permission.

research studies, are used as a basis for a brief introduction to how different components affect and can affect the pelletizing properties of different biomaterials.

It is clear that there is a wide difference between how types of substances affect the pelletizing properties. As shown in Figs. 3.4 and 3.5, the most common component within the biomasses of the polysaccharide, cellulose, is not affecting the process very much. The proportion of cellulose in wood is rarely discussed in pellet manufacturing. One reason for this is that cellulose, unlike other components, does not reach its softening temperature during pelletization. It is known that the softening temperature of cellulose is unchanged by water sorption, and both wet and dry celluloses have glass transition temperatures above 220°C (Goring, 2018), which is well above the temperature that occurs within a pellet press. It has also been shown that cellulose is not so dependent on high pressure and rather independent of moisture during pelletization (Frodeson *et al.*, 2019a). The fact that pure cellulose has good bonding properties is well known from other industrial applications, such as the paper industry, and pelletizing pure cellulose generates a high quality of cellulose pellets. Cellulose pelletizing does occur within applications other than fuel pellet production. However, pelletizing pure cellulose has its difficulties and the author's experience is that cellulose powder is easier to pelletize compared to fibers, when operation properties are found, and pure cellulose pellets are also very durable and hard. However, the effect of cellulose on the pelleting process of wood or herbs is probably low and, as seen in Fig. 3.5, adding 10% cellulose powder to wood does not significantly affect the backpressure or the hardness, which is probably the reason why it is rarely discussed in research contexts.

The hemicelluloses seem to affect the pelletization process even more. Stelte *et al.* (2011a) assumed that the glass transition temperature for hemicelluloses in wood is around room temperature but stated that further studies are required to prove this (Stelte *et al.*, 2012a). In many research studies, the hemicelluloses are often named as one parameter and not divided into separated substances. Based on experience from pelletizing five different hemicelluloses, two mannans and three xylans, is it clear that xylans generate more die friction, increase energy for compression, are more sensitive to moisture, and contribute to strong bondings within the pellets. Mannan does the opposite in all parameters, decreases compression need, generates low friction, is rather independent of water, and has a low impact on bonding properties (see Figs. 3.4 and 3.5). Larsson *et al.* (2021) tested to model and predict the pelletizing properties based on the chemical composition of biomasses. Even if their study did not identify a difference between the hemicelluloses mannan and xylan, they found that these hemicelluloses are important predictors for target moisture content in a feedstock (Larsson *et al.*, 2021). However, it is clear that xylan becomes fluent and changes its structure during the physical conditions that occur under pelletization. None of the studies on pure substances (Frodeson *et al.*, 2018; Frodeson, 2019; Frodeson *et al.*, 2019a, 2019b) have shown that mannan has this behavior. It has been shown that xylan changes its softening temperature with increased moisture, and Goring (2018) verified that birch xylan at 23% moisture had a softening temperature of 54°C (Goring, 2018). The difference between xylan and mannan can explain why hardwood, with more xylan in the hemicellulose, has been shown to generate pellets with higher durability in comparison to softwood (Agar *et al.*, 2018).

Within the group of other polysaccharides, a common substance used in pellet production is starch, mostly used as an additive (Kuokkanen *et al.*, 2011; Tarasov *et al.*, 2013; Larsson *et al.*, 2015). Even if starch is a commonly used additive, there is still a paucity of knowledge regarding exactly how it affects the densification process. Normally, these additive starches are waste products from other industries and are not primarily designed to be optimized for the mechanical conditions of pelleting. These waste products work well, and several studies have shown that even adding only a small amount of starch increases durability and decreases the number of fines (Kuokkanen *et al.*, 2011; Ståhl *et al.*, 2012; Tarasov *et al.*, 2013; Samuelsson *et al.*, 2014; Larsson *et al.*, 2015; Ståhl *et al.*, 2019). However, in relation to inlet power supply and bulk density, the results are not clear as to whether starch is the most important parameter or if there are other factors, such as moisture content, that affect the results more. As shown in Fig. 3.5, there is a difference between potato and wheat starch in relation to both friction and bonding properties. Therefore, even if the quality of the pellets increases by using starches, there are variations between types of starch that should be investigated further. For example, the difference between the stiffer amylose and the more flexible amylopectin (Fengel and Wegener, 1989) and their roles during pelletizing has not been studied and should be clarified. However, although starch has shown potential to improve pelletizing properties, Fig. 3.5 shows that there are other substances that have better properties. An interesting component that has shown effective pelleting properties is galactan, and as Fig. 3.5 shows, arbinogalactan from larch trees both increases hardness and has a lubricating effect, and at the same time, galactan seems to decrease the energy need for compression meanings that it has good potential to booth increase pellet quality and at the same time decrease the need of energy for production (see Fig. 3.4).

When it comes to proteins and pectin, it is difficult to state their roles based on the current study. As Fig. 3.5 shows, it is not possible to state whether they lubricate within the die or strengthen the bonding properties. However, studies on pure substances have shown that both galactan and protein become fluent during the pelletizing process and that these properties are important for the creation of strong bonds.

In relation to pelletizing research, the most discussed component is lignin, which has been identified as the most important binding agent in pellets (Whittaker and Shield, 2017) based on its ability to become fluent and create solid bridges when its glass transition temperature has been reached (Kaliyan and Morey, 2010; Wilson, 2010; Stelte *et al.*, 2011a and b). It has been found that, if the pelletizing temperature is below the glass transition temperature, the polymer chains are anchored to each other by van der Waals forces, hydrogen bonds, or other attractive forces (Goring, 2018). When the glass transition temperature is approached, the chains acquire sufficient energy for large-scale movement with respect to each other (Goring, 2018) and stronger bondings are created. In relation to lignin, studies have found that pellets formed above the glass transition temperature have significantly higher compression strength, greater unit density, and expand less in length after pelletization (Stelte *et al.*, 2012a; Whittaker and Shield, 2017). However, dry lignin's glass transition temperature varied from 127–193°C and it softened at 72–128°C, which is higher than the softening temperature of birch xylan that was 54°C at a moisture content of 23% and 40–50°C for waxes (Stelte *et al.*, 2011a). In an earlier study by Frodeson *et al.* (2018), it was shown that xylan, galactan, tannin, and resin all became fluent during pelletization and that xylan, galactan, and tannin created hard pellets after cooling (Frodeson *et al.*, 2018). However, waxes are so fluent that it is not possible to pelletize them as pure substances at ordinary die temperatures of around 100°C (Frodeson *et al.*, 2018). As shown in Fig. 3.5, xylan, galactan, tannin, and resin generate hard pellets, and the increase in hardness is much higher than with lignin. It is well known that extractives affect pellet properties but do not always create strong bondings. The extractives affect pelletability in two ways. The first effect is that high amounts of oleophilic low molecular weight hydrocarbon solutions, such as fatty acids, oil, and waxes, decrease pellet strength (Finell *et al.*, 2009; Nielsen *et al.*, 2010; Stelte *et al.*, 2011a, 2012a). Possible explanations are that either the extractives block and consequently reduce the number of binding sites on the surface (Back Ernst, 1987; Stehr and Johansson, 2000; Samuelsson *et al.*, 2009) or that the inter-particle binding is limited to weak van der Waals interactions and fiber interlocking (Stelte *et al.*, 2012a). The other effect is that extractives act as a lubricating agent inside the die channels when they migrate to the pellet surface at elevated temperatures (Stelte *et al.*, 2012a; Castellano *et al.*, 2015), leading to decreased friction and the current effect (Samuelsson *et al.*, 2009; Filbakk *et al.*, 2011; Stelte *et al.*, 2012a). However, as shown in Fig. 3.5, both tannin and resin increase bonding capacity while tannin increases friction and resin follows the lubricating theory named above.

Therefore, even though it is uncertain that any of these substances play a greater role than lignin, it can be stated that there are many substances in the chemical composition of biomasses that have significant effects on pelleting properties. This means that future research needs to be clearer in reporting chemical composition when new feedstocks are being tested.

3.4. Summary

The purpose of this chapter was to give a brief introduction to the chemical composition of biomasses and how these different components can affect the pelleting characteristics of wood and herbaceous materials. It is difficult to identify exactly what happens both inside the die and within a mixture of these components as they are found in nature, and there is a paucity of knowledge within these areas.

One important parameter is where different types of substances are located within the biomass structure. If we study the cell walls in wood fiber, lignin is the main substance within the middle lamella, while xylan is bound within a matrix of cellulose found in the secondary cell walls. The fact that xylan is bonded deeper within the cell structure means that its potential to reach the particle surface is limited in comparison to lignin. Therefore, lignin's potential to generate strong bonds is probably higher because it is more easily accessible than, e.g., xylan. Nevertheless, this knowledge can explain some studies correlated to particle sizes and grinding effects on pelletability. It has been seen that an increase in particle size significantly decreases both attractive forces and mechanical durability (Kaliyan and Morey, 2009; Dhamodaran and Afzal, 2012; Castellano *et al.*, 2015) and that grinding leads to fractured surfaces that end up in the fiber wall and expose cellulose and hemicellulose (Näslund, 2003). Thus, if a biomass contains a high amount of xylan, it is likely that its properties to form strong bonds will increase if they are ground harder in comparison with materials containing large amounts of the glucomannan in hemicellulose. This knowledge increases the understanding of biomasses that would be ground into small particles and, thus, a more energy-efficient production process can be designed.

With greater knowledge about how chemical composition affects pellet properties in wood and herbaceous materials, the pellet industry can be more flexible in adding new raw materials without risks to production, which in turn will increase the use of more waste products and increase the availability of reliable biomasses.

References

Abbaszadeh, A., MacNaughtan, W., Sworn, G., & Foster, T. J. (2016). New insights into xanthan synergistic interactions with konjac glucomannan: A novel interaction mechanism proposal. *Carbohydr. Polym. 144*, 168–177.

Agar, D. A., Rudolfsson, M., Kalén, G., Campargue, M., Da Silva Perez, D., & Larsson, S. H. (2018). A systematic study of ring-die pellet production from forest and agricultural biomass. *Fuel Process. Technol.*, *180*, 47–55.

Arifkhodzhaev, A. O. (2000). Galactans and galactan-containing polysaccharides of higher plants. *Chem. Nat. Compd.*, *36*(3), 229–244.

Back Ernst, L. (1987). The bonding mechanism in hardboard manufacture review report. *Holzforschung — Int. J. Biol. Chem. Phys. Technol. Wood*, *41*, 247.

Bakker, R. R., & Elbersen, H. W. (2005, October). Managing ash content and quality in herbaceous biomass: an analysis from plant to product. In *14th European Biomass Conference* (Vol. 17, p. 21).

Barsett, H., Ebringerová, A., Harding, S. E., Heinze, T., Hromádková, Z., Muzzarelli, C., Muzzraelli, R. A. A., Paulsen, B. S., & ElSeoud, O. A. (2005). *Polysaccharides I: Structure, Characterisation and Use.* Berlin, Heidelberg New York: Springer Science & Business Media.

Berglund, J., Angles d'Ortoli, T., Vilaplana, F., Widmalm, G., Bergenstråhle-Wohlert, M., Lawoko, M., Henriksson, G., Lindström, M., & Wohlert, J. (2016). A molecular dynamics study of the effect of glycosidic linkage type in the hemicellulose backbone on the molecular chain flexibility. *Plant J.*, *88*(1), 56–70.

Bhattacharya, S. & Abdulsalam, P. (2002). Low greenhouse gas biomass options for cooking in the developing countries. *Biomass Bioenergy*, *22*(4), 305–317.

Bi, R., Berglund, J., Vilaplana, F., McKee, L. S., & Henriksson, G. (2016). The degree of acetylation affects the microbial degradability of mannans. *Polym. Degrad. Stab.*, *133*, 36–46.

Boerjan, W., Ralph, J., & Baucher, M. (2003). Lignin biosynthesis. *Ann. Rev. Plant Biol.*, *54*(1), 519–546.

Castellano, J. M., Gómez, M., Fernández, M., Esteban, L. S., & Carrasco, J. E. (2015). Study on the effects of raw materials composition and pelletization conditions on the quality and properties of pellets obtained from different woody and non woody biomasses. *Fuel*, *139*, 629–636.

Dhamodaran, A. & Afzal, M. T. (2012). Compression and springback properties of hardwood and softwood pellets. *BioResources*, *7*(3), 4362–4376.

EC (2012). *Innovating for Sustainable Growth: A Bioeconomy for Europe.* Brussels, Belgium: European Commission.

Ek, M., Gellerstedt, G., & Henriksson, G. (2009). *Wood Chemistry and Biotechnology.* Berlin: Walter de Gruyter.

Fengel, D., & Wegener, G. (1989). *Wood: Chemistry, Ultrastructure, Reactions*, 2nd edn. Berlin, New York: Walter de Gruyter.

Filbakk, T., Skjevrak, G., Høibø, O., Dibdiakova, J., & Jirjis, R. (2011). The influence of storage and drying methods for Scots pine raw material on mechanical pellet properties and production parameters. *Fuel Process. Technol.*, *92*(5), 871–878.

Finell, M., Arshadi, M., Gref, R., Scherzer, T., Knolle, W., & Lestander, T. (2009). Laboratory-scale production of biofuel pellets from electron beam treated Scots pine (*Pinus silvestris* L.) sawdust. *Radiat. Phys. Chem.*, *78*(4), 281–287.

Frodeson, S. (2019). Towards understanding the pelletizing process of biomass: Perspectives on energy efficiency and pelletability of pure substances. PhD, Karlstads universitet.

Frodeson, S., Anukam, A. I., Berghel, J., Ståhl, M., Lasanthi Kudahettige Nilsson, R., Henriksson, G., & Bosede Aladejana, E. (2021). Densification of wood — Influence on mechanical and chemical properties when 11 naturally occurring substances in wood are mixed with beech and pine. *Energies*, *14*(18), 5895.

Frodeson, S., Henriksson, G., & Berghel, J. (2018). Pelletizing pure biomass substances to investigate the mechanical properties and bonding mechanisms. *BioResources*, *13*(1), 1202–1222.

Frodeson, S., Henriksson, G., & Berghel, J. (2019a). "Effects of moisture content during densification of biomass pellets, focusing on polysaccharide substances. *Biomass Bioenergy*, *122*, 322–330.

Frodeson, S., Lindén, P., Henriksson, G., & Berghel, J. (2019b). Compression of biomass substances — A study on springback effects and color formation in pellet manufacture. *Appl. Sci.*, *9*(20), 4302.

Goring, D. A. I. (2018). Thermal softening, adhesive properties and glass transitions in lignin, hemicellulose and cellulose. In Consolidation of the Paper Web, Trans. of the IIIrd Fund. Res. Symp. Cambridge, 1965, FRC, Manchester, 2018.

Holm, J. K., Henriksen, U. B., Hustad, J. E., & Sørensen, L. H. (2006). Toward an understanding of controlling parameters in softwood and hardwood pellets production. *Energy Fuels*, *20*(6), 2686–2694.

Janssen, E. B. R. & Rutz, D. (2012). *Bioenergy for Sustainable Development in Africa*. London, New York: Springer Verlag.

Kaliyan, N. & Morey, R. V. (2009). Factors affecting strength and durability of densified biomass products. *Biomass Bioenergy*, *33*(3), 337–359.

Kaliyan, N. & Morey, R. V. (2010). Natural binders and solid bridge type binding mechanisms in briquettes and pellets made from corn stover and switchgrass. *Bioresour. Technol.*, *101*(3), 1082–1090.

Kamanula, J., Sileshi, G. W., Belmain, S. R., Sola, P., Mvumi, B. M., Nyirenda, G. K. C., Nyirenda, S. P., & Stevenson, P. C. (2010). Farmers' insect pest management practices and pesticidal plant use in the protection of stored maize and beans in Southern Africa. *Int. J. Pest Manage.*, *57*(1), 41–49.

Kisiangani, E. & Masters, L. (2011). *Natural Resources Governance in Southern Africa*. Africa Institute of South Africa: EBSCOhost.

Kudahettige-Nilsson, R. L., Ullsten, H., & Henriksson, G. (2018). Plastic composites made from glycerol, citric acid, and forest components. *BioResources, 13*(3), 6600–6612.

Kulasinski, K., Salmén, L., Derome, D., & Carmeliet, J. (2016). Moisture adsorption of glucomannan and xylan hemicelluloses. *Cellulose, 23*(3), 1629–1637.

Kuokkanen, M. J., Vilppo, T., Kuokkanen, T., Stoor, T., & Niinimäki, J. (2011). Additives in wood pellet production — A pilot-scale study of binding agent usage. *BioResources, 6*(4), 4331–4355.

Larsson, S., Agar, D., Rudolfsson, M., Da Silva Perez, D., Campargue, M., Kalen, G., & Thyrel, M. (2021). Using macromolecular composition to predict optimal process settings in ring-die biomass pellet production. *Fuel, 283*, 9.

Larsson, S., Lockneus, O., Xiong, S., & Samuelsson, R. (2015). Cassava stem powder as an additive in biomass fuel pellet production. *Energy Fuels, 29*(9), 5902–5908.

Mani, S., Tabil, L., & Sokhansanj, S. (2003). Compaction of biomass grinds-an overview of compaction of biomass grinds. *Powder Handling Process., 15*(3), 160–168.

Mani, S., Tabil, L. G., & Sokhansanj, S. (2004). Evaluation of compaction equations applied to four mass species. *Can. Biosyst. Eng., 46*(3), 55–61.

Mani, S., Tabil, L. G., & Sokhansanj, S. (2006). Effects of compressive force, particle size and moisture content on mechanical properties of biomass pellets from grasses. *Biomass Bioenergy, 30*(7), 648–654.

Matakala, P. W., Kokwe, M., & Statz, J. (2015). Zambia national strategy to reduce emissions from deforestation and forest degradation (REDD+). *Forestry Department, Ministry of Lands Natural Resources and Environmental Protection, FAO, UNDP, and UNEP, Government of the Republic of Zambia, Zambia.*

Näslund, M. (2003). *Teknik och Råvaror för Ökad Produktion av Bränslepellets* (*Technology and Raw Materials for Increasing Pellet Production*; Energidalen i Sollefteå AB: Sollefteå, Sweden. Swedish, with summary in English). There is no corresponding record for this reference.

Nielsen, N. P. K. (2009). Importance of raw material properties in wood pellet production — Effects of differences in wood properties for the energy requirements of pelletizing and the pellet quality. PhD PhD, University of Copenhagen.

Nielsen, N. P. K., Gardner, D. J., & Felby, C. (2010). Effect of extractives and storage on the pelletizing process of sawdust. *Fuel, 89*(1), 94–98.

Nielsen, N. P. K., Gardner, D. J., Poulsen, T., & Felby, C. (2009). Importance of temperature, moisture content, and species for the conversion process of wood residues into fuel pellets. *Wood Fiber Sci., 41*(4), 414.

Pennise, D. M., Smith, K. R., Kithinji, J. P., Rezende, M. E., Raad, T. J., Zhang, J., & Fan, C. (2001). Emissions of greenhouse gases and other airborne pollutants from charcoal making in Kenya and Brazil. *Journal of Geophysical Research: Atmospheres, 106*(D20), 24143–24155.

Peša, I. (2017). Sawdust pellets, micro gasifying cook stoves and charcoal in urban Zambia: Understanding the value chain dynamics of improved cook stove initiatives. *Sustainable Energy Technol. Assess.*, *22*, 171–176.

Poddar, S., Kamruzzaman, M., Sujan, S. M. A., Hossain, M., Jamal, M. S., Gafur, M. A., & Khanam, M. (2014). Effect of compression pressure on lignocellulosic biomass pellet to improve fuel properties: Higher heating value. *Fuel*, *131*, 43–48.

Pooja, N. S. & Padmaja, G. (2015). Enhancing the enzymatic saccharification of agricultural and processing residues of cassava through pretreatment techniques. *Waste Biomass Valorization*, *6*(3), 303–315.

Rabemanolontsoa, H. & Saka, S. (2013). Comparative study on chemical composition of various biomass species. *RSC Adv.*, *3*(12), 3946–3956.

Ramírez-Gómez, Á. (2016). Research needs on biomass characterization to prevent handling problems and hazards in industry. *Part. Sci. Technol.*, *34*(4), 432–441.

Roffael, E. (2016). Significance of wood extractives for wood bonding. *Appl. Microbiol. Biotechnol.*, *100*(4), 1589–1596.

Rumpf, H. (1962). The strength of granules and agglomerates. In *Agglomeration-Proceedings of the First International Symposium on Agglomeration, Philadelphia*, pp. 379–418.

Saha, B. C. (2003). Hemicellulose bioconversion. *J. Ind. Microbiol. Biotechnol.*, *30*(5), 279–291.

Samuelsson, R., Finell, M., Arshadi, M., Hedman, B., & Subirana, J. (2014). Inblandning av stärkelse och lignosulfonat i pellets vid Bioenergi i Luleå AB. Swedish University of Agricultural Sciences, Umeå, Report, 29, pp. 1–18.

Samuelsson, R., Thyrel, M., Sjöström, M., & Lestander, T. A. (2009). Effect of biomaterial characteristics on pelletizing properties and biofuel pellet quality. *Fuel Process. Technol.*, *90*(9), 1129–1134.

Seung, D. (2020). Amylose in starch: Towards an understanding of biosynthesis, structure and function. *New Phytol.* *228*(5), 1490–1504.

Shane, A., Gheewala, S. H., Fungtammasan, B., Silalertruksa, T., Bonnet, S., & Phiri, S. (2016). Bioenergy resource assessment for Zambia. *Renewable and Sustainable Energy Rev.*, *53*, 93–104.

Souza, G. M., Ballester, M. V. R., de Brito Cruz, C. H., Chum, H., Dale, B., Dale, V. H., Fernandes, E. C. M., Foust, T., Karp, A., Lynd, L., Maciel Filho, R., Milanez, A., Nigro, F., Osseweijer, P., Verdade, L. M., Victoria, R. L., & Van der Wielen, L. (2017). The role of bioenergy in a climate-changing world. *Environ. Dev.*, *23*, 57–64.

Ståhl, M., Berghel, J., Frodeson, S., Granström, K., & Renström, R. (2012). Effects on pellet properties and energy use when starch is added in the wood-fuel pelletizing process. *Energy Fuels*, *26*(3), 1937–1945.

Ståhl, M., Frodeson, S., Berghel, J., & Olsson, S. (2019). Using secondary pea starch in full-scale wood fuel pellet production decreases the use of steam conditioning. World Sustainable Energy Days/European Pellet Conference. Wels, 27 February to 1 March.

Stehr, M. & Johansson, I. (2000). Weak boundary layers on wood surfaces. *J. Adhes. Sci. Technol.*, *14*(10), 1211–1224.

Stelte, W., Clemons, C., Holm, J. K., Ahrenfeldt, J., Henriksen, U. B., & Sanadi, A. R. (2011a). Thermal transitions of the amorphous polymers in wheat straw. *Ind. Crops Prod.*, *34*(1), 1053–1056.

Stelte, W., Clemons, C., Holm, J. K., Ahrenfeldt, J., Henriksen, U. B., & Sanadi, A. R. (2012a). Fuel pellets from wheat straw: The effect of lignin glass transition and surface waxes on pelletizing properties. *BioEnergy Res.*, *5*(2), 450–458.

Stelte, W., Holm, J. K., Sanadi, A. R., Barsberg, S., Ahrenfeldt, J., & Henriksen, U. B. (2011b). Fuel pellets from biomass: The importance of the pelletizing pressure and its dependency on the processing conditions. *Fuel*, *90*(11), 3285–3290.

Stelte, W., Holm, J. K., Sanadi, A. R., Barsberg, S., Ahrenfeldt, J., & Henriksen, U. B. (2011c). A study of bonding and failure mechanisms in fuel pellets from different biomass resources. *Biomass Bioenergy*, *35*(2), 910–918.

Stelte, W., Sanadi, A. R., Shang, L., Holm, J. K., Ahrenfeldt, J. & Henriksen, U. B. (2012b). Recent developments in biomass pelletization — A review. *BioResources*, *7*(3), 4451–4490.

Tarasov, D., Shahi, C., & Leitch, M. (2013). Effect of additives on wood pellet physical and thermal characteristics: A review. *ISRN Forestry*, *2013*, 6.

Vassilev, S. V., Baxter, D., Andersen, L. K., & Vassileva, C. G. (2010). An overview of the chemical composition of biomass. *Fuel*, *89*(5), 913–933.

Whittaker, C. & Shield, I. (2017). Factors affecting wood, energy grass and straw pellet durability — A review. *Renewable Sustainable Energy Rev.*, *71*, 1–11.

WHO (2016). *Burning Opportunity: Clean Household Energy for Health, Sustainable Development, and Wellbeing of Women and Children.* Geneva, Switzerland: World Health Organization.

Wilson, T. O. (2010). *Factors Affecting Wood Pellet Durability.* Pennsylvania: The Pennsylvania State University.

World Energy (2016). *World Energy Resources.* London: World Energy Council.

https://doi.org/10.1142/9781800613799_0004

Chapter 4

Blending and Densification: Significance and Quality

Stefan Frodeson*,‡ and Jaya Shankar Tumuluru†,§

**Environmental and Energy Systems*
Department of Engineering and Chemical Science
Karlstad University, Karlstad, Sweden
†Southwestern Cotton Ginning Research Laboratory
United States Department of Agriculture
Agricultural Research Service
Las Cruces, New Mexico, USA
‡stefan.frodeson@kau.se
§jayashankar.tumuluru@usda.gov

Abstract

Future pellet plants will need to be more flexible regarding the need to vary the feedstock species and increase the areas for biomass. By blending biomass species, pellets can be a solution that meets new desirable specifications for different conversion pathways, such as biochemical, thermochemical, and biopower. In this chapter, the authors address opportunities for blending and densification to homogenize the chemical composition of biomaterials and improve the potential for other processing and transformation methods. The advantages of blending biomasses are many: increasing the potential supply of biomass, creating new markets, reducing raw material costs, and improving biomass flow and pelletizing properties. Important densification parameters to keep in mind are highlighted, and five different studies are summarized; these include investigations from a single pellet press study to briquette applications. Some conclusions are that blending woody and herbaceous biomasses yields products with uniform product quality in terms of physical properties and chemical composition. Furthermore, blending pine and switchgrass yields a compressed product with higher density and durability while reducing energy consumption.

Keywords: Biomass blends, feedstock cost, co-pelleting, co-briquetting, blend quality, press channel length, blend pellet and briquette quality

4.1. Why Develop Blends for Future Pellet Production?

This chapter discusses feedstock solutions related to blending with one main question: *How can the biofuel densification industry secure a broad raw material base in a future circular bioeconomy society?* In the future bioeconomy, both demand and the economical transport of biomaterials will increase; the need to develop new products and applications will also increase. By densifying woody and herbaceous biomass blends, we can improve their physical properties and chemical compositions (Yancey *et al.*, 2013). In addition, by blending, we can improve fuel pellets with new feedstock solutions (Henriksson *et al.*, 2019), reduce the effect of moisture during pelletization (Tumuluru *et al.*, 2011), and maintain the nutrient content of feed by blending the ingredients of biomass sources (Tumuluru *et al.*, 2011), as well as improve the feedstock specifications (Edmunds *et al.*, 2018) and overcome limitations in cost and quality (Ray *et al.*, 2017). As such, the advantages of biomass blending are as follows: (1) an increase in the potential biomass supply, (2) feedstock cost reduction, and (3) improvement in biomass flow and pelleting characteristics (Yancey *et al.*, 2013; Tumuluru, 2021; Crawford *et al.*, 2015). However, perhaps the most important parameter in this blending process is the potential to overcome the challenges associated with feedstock quality, variability, supply, and cost (Ray *et al.*, 2017; Kenney *et al.*, 2013; Thompson *et al.*, 2013). These challenges include variabilities in the physical properties of biomass in terms of particle size, moisture, and density, in addition to the chemical properties (Tumuluru *et al.*, 2011; Frodeson *et al.*, 2018; Frodeson *et al.*, 2021).

Thus, learning more about blending and co-pelletizing is important to improve feedstock solutions, decrease cost, and increase

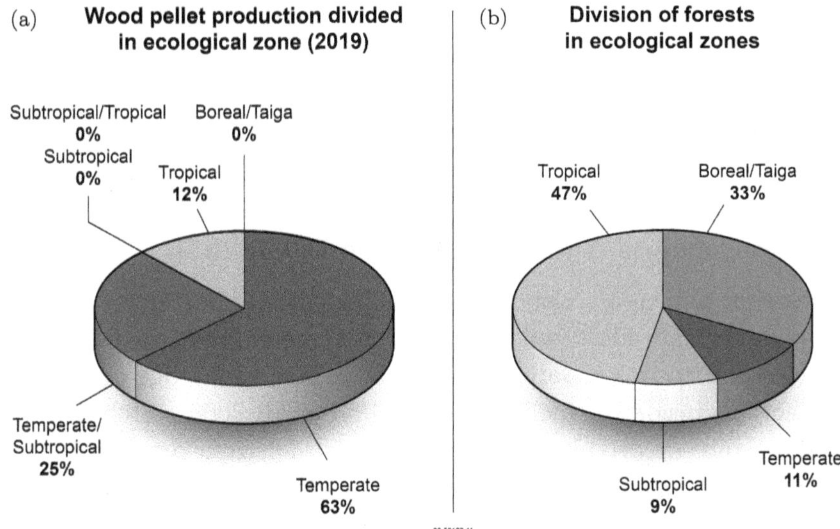

Figure 4.1. (a) World fuel pellet production from 2019, generally divided by ecological zones; (b) variation on the total mix of forests, divided by ecological zones.

quality. Beyond that, if fuel pellet producers seek to increase and expand into new markets and play an even more important role in the transition to more sustainable energy production, a variation in feedstock species is necessary. World fuel pellet production in 2019 was approximately 41 Mt (FAO, 2021). Suppose we divide production from 2019 from a perspective of ecological zones. In that case, we can state that over 60% of today's fuel pellet production comes from pure temperate forests — around 25% of which are located in various temperate and subtropical zones, as shown in Fig. 4.1. Only 12% of pellet production is located in pure tropical forests, meaning that most knowledge regarding industrial fuel pelletization is related to temperate forests — specifically coniferous and deciduous forests. Forests with relatively small tree species variations are much more suitable for forestry, where large quantities of specific raw materials are harvested. However, suppose pellets for biofuels are to play an even greater role in the transition from fossil-based energy to a renewable energy solution. In that case, the pellet industry must utilize biomasses and wastes from all types of ecological zones, of which tropical forests make up 50% of all available forests globally, compared to 11% for temperate forests. Since tropical forests provide a wide range of biodiversity, the further expansion of pellet production in countries located in tropical zones must involve an increased variability of biomass species through blending.

4.2. Parameters Affecting Blending Possibilities

If we want to increase blending possibilities, it is important to look at parameters that affect pelletability — parameters that the feedstock species and technical setup can define. Exact knowledge regarding the actions inside the die is limited (Mani *et al.*, 2006a; Stelte *et al.*, 2012a), but the combination of feedstock species, pretreatments, densification steps, and post-treatments directly affects the result. When the right conditions occur, strong bonds are created between the particles in the pellets, even if the exact nature between these bonds is not known with any certainty (Mani *et al.*, 2006a; Rumpf, 1962; Mani *et al.*, 2003; Kaliyan and Morey, 2009, 2010; Stelte *et al.*, 2012a; Poddar *et al.*, 2014; Ramírez-Gómez, 2016; Holm *et al.*, 2006; Anukam *et al.*, 2021). Explanations for what occurs are derived from the agglomeration field (Stelte *et al.*, 2012a; Rumpf, 1962; Kaliyan and Morey, 2009; Križan, 2015; Pietsch, 1991), which categorizes the binding mechanisms into (1) solid bridges; (2) adhesion and cohesion forces; (3) surface tension and capillary pressure; (4) attraction forces; and (5) interlocking bonds (Križan, 2015; Rumpf, 1962). However, even if these explanations help us understand what takes place inside the die, these theories are difficult to apply to increase new biomasses and blends in relation to the production parameters.

If we examine the technique behind pelletization, there are several specific parameters such as die dimension (e.g., diameter, cone design, and active press-channel length (PCL)), die or roller wheel speed, die temperature, and pressure (e.g., the gap between the roller and die), which are all important parameters to study when new material and blends are tested and studied. The densification process can be divided into three subprocesses (Nielsen, 2009): (1) the compression stage; (2) the flow stage; and (3) the friction stage, where the friction is important for two reasons: to generate the right temperature and create backpressure sufficient to generate pressure from the rollers (Nielsen, 2009; Frodeson, 2019). Both the correct backpressure and die temperature are necessary for the production of high-quality pellets (Nielsen, 2009), where the die temperature often reaches from 100°C to 130°C (Nielsen, 2009) and the pressure reaches between 210 MPa and 450 MPa (Nielsen, 2009; Seki *et al.*, 2013). This is the case when the parameters within the die channel are optimized. As such, the die channel plays an active role where friction occurs, as well as an inactive role that is necessary to ensure the mechanical strength of the die. For pellet production, the die ratio (L/D) is mentioned as a data point for pellet production, where D is the diameter and L represents the PCL. However, it must be remembered that two different lengths are involved. In addition, the pressway is an important parameter for pellet production, including the cone height beyond the PCL. Lavergne *et al.* (2021) tested 10 different raw materials with different moisture content values and die designs related to PCL. They found that increased PCL positively affected all feedstock assortments related to bulk density, while the effect on durability varied for Scots pine bark, poplar, and wheat straw. No significant impact was found between increased PCL and durability, but increased PCL had a distinctive correlation with other biomasses (Lavergne *et al.*, 2021). As such, die design is important and is chosen based on a specific material or mixture. This leads to a limitation in possibilities to blend and variate since pellet producers must strive for a feedstock with as small a chemical variation as possible. However, as mentioned above, this is not a limitation for many pellet mills because they are often located within a temperate forest next to a sawmill, which focuses on a specific species. However, suppose the pellet industry seeks to increase production with new raw materials. In that case, they need to include seasonal biomasses, establish themselves in more tropical parts of the world, or truly strive to limit the cost of feedstocks. This limitation can lead to high transportation and raw materials costs. Therefore, by blending, it is possible to increase the number of species in the feedstock; however, it is important to identify how added, and blended materials affect the biomass flow within the die and the parameters that affect these relationships.

One important parameter is the moisture content, and different species have different levels of optimal moisture during pelletizing — generally, 6–12% in woody biomass (Filbakk *et al.*, 2011; Obernberger and Thek, 2004; Li *et al.*, 2015; Nguyen *et al.*, 2015; Huang *et al.*, 2017; Serrano *et al.*, 2011) and up to 20% in agricultural biomasses such as barley (Serrano *et al.*, 2011). The freshness of the feedstock is also a factor — fresh feedstock needs less moisture content in comparison with stored feedstock (Filbakk *et al.*, 2011; Arshadi *et al.*, 2008). Thus, water affects pelletization and is one important parameter when blends are chosen; however, the effects of water can be quite ambiguous. It is also important to remember that the pellet industry mostly uses wet raw material as feedstock, which must be dried before pelletization. The drying process is associated with major costs, and approximately 25% of the total pellet production cost can be related to this process (Monterio *et al.*, 2012; Mani *et al.*, 2006b; Obernberger and Thek, 2004; Uasuf and Becker, 2011). Therefore, if pellet production through blending can be achieved at a higher moisture content, there is great operational and economic potential. However, in relation to the bonding properties of pellets, water on surfaces can act like bridges between the particle sites and cause cohesive forces that increase binding strength (Kaliyan and Morey, 2009), creating binding forces due to capillary pressure that disappear once the water evaporates (Rumpf, 1962; Samuelsson *et al.*, 2012) and act as a plasticizer and increase the molecular mobility of the amorphous polymers, which increases the flowability of hemicelluloses and extractives with a low glass transition temperature (Stelte *et al.*, 2011a, 2011b; Puig-Arnavat, 2016). Thus, it is sometimes difficult to find the optimal moisture content, and the operating window related to water can be rather small since both low and high moisture content will negatively affect the pellet (Stelte *et al.*, 2012a; Nielsen, 2009; Filbakk *et al.*, 2011; Stelte *et al.*, 2011a; Puig-Arnavat, 2016; Back, 1987; Carone *et al.*, 2011; Rhén, 2005). Since the incompressibility of water affects the agglomeration of the particles (Rumpf, 1962), excessive internal pressure from steam is generated inside the pellets (Filbakk *et al.*, 2011; Larsson *et al.*, 2008), and water is absorbed by the hydrogen bonds, leading to occupied locations of particle–particle binding, which therefore reduces the number of hydrogen bonds (Nielsen, 2009; Back, 1987; Larsson *et al.*, 2008). It is also clear that the amount of water affects the friction (Samuelsson *et al.*, 2012) since water acts as a lubricating media. Therefore, as optimal moisture content is an important parameter that gives a rather small operating window, blending can be a solution to expand the window; however, it is important to clarify the types of biomasses that have a wide range of acceptable moisture content.

In addition, particles and their sizes are often mentioned as one parameter affecting pelletability, but the recommendation is not clear. Studies have found that particle sizes that are often below 5 mm (Stelte *et al.*, 2012a) should vary depending on the diameter of the pellets — around 4 mm for a 6 mm pellet (Thek and Obernberger, 2010) — or that those below 8 mm are probably not necessary (Bergström, 2008). However, increases in particle size significantly decrease the attractive forces (Kaliyan and Morey, 2009; Dhamodaran and Afzal, 2012). In addition, since smaller milling screen sizes positively affect mechanical durability (Castellano *et al.*, 2015), one important effect related to particle sizes and grinding is that it can lead to fractured surfaces in the fibers and exposed substances such as carbohydrates and lignin. As such, a small particle size leads to increased friction between the particles and the die channel (Stelte *et al.*, 2011b; Castellano *et al.*, 2015; Mani *et al.*, 2006a) and thus, increased die temperature.

Press ratio (L/D), moisture content, and particle size are technique and pretreatment solutions' parameters. When we pelletize lignocellulosic biomasses, which are mainly composed of carbohydrate polymers (e.g., cellulose and hemicellulose) and aromatic polymers (e.g., lignin), the biomass can be divided into two main types: (1) *herbs*, which are the plants that grow over soil that are non-woody and often wither and die after one or two seasons; and (2) *woody plants*, including bushes and trees with stiff stems, branches, and permanent over soil growth (Ek *et al.*, 2009). From a pelletizing perspective, the "extractives" are important. The proportion of extractives varies greatly, not only between different biomasses but also within a specific species, such as different parts of a tree. Furthermore, the term "extractive" covers many other low molecular weight organic substances (Frodeson, 2019). Therefore, the chemical composition is varied. If different biomasses are blended, it is important to understand how the composition will affect bonding properties and friction within the die.

Lignin has been identified as the most important binding agent in pellets (Whittaker and Shield, 2017) based on its possibility to become fluent and create solid bridges when its glass transition temperature has been reached (Kaliyan and Morey, 2010; Wilson, 2010; Stelte *et al.*, 2011c; Stelte *et al.*, 2012b). However, the glass transition temperature for waxes is lower than that for lignin, at approximately 40–50°C (Stelte *et al.*, 2011a), meaning that waxes flow and create a coating layer more quickly than lignin (Stelte *et al.*, 2012b). In addition, hemicelluloses — especially xylan and galactan — have been shown to become fluent at temperatures far below those at which lignin will. They also generate a strong bond (Frodeson *et al.*, 2018; Frodeson *et al.*, 2019a), while glucomannan in the hemicellulose seems to be inhibiting (Frodeson *et al.*, 2019b). Beyond affecting the bonding properties, the relationship between the different components that contribute to friction maybe even more important. One important parameter that is affected by a lubricating media is the need for energy during densification. A lubricating media decreases the need for energy; however, it also affects backpressure, which means that it will also impact the bulk density and lower the die temperature. Extractives are well known as components that act as lubricants (Castellano *et al.*, 2015; Stelte *et al.*, 2011c); however, it is important to keep in mind that the term "extractives" includes a number of different substances and lubrication media (Frodeson *et al.*, 2021). Therefore, it is important to understand how different species affect the densification part of the pelletization process for a successful blending solution.

4.3. Difficulties with Storage of Biomass Blends

A pelletized product can be used in various contexts; in addition to energy, the application areas are raw materials for biorefinery, such as feeding into extruders. This means that a blend of different biomass raw materials helps meet the desired specifications for different conversion pathways, such as biochemical, thermochemical, and biopower. The blending of these biomasses, therefore, has the potential to homogenize their chemical compositions and improve downstream processing and conversion. For example, agricultural biomass with a high ash content has variable physical properties (e.g., particle size) and a variable chemical composition (e.g., proximate and ultimate composition, as well as ash content) can be blended with woody biomass in different ratios to meet the desired physical properties, chemical composition, ash content, and ash speciation. There are studies indicating that the blending of biomass feedstocks helps improve both the thermochemical and thermomechanical conversion. Blending also helps reduce feedstock costs. Ray *et al.* (2017) research indicated that blending of various woody, herbaceous, and municipal solid waste biomasses helps overcome the cost and quality limitations of using lignocellulosic biomass for biofuels production. A study conducted by Edmunds *et al.* (2018) on blending pine with switchgrass indicated that the feedstock specifications are improved for thermochemical conversion. Another study by Yancey *et al.* (2013) on the pre-processing of various woody and herbaceous biomass blends indicated that the variability in physical properties and chemical composition is reduced. A study by Mahadevan *et al.* (2016) using the blends of switchgrass and southern pinewood resulted in bio-oils with low acidity and viscosity.

Despite these advantages, blending biomass has numerous challenges, such as feeding, handling, storage, and transportation issues. Some of these challenges are as follows: (a) variability in particle size, shape, distribution, bulk, and particle density, as well as the rheological properties, which create particle entrainment and classification resulting in feeding challenges; (b) variability in particle size and bulk density, resulting in particle segregation; and (c) variability and low density, creating feeding and handling challenges resulting in the plugging of conveyors, bridging of the drop chutes, and other conversion inefficiencies. These challenges can be overcome by densification of biomass using a pellet mill, briquette press, screw extruder, roller press, or agglomerator (Tumuluru *et al.*, 2011). The major advantage of densification using these systems is that it creates a commodity-type product with uniform size, shape, and higher bulk density, typically three to five times higher than the raw biomass. These densified products have improved handling and conveyance efficiency in bioenergy infeed supply systems. In addition, the densified products have better control over the particle size distribution of the product stream for improved feedstock uniformity and density, thereby enhancing the deconstruction of the structural components of the biomass and improving the biochemical and thermochemical conversions. With increased density, biomass transportation costs can be decreased significantly.

4.4. Research Needed to Expand the Use of Blends

Several papers and reviews related to biomass blends have been published. What is important is that these results are comparable, and a recently conducted study by Berghel *et al.* (2022) has shown that it is difficult to compare results from single pellet press studies (Berghel *et al.*, 2022). Therefore, research to increase blends or additives should verify densification properties related to the die, such as the press length, energy used for compression, or back pressure, to evaluate the degree of pelletability. Today, most of the ongoing worldwide pellet research is applied based on studies of different pellet presses and associated equipment. These tests are generally produced in the full-scale application, bench scales, or single pellet presses.

Research conducted in full-scale production can utilize large amounts of material under stable operating conditions, which create meaningful opportunities for tests with specific dies or varied preconditions. However, one drawback is that production might have to be stopped while tests are conducted, meaning a loss in production income. In addition, switching the die to test new blend solutions and materials is also difficult. Research conducted in a laboratory bench press or pilot press (10–500 kg/h) resembles that of full-scale production. The advantage of bench-scale production is that less material must be tested and that tests on blends and new materials are rather easily conducted and provide a simpler understanding of the operating properties, such as optimal moisture content and press length. Maintaining more control over the data and operating parameters, including testing with different press ratios, is also easier. However, it is difficult to study the parameters within the die, even at the laboratory scale. As such, researchers must use test apparatus that are even smaller in size, referred to as single pellet presses (SPPs). SPPs are used in research laboratories worldwide since they are rather easy and cheap to study pelletability and material-oriented issues. SPPs are also easy to control where parameters such as compression energy and backpressure can be studied. One primary drawback, however, is that almost all SPP studies are based on batches. Since industrial production is a continuous process, it is difficult to transfer the SPP results to full-scale production. However, there is much that can be learned in SPP production from both a material and process point of view — especially when it comes to new material and blending solutions. Independent of the scale of tests, the purpose of testing new material and/or blends is to discover how this new material can be used from a pelleting perspective.

4.5. Pellet Production via Blending Can Make a Difference

Blending related to pelletization can thus have different purposes, ranging from homogenizing materials to broadening raw material bases for pellet producers. In this chapter, we will highlight the potential that pellet production can offer to move toward a more sustainable society.

According to Janssen and Rutz (2012), the biomass in Africa is used mainly as firewood and charcoal. For example, in Lusaka, Zambia, 85% of urban households use charcoal for food cooking, as observed in Fig. 4.2. The estimated consumption of charcoal per household per year is 1.3 t, the production of which requires approximately 8 t of wood (Gumbo *et al.*, 2013). During charcoal production, 70–90% of the energy is lost in primitive charcoal piles

Figure 4.2. Picture from a charcoal market in Lusaka (courtesy of Jonas Berghel).

(Janssen and Rutz, 2012; Pennise, 2001). Beyond this high energy loss, the teardown of wood for charcoal also leads to deforestation, which in Zambia is around 270,000 hectares per year — one of the 10 highest deforestation rates in the world (Matakala *et al.*, 2015). Beyond that, the combustion of charcoal often occurs in stoves with an efficiency rate of approximately 12–27% (Bhattacharya and Abdulsalam, 2002). Indoor air pollution generated from particulate matter and carbon monoxide associated with these conventional cooking methods causes 4.5 million premature deaths every year (World Health Organization, 2016). The increasing population and energy needs do not make the situation sustainable.

A sustainable solution is possible when the cooking method neither uses charcoal nor generates poisonous gases, as seen in Fig. 4.3. Today, cooking food and generating heat from pellet stoves designed for cooking is possible. Pellet cooking stoves are capable of optimal

Figure 4.3. Picture of a pellet cooking stove that can replace charcoal stoves and generate a more sustainable food cooking solution in Africa (courtesy of Jonas Berghel).

combustion (Bhattacharya and Abdulsalam, 2002; Peša, 2017; Souza *et al.*, 2017); however, even if the pellet cooking stove can solve the problems associated with charcoal, pellets must be available, and the pellet production should be based on available waste and biomasses not used today. As such, blends and blending possibilities, such as feedstocks related to sustainable pellet production in Africa, have become the best solution.

4.6. Examples of Blending Studies in Different Scales

In this section, we will highlight different studies to show some examples of how blends can be evaluated and related to studies on different scales, from gram level to full-scale activities, as well as different purposes. We also show how results can be achieved and how experimental results can be evaluated to blend.

4.6.1. *Study 1: A Way of Thinking Around Blending Related to a Study on a Single Pellet Press*

Within an SPP study, 12 waste biomasses, typically available in Zambia, were used as raw material in a pellet production facility (Henriksson *et al.*, 2019). These 12 biomasses tested were as follows: (a) bamboo, (b) cassava peel, (c) cassava stem, (d) eucalyptus, (e) gliricidia, (f) peanut shell, (g) lantana camara, (h) miombo seed capsules, (i) pigeon pea, (j) pine, (k) sicklebush, and (l) tephrosia. One way to find a good blend is to find a mixture of materials with similar pelletizing properties. The press ratio (L/D) is impossible to evaluate within the most common SPP studies since the pellets are pressed batch-wise and not within a continuous possibility. In the following, therefore, an example is presented of how different raw materials and their possibilities can be tested and analyzed using compressed energy and backpressure. Materials with similar compression and friction energy ought to have related behavior in a full production pellet mill and thus could be used with the same press length. As shown in Fig. 4.4, miombo (Miom), peanut shell (PeaS), pigeon pea (PigP), and sicklebush (Sick) have similar compression and friction energies at a certain MC and can be an effective starting point for a blending mix. A producer could probably use these materials as single raw materials or in a blend, without modifying the process conditions. The knowledge of suitable press length in pellet production allows for economic and time-efficient production when using these biomasses.

Furthermore, two interesting groups of materials can be identified to function as combinations of raw material bases to broaden raw material availability and create sustainable pellet production: Group 1, those consisting of equal pellet abilities; and Group 2, the independent water group. Within Group 1, all the materials possessing equivalent pellet properties, such as Miom, PeaS, PigP, and Sick, are found. All these materials should have the ability to be used as a varied raw material stream, independent of the amount of each material, provided particle sizes are equal. The advantage of this group is that the same die press length could be used. Group 2 requires better control of the mixtures of raw materials. With this

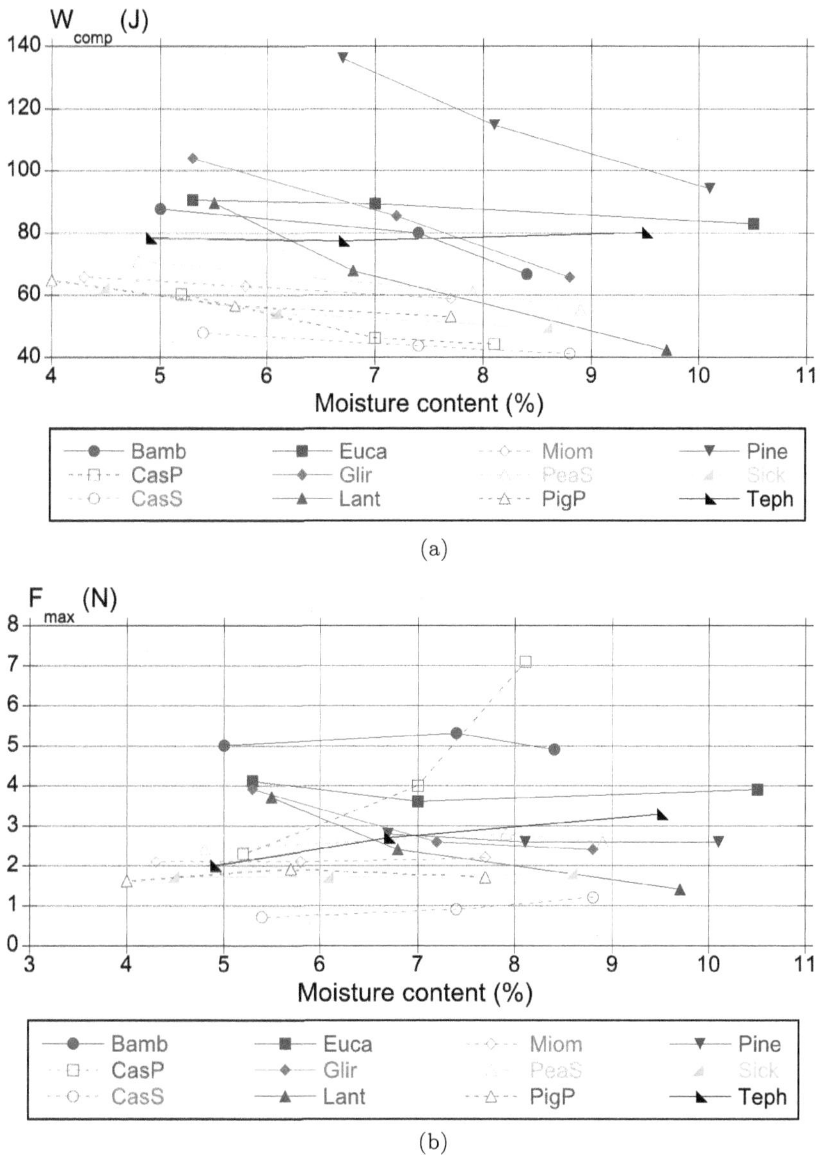

Figure 4.4. Compression work (a) and maximal force (b) needed for 12 different biomasses from Zambia pelletized in a single pellet press. Results are published in (Henriksson *et al.*, 2019) and presented with permission.

group, the materials, such as eucalyptus (Euca), miombo (Miom), peanut shell (PeaS), pigeon pea (PigP), and sicklebush (Sick), are independent of moisture content. In case these materials are mixed with a specific fixed share split, the mixture should have a rather wide optimum for MC, simplifying the production properties. More results and discussions can be found in the studies conducted by Henriksson *et al.* (2019).

Therefore, even if the materials are not blended, it is possible to show and evaluate the possibilities of them being utilized as either blended feedstock or a broader raw materials base, as they have the same pellet abilities. However, the study presented in the following shows examples of a study based on a laboratory-scale unit and how pine and switchgrass can be evaluated through pellet properties and tested before they can be pelletized with high moisture content.

4.6.2. *Study 2: Blending and Pelleting of Pine and Switchgrass in a Laboratory-Scale Flat Die Pellet Mill*

Tumuluru (2019) studied the pelleting characteristics of pine and switchgrass blends by using a flat die pellet mill (see Fig. 4.5) and a high-moisture pelleting process to understand the impact of the L/D ratio of the pellet die and blend moisture content on pellet quality in terms of density and durability. The blend within this study was switchgrass, *P. virgatum* L., which was field-grown and harvested in Vonore, TN, USA, and is further processed in a tub grinder by Genera Energy, and 2-in. (inch) top pine residue samples, harvested from forest stands near Auburn, AL, USA. These residues were then dried, and the 2-in. top pine residue and switchgrass were size-reduced in a hammer mill fitted with a 3/16-in. (inch) (4.76 mm) screen. These grinding tests were conducted at Herty Advanced Biomaterials in Savannah, GA, USA.

In this study, three-blend ratios (e.g., 50% 2-in. top pine residue +50% switchgrass, 75% 2-in. top pine residue +25% switchgrass, and 25% 2-in. top pine residue +75% switchgrass), three L/D ratios (1.5, 2.0, and 2.6), and three-blend moistures (20, 25, and 30% w.b.)

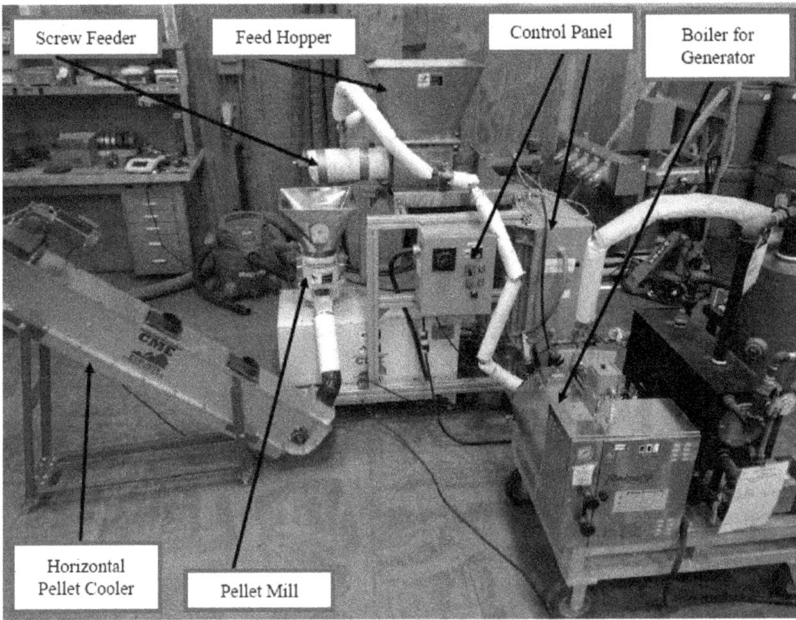

Figure 4.5. Flat die pellet mill used for the pelleting studies (Tumuluru, 2019).

were tested. A laboratory-scale ECO-10 flat die pellet mill was used to perform the pelleting tests. Figure 4.6 shows examples of the pellets produced in different process conditions.

Some of the major observations of blending 25% switchgrass +75% 2-in. top pine residue and 50% switchgrass +50% 2-in. top pine residue helped achieve the requisite bulk density and durability (>550 kg/m^3 and >95%) conditions as shown in Fig. 4.7. The pelleting process conditions of an L/D ratio of 2.6 and a blend moisture content of 20% (w.b.) resulted in bulk density (>550 kg/m^3) and durability (>95%). The results indicated that for the 50% 2-in. pine top residue +50% switchgrass, the moisture content of the pellet decreased with a corresponding increase in L/D ratio and decrease in blend moisture content. The bulk density of the blend pellets was lower at a higher blend moisture content and lower L/D ratio, whereas a higher L/D ratio (2.6) and lower blend moisture content (20–22%, w.b.) increased the bulk densities of >580 kg/m^3. The L/D ratio and blend moisture content significantly affected the

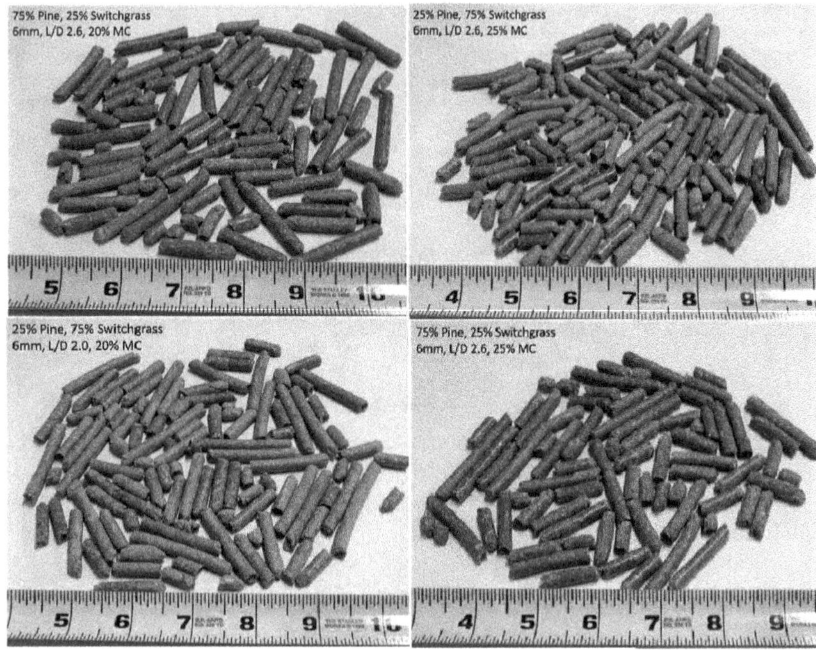

Figure 4.6. Blend pellets made from 2-in. top pine + switchgrass blends at different moisture contents and L/D ratios of the pellet die (Tumuluru, 2019).

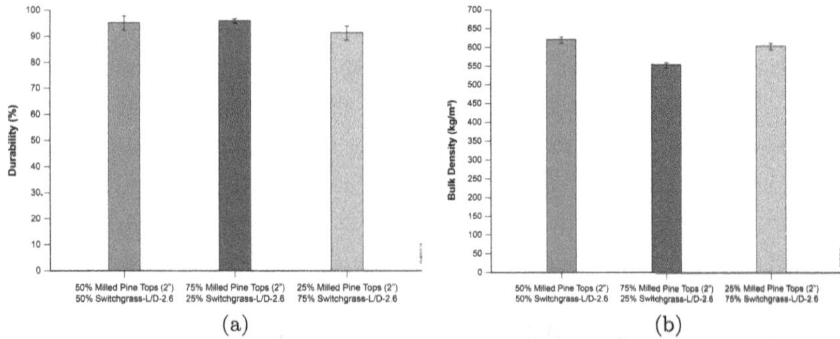

Figure 4.7. Durability (a) and bulk density (b) from pellets produced using a switchgrass, pelletized at 20% moisture content and an L/D ratio of 2.6.

pellet durability. A lower to medium moisture content of 20–25% (w.b) and a higher L/D ratio of 2.6 resulted in durability values of >94%. In addition, the pelleting energy consumption decreased at a lower L/D ratio. Higher pelleting energy of >140 kWh/t was observed at an L/D ratio in the pellet die (e.g., approximately 2.4–2.6) and a lower moisture content of 20–22% (w.b.), whereas lowering the L/D ratio to 1.5 at the same moisture content resulted in lower pelleting energy of <102 kWh/t.

In the case of blend ratio, the 75% 2-in. milled pine top residue and 25% milled switchgrass revealed a similar trend in terms of the observed moisture content, where a higher L/D ratio and a blended moisture content resulted in higher moisture loss of approximately 5–7% (w.b) during pelleting. The bulk density of the blended pellet increased with an increase in the L/D ratio of the pellet die and a decrease in blend moisture content. The highest bulk density of >540 kg/m^3 and lowest bulk density values of <364 kg/m^3 were observed. In the case of durability, the pellets produced had durability values of about 98%. Regarding pelleting energy, a lower value of <78 kWh/t was observed for the L/D ratio of 2.2–2.6 at a lower blend moisture content of 20% (w.b.).

In the case of blend ratio for the 25% 2-in. milled pine top residue +75% milled switchgrass, a lower blend moisture content produced pellets with the lowest moisture content. For example, a moisture loss of approximately 8–10% (w.b) was seen at a 30% (w.b.) initial blend moisture content, whereas at a 20% (w.b.) initial moisture content, the loss of moisture observed during pelleting was only approximately 6–7% (w.b.). The bulk density higher at a higher L/D ratio and at lower blend moisture content. The maximum bulk density observed at 20% (w.b) moisture content, and an L/D ratio of 2.6 was 580 kg/m^3. In the case of durability, the maximum observed values were >92% at an L/D ratio of 2.6 and a blend moisture content of 20% (w.b.). The lower L/D ratio of 1.5, lower durability values of 76–78% were observed at different blend moisture contents of 20–30% (w.b.). The pelleting energy observed was 80 kWh/t at an L/D ratio of 1.5, whereas at a higher L/D ratio of 2.6 and a lower moisture content of 20% (w.b.), the pelleting energy was >180 kWh/t.

To summarize, this laboratory study, based on a flat die mill, shows that, by blending pine with switchgrass, it is possible that both a higher moisture content than usual can be used and high durability can be reached. Even if durability standards used within fuel pellets are not always reached, it must be remembered that fuel pellets represent only one field of end products where pellets are used. However, this study was based on a flat die solution, and another step is a 1-t ring die, which was used in the next study.

4.6.3. *Study 3: Pilot-Scale Pellet Mill Studies on Pine and Switchgrass Blends*

A pilot-scale ring-die pellet mill at Idaho National Laboratory with a throughput capacity of 1 t/h, as shown in Fig. 4.8, was used for both the preliminary testing of switchgrass (SG)-southern yellow pine (SYP) blends and the 2.5-t pelleting demonstration (Tumuluru *et al.*, 2022). The focus of the preliminary pelleting study was to understand how the pellet die L/D ratios (5, 7 and 9), blend ratios and blend moisture contents impact the quality of the pellets. Based on the preliminary pelleting studies, we selected the appropriate

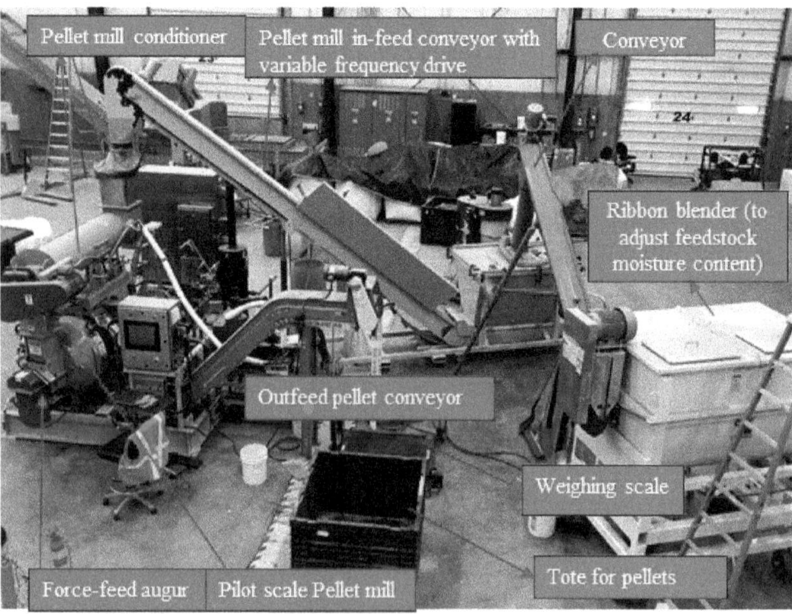

Figure 4.8. Pilot-scale pellet mill used in the study.

combination of parameters and demonstrated the pelleting of 2.5 tons of 2-in. (inch) and 6-in. (inch) SYP residues blended with switchgrass at a ratio of 60:40 and 50:50. The compression or L/D ratio of the pellet die used for the demonstration studies is 9. Figure 4.9 shows the raw material and pellet photos of pine, switchgrass, and their blends based on the preliminary pelleting studies. The pellets produced using pine, switchgrass, and their blends were measured for their physical properties, such as pellet moisture content, unit, bulk, tapped density, durability, and the energy consumption of the process. The mean particle size and bulk density of switchgrass (SG) and southern yellow pine (SYP) ground in a hammer mill fitted with a 1/4-inch (6.35 mm) screen before pelleting is 1.25 and 0.91 mm and 104 and 190 kg/m^3.

Pelleting results based on the preliminary tests indicated a moisture loss of approximately 5–10% (w.b.) was observed during the pelleting process for 100% pure pine pellets. The unit, bulk, and tapped densities at different moisture levels and L/D ratios

Figure 4.9. Raw materials and pellets are produced using Southern Yellow Pine (SYP) and Switchgrass (SG) blends. (a) SYP tops ground to 6.35 mm screen size in a hammer mill; (b) SG biomass ground to 6.35 mm screen size in a hammer mill; (c) 100% SYP pellets; (d) 100% SG pellets; (e) 75% SYP and 25% SG pellets; and (f) 75% SG and 25% SYP pellets (Tumuluru *et al.*, 2022).

were in the range of 1096–1169, 471–569, and 518–616 kg/m^3 and the durability was in the range of 89–95%. In the case of the 75% pine and 25% switchgrass pellets, a moisture loss of 6–8% (w.b.) was observed at a higher L/D ratio of 9 and at 25% (w.b.) blend moisture content. The unit, bulk, and tapped densities of 799–1117, 450–573, and 499–631 kg/m^3, and the durability in the range of 85–93% was observed. In the case of the 50% pine pellets, a moisture loss of 6–10% (w.b.) was observed at a higher L/D ratio of 9 and a higher blend moisture content of 25% (w.b.). The unit, bulk, and tapped densities were 1087–1115, 516–572, and 569–623 kg/m^3, respectively. Overall, the durability values were in the range of 85–95%, and a higher L/D ratio resulted in higher durability of 95%. For the 25% pine and 75% switchgrass pellets, moisture loss of about 5–9% (w.b.) was observed. The unit, bulk, and tapped densities were 889–1146, 465–607, and 510–654 kg/m^3, respectively. The lowest durability of 83% was observed at an L/D ratio of 5 and a blend moisture content of 20% (w.b.). In contrast, the highest durability of 95% was observed at an L/D ratio of 9 and at about 20% (w.b.) blend moisture content. In the case of the 100% switchgrass pellets, the moisture loss was similar to 100% pine pellets (e.g., about 5–10% w.b.). The unit, bulk, and tapped densities were between 991–1156, 499–552, and 546–621 kg/m^3, respectively, while the durability was between 85 and 95%.

The high moisture pellets were further dried at 70°C for 2–3 h in a laboratory oven, reducing the pellet's moisture content to less than 10% (w.b.). Drying the pellets reduced the unit and bulk density of the pellets produced. Pellet durability increased slightly after low-temperature drying of the pellets. The blending of pine with switchgrass reduced the energy consumption of the pelleting process. Compared with the 100% pine, the 100% SG pellets consumed more energy. For example, at an L/D ratio of 9 and 20% moisture content, the energy consumption of the 100% SG pellets was 123 kWh/t, whereas the 100% pine only consumed 105 kWh/t of energy. Also higher blend moisture content and lower L/D ratio increased the pelleting energy consumption. In the blended feedstocks, higher amounts of pine (75%) reduced the energy consumption.

4.6.4. *Study 4: 2.5-t Demonstration of High Moisture Pelleting*

Following the preliminary pilot-scale studies (Studies 2 and 3) on SYP, SG, and their blends, the process was demonstrated at a 2.5-t scale (Tumuluru *et al.*, 2022). An L/D ratio of 9 and moisture content of about 20% (w.b.) was selected since these pelleting conditions resulted in higher pellet quality in terms of bulk density and durability at lower energy consumption. SYP residue (e.g., 2- and 6-in. top) and SG were blended at two different ratios (e.g., 40:60 and 50:50). The pelleting properties indicated that they were not significantly different for the two types of pine residues tested. The pellet's physical properties, chemical composition, and energy consumption for the demonstration tests are given in Table 4.1. In the case of 6-in. (152.4 mm) pine residues, the 60% blend resulted in a higher unit density but the bulk and tapped density and durability were not significantly different. The energy consumption was slightly higher for the 50% blend of 6-in. (152.4 mm) SYP at 98 kWh/t compared to the 60% blend (89 kWh/t). In the case of 2-in. (50.8 mm) pine residue, a 60% blend produced pellets with a lower unit, bulk, and tapped densities, as well as durability, but the specific energy consumption of the process was lower, at 87 kWh/t than that of the 50% blend (95 kWh/t). Among the four tested blends, 2-in. (50.8 mm) SYP-60 produced lower quality pellets, whereas 6-in. (152.4 mm) SYP at a 50% blend ratio produced comparably higher quality pellets. A higher pine content led to lower pelleting energy consumption. Changes in pine blend ratio and stem diameter had no significant effect on the elemental composition, heating values, and ash content of the pellets (Table 4.1). This study has indicated that blending of southern yellow pine and switchgrass has led to the homogenization of the chemical properties.

4.6.5. *Study 5: Briquetting Studies on Pine, SG, and Corn Stover*

The fifth study is related to briquetting, and the main focus of this research was to understand how blends of lodgepole pine, SG, and

Table 4.1. Average physical-chemical properties of SG and SYP blended pellets immediately after pelleting during the 2.5-t demonstration (Tumuluru *et al.*, 2022).

Pellet properties/ Sample ID	6-in. (152.4 mm) SYP-50	6-in. (152.4 mm) SYP-60	2-in. (50.8 mm) SYP-50	2-in. (50.8 mm) SYP-60
Unit density (kg/m^3)	944 ± 84	1016 ± 37	1083 ± 48	965 ± 119
Bulk density (kg/m^3)	561 ± 1	557 ± 7	586 ± 1	474 ± 5
Tapped density (kg/m^3)	604 ± 3	599 ± 3	629 ± 3	510 ± 3
Durability (%)	93 ± 0	92 ± 0	91 ± 2	80 ± 2
Carbon (%)	47 ± 0	47 ± 0	47 ± 0	47 ± 0
Hydrogen (%)	7 ± 0	7 ± 0	6 ± 0	6 ± 0
Nitrogen (%)	0 ± 0	0 ± 0	1 ± 0	0 ± 0
Oxygen (%)	46 ± 0	47 ± 0	46 ± 0	47 ± 0
Ash (%)	2 ± 0	2 ± 0	2 ± 0	2 ± 0
HHV (J/g)	$19{,}851 \pm 9$	$19{,}729 \pm 5$	$19{,}899 \pm 10$	$19{,}821 \pm 8$
SEC (kWh/t)	98	89	95	87

Note: SYP — Southern yellow pine tops; SYP and SG were ground to 6.35 mm hammer screen size and blended at 60:40 or 50:50 ratio; SEC — Specific energy consumption; HHV — Higher heating value.

corn stover impact the physical properties, chemical composition, and caloric value of the produced briquettes, as well as how feedstock properties such as moisture content and grind size impact their quality (Tumuluru and Fillerup, 2020). Many researchers have found that feedstock properties, such as moisture content and particle size, significantly impact particles' interlocking and binding mechanisms during compression in the briquetting process. A review article published by Tumuluru *et al.* (2011) on biomass densification systems suggested that feedstock properties (such as grind size and moisture), biomass composition (such as lignin, protein, and starch), and densification process variables (such as compression pressure and pellet die dimensions) have a great impact on the physical properties and chemical composition of densified products. Keeping this in view, the specific objectives of this study were to (1) understand how binary and ternary blend ratios of SG, lodgepole pine, and corn stover impact the physical properties, chemical composition,

HHV, and specific energy consumption of the briquetting process; (2) understand how process variables, such as blend moisture content value (i.e., 12%, 15%, and 18% w.b.) and hammer mill screen size (i.e., 4.8 mm and 12.7 mm), impact the briquetting process; (3) develop regression models to predict briquette properties and specific energy consumption; and (4) test the impact that a larger hammer mill grind size of 19.05 mm (3/4-in.) has on briquette quality and energy consumption at 15% (w.b.) moisture content (Tumuluru and Fillerup, 2020).

A pilot-scale heavy-duty hydraulic briquette press (WEIMA C150 model, Fort Mill, SC, USA) was used in this study. A three-phase 460 V is required to run this machine (Tumuluru, 2019a; Tumuluru and Fillerup, 2020). The briquette press is provided with a hopper of 1.1 m^3 volume. This press has a throughput of approximately 50–80 kg of briquettes per hour. The complete details of the briquette press configuration are discussed by Tumuluru and Fillerup (2020). Three feedstocks (e.g., lodgepole pine, switchgrass (SG), and corn stover) were size-reduced in a hammer mill fitted with 4.8-, 12.7-, and 19.05-mm screens and then reconditioned to three different moisture content levels (e.g., 12%, 15%, and 18% (w.b.)). The blending of the three feedstocks and corresponding moisture adjustments were carried out in a ribbon blender (RB 500, Colorado Mill Equipment, CO, USA). Mixing these blends with water was carried out for about 30 min to achieve a homogenous blend and allow the moisture to distribute evenly. The blended and moisture-adjusted feedstock was stored overnight at a room temperature of 20°C to allow the moisture to equilibrate (Tumuluru and Fillerup, 2020). The briquetting process in the press is carried out in four different stages. In the first stage of briquetting, the biomass was prepressed using the augur provided in the hopper. This material was further prepressed in a second stage with a load pusher before being formed into a briquette, in the third stage pressure is applied using a hydraulic ram which results in briquettes (Tumuluru and Fillerup, 2020). Finally, the briquettes were discharged from the press. The briquettes produced have a diameter of about 2 in. (50.8 mm), which were further stored for five days at room temperature of about 20°C to allow for a stable density (Tumuluru and Fillerup, 2020). The briquette properties such as unit, bulk, tapped density, and chemical properties were measured after five days of storage. Two different blend ratios were tested: 1:1 and 1:1:1.

4.6.5.1. *Two-blend studies (1:1 ratio)*

Figure 4.10 indicates the various briquettes produced with a 50:50 blend ratio of corn stover:SG; SG:lodgepole pine; and lodgepole pine:corn stover (Tumuluru and Fillerup, 2020). The unit densities of the briquettes made with the blend ratios of pairs of 50:50 SG, lodgepole pine, and/or corn stover indicated that a lower moisture content value of 12% (w.b.) resulted in unit density values ranging from 775 to 825 kg/m^3 for corn stover and SG blends at both screen sizes (e.g., 4.8 and 12.7 mm), whereas unit density values were reduced at a moisture content of 18% (w.b.). Blending lodgepole pine with other feedstocks, such as corn stover and SG, at a lower moisture content value of 12% (w.b.) did not significantly impact the unit density values measured; however, the medium and higher blend moisture contents of 15% and 18% (w.b.) resulted in lower unit density values. In the case of bulk density and a lower blend moisture content of 12% (w.b.), bulk densities were in the range of 425–475 kg/m^3 for both hammer mill screen sizes at all three of the blend ratios tested. Increasing the blend moisture content to 18% (w.b.) for the 4.8 mm grind reduced bulk density to approximately 380–410 kg/m^3, whereas increasing the screen size to 12.7 mm at the higher moisture content value of 18% (w.b.) further reduced the bulk density to approximately 340–380 kg/m^3. The results also indicated that the bulk density values increased when lodgepole pine was blended with the other two herbaceous feedstocks (i.e., SG and corn stover). The durability data indicated that a smaller screen size of 4.8 mm resulted in lower durability values than a 12.7-mm grind size. In the case of the SG and corn stover blend, a lower to medium moisture content value of 12% and 15% (w.b.), as well as a smaller screen size grind, reduced durability values to approximately 92–93%, whereas increasing the moisture content value to 18% (w.b.) increased durability values to >95%. The blending of lodgepole pine

(a)

(b)

Figure 4.10. Briquettes produced at different blend ratios (1:1) and in different process conditions (Tumuluru and Fillerup, 2020): (a) lodgepole pine:corn stover (15%, w.b. moisture content and 12.7 mm hammer mill screen size); (b) lodgepole pine:switchgrass (15%, w.b. moisture content and 12.7 mm hammer mill screen size).

and SG did not improve durability values, which were observed to be 89–91%, whereas blending lodgepole pine with corn stover improved durability values from 92% to 93.3%. By increasing the hammer mill screen size to 12.7 mm, the SG/corn stover and lodgepole pine/corn stover blends resulted in durability rating values of >95% at all three moisture content values tested. In the case of the lodgepole/SG blend, a higher moisture content value of 18% (w.b.) and a 12.7-mm hammer mill screen size resulted in durability values of >95%. Finally, blending lodgepole pine with SG at a higher moisture content value of 18% (w.b.) for the 4.8 mm hammer mill screen size grind resulted in durability rating values lower than 90%.

4.6.5.2. *Three-blend studies (1:1:1)*

Within the three-blend tests, the unit density of the briquettes was in the range of 725–800 kg/m^3. The highest of these, at about 800 kg/m^3, was observed for the 12.7-mm hammer mill grind at a moisture content value of 12% (w.b.). The 4.8-mm hammer mill grind also produced blend briquettes with unit density values in the range of 725–786 kg/m^3 for all three of the moisture content values tested. The lower moisture content values generally resulted in higher unit density values. Increasing moisture content values for both hammer mill screen sizes reduced the unit density values, but the change in hammer mill screen size only had a marginal effect. The bulk density of the briquettes made using lodgepole pine/SG/corn stover blends was in the range of 414–455 kg/m^3 at the different blend moisture content values tested. Increasing the blend moisture content value impacted the bulk density of the briquettes produced, whereas increasing the screen size did not significantly change the bulk density values. For both the 4.8- and 12.7-mm hammer mill grind sizes, the lowest bulk density values were observed at a blend moisture content value of 18% (w.b.), at 414 kg/m^3, whereas the highest value of approximately 455 kg/m^3 was observed for the 12% and 15% blend moisture content values with the 4.8-mm hammer mill grind. There was no significant difference in bulk density values for the briquettes made using the 12.7-mm hammer mill grind, where the measured values were in the range of 447–455 kg/m^3

at 12% and 15% (w.b.) blend moisture values. The bulk density of the two-blend ratio briquettes resulted in slightly higher values than the three-blend ratio. Durability rating values decreased with a corresponding decrease in moisture and hammer mill screen size. The lowest measured value on the blends tested was 91.5% for the 4.8-mm grind at a blend moisture content of 12% (w.b.). Increasing the screen size to the 12.7-mm hammer mill grind size increased the durability rating to >95%. Moisture content positively impacted the durability rating values for both hammer mill grind sizes tested. Increasing the blend moisture content to 18% (w.b.) increased the durability rating values. The highest durability rating value achieved, 97.64%, was observed at a moisture content of 18% (w.b.) for the 12.7-mm hammer mill grind size. The results indicate that increasing the hammer mill screen size and moisture content value of the blend increased the durability rating values.

The energy consumption of the briquetting process for the three-blend feedstock was influenced by both hammer mill screen size and blend moisture content. For the 4.8-mm hammer mill screen size, the briquetting energy consumption was 82–88 kWh/t. At this screen size, the energy consumption marginally decreased with a corresponding increase in blend moisture content. In the case of the 12.7-mm hammer mill size, an increase in moisture content from 12% to 15% (w.b.) increased energy consumption values, where the highest value of approximately 97 kWh/t was observed at blend moisture of 18% (w.b.). The lowest value of approximately 79 kWh/t was observed for blend moisture of 12% (w.b.). It was concluded from these results that for the smaller hammer mill grind size of 4.8-mm, medium to high blend moisture content (i.e., 15% and 18%, w.b.) was more desirable. In the case of a larger 12.7-mm hammer mill screen grind size, a low to medium blend moisture content was desirable to reduce the energy consumption of the briquetting process.

4.7. Summary

Blending is a common method; different types of biomasses are mixed to improve their physical properties and chemical composition. Biomass blending helps overcome feedstock challenges, such as quality, variability, supply, and cost. The major advantages of biomass blending are an increase in potential biomass supply for a given biorefinery area, feedstock cost reduction, and improvements in biomass flow and pelleting characteristics. The challenges associated with blending, such as variability in particle size, moisture, and density, can be overcome by densification using mechanical systems, such as a pellet mill, briquette press, extruder, or roller press. The densification process is critical for blends to produce a feedstock material suitable as a commodity product. Densification helps overcome the variability issues of the physical properties, such as particle size, shape, density, and moisture. The blends after densification can have improved handling and conveyance efficiencies throughout the supply system and biorefinery infeed and feedstock uniformity and density. Various pelleting and briquetting studies have indicated that blending woody and herbaceous biomasses leads to homogenization of the biomass and yields products with uniform product quality in terms of physical properties and chemical composition. Studies have also identified new markets where pellets can make a significant difference in terms of the move towards a more sustainable society and where blending and variated feedstocks are the best solution. The blending and pelleting or briquetting of pine and switchgrass have resulted in higher density and durability and reduced the process energy consumption.

References

Anukam, A., Berghel, J., Henrikson, G., Frodeson, S., & Ståhl, M. (2021). A review of the mechanism of bonding in densified biomass pellets. *Renew. Sust. Energ. Rev.*, *148*, 111249.

Arshadi, M., Gref, R., Geladi, P., Dahlqvist, S.-A., & Lestander, T. (2008). The influence of raw material characteristics on the industrial pelletizing process and pellet quality. *Fuel Process. Technol.*, *89*(12), 1442–1447.

Back, E. L. (1987). The bonding mechanism in hardboard manufacture review report. *Holzforschung*, *41*(4), 247–258. https://doi.org/10.1515/hfsg.1987.41.4.247.

Berghel, J., Ståhl, M., Frodeson, S., Pichler, W., & Weigl-Kuska, M. (2022). A comparison of relevant data and results from single pellet press research is mission impossible: A review. *Bioresour. Technol. Rep.*, *18*, 101054.

Bergström, D., Israelsson, S., Öhman, M., Dahlqvist, S.-A., Gref, R., Boman, C., & Wästerlund, I. (2008). Effects of raw material particle size distribution on the characteristics of Scots pine sawdust fuel pellets. *Fuel Process. Technol.*, *89*(12), 1324–1329.

Bhattacharya, S., & Abdulsalam, P. (2002). Low greenhouse gas biomass options for cooking in the developing countries. *Biomass Bioenerg.*, *22*(4), 305–317.

Carone, M. T., Pantaleo, A., & Pellerano, A. (2011). Influence of process parameters and biomass characteristics on the durability of pellets from the pruning residues of *Olea europaea L. Biomass Bioenerg.*, *35*(1), 402–410.

Castellano, J. M., Gómez, M., Fernández, M., Esteban, L. S., & Carrasco, J. E. (2015). Study on the effects of raw materials composition and pelletization conditions on the quality and properties of pellets obtained from different woody and non woody biomasses. *Fuel*, *139*, 629–636.

Crawford, N. C., Ray, A. E., Yancey, N. A., & Nagle, N. (2015). Evaluating the pelletization of 'pure' and blended lignocellulosic biomass feedstocks. *Fuel Process. Technol.*, *140*, 46–56.

Dhamodaran, A., & Afzal, M. T. (2012). Compression and springback properties of hardwood and softwood pellets. *BioResearch*, *7*(3), 4362–4376.

Edmunds, C. W., Reyes Molina, E. A., André, N., Hamilton, C., Park, S., Fasina, O., Adhikari, S., Kelley, S. S., Tumuluru, J. S., Rials, T. G., & Labbé, N. (2018). Blended feedstocks for thermochemical conversion: Biomass characterization and bio-oil production from switchgrass-pine residues blends. *Front. Energy Res.*, *6*, 79.

Ek, M., Gellerstedt, G., & Henriksson, G. (Eds.). (2009). *Wood Chemistry and Wood Biotechnology.* Vol. 1. Berlin, Germany: Walter de Gruyter GmbH.

FAO (2021). *FAOSTAT.* Food and Agriculture Organization of the United Nations. Retrieved from December 21 from 2021 https://www.fao.org/faostat/en/#data/FO.

Filbakk, T., Skjevrak, G., Høibø, O., Dibdiakova, J., & Jirjis, R. (2011). The influence of storage and drying methods for Scots pine raw material on mechanical pellet properties and production parameters. *Fuel Process. Technol.*, *92*(5), 871–878.

Frodeson, S., Anukam, A. I., Berghel, J., Ståhl, M., Lasanthi Kudahettige Nilsson, R., Henriksson, G., & Bosede Aladejana, E. (2021). Densification of wood — Influence on mechanical and chemical properties when 11 naturally occurring substances in wood are mixed with beech and pine. *Energies*, *14*(18), 5895.

Frodeson, S., Lindén, P., Henriksson, G., & Berghel, J. (2019a). Compression of biomass substances — A study on springback effects and color formation in pellet manufacture. *Appl. Sci.*, *9*(20), 4302.

Frodeson, S., Henriksson, G., & Berghel, J. (2019b). Effects of moisture content during densification of biomass pellets, focusing on polysaccharide substances. *Biomass Bioenerg.*, *12*, 322–330.

Frodeson, S., Henriksson, G., & Berghel, J. (2018). Pelletizing pure biomass substances to investigate the mechanical properties and bonding mechanisms. *BioResearch*, *13*(1), 1202–1222.

Frodeson, S. (2019). Towards understanding the pelletizing process of biomass: Perspectives on energy efficiency and pelletability of pure substances. Ph.D. Thesis, Karlstad University, Karlstad, Sweden.

Gumbo, D. J., Moombe, K. B., Kandulu, M. M., Kabwe, G., Ojanen, M., Ndhlovu, E., & Sunderland, T. C. H. (2013). Dynamics of the charcoal and indigenous timber trade in Zambia: A scoping study in Eastern, Northern and Northwestern provinces, Center for International Forestry Research (CIFOR) Occasional Paper 86, Bogor, Indonesia. Retrieved from https://www.cifor.org/publications/pdf_files/OccPapers/OP-86.pdf.

Henriksson, L., Frodeson, S., Berghel, J., Andersson, S., & Ohlson, M. (2019). Bioresources for sustainable pellet production in Zambia: Twelve biomasses pelletized at different moisture contents. *BioResearch*, *14*(2), 2550–2575.

Holm, J. K., Henriksen, U. B., Hustad, J. E., & Sørensen, L. H. (2006). Toward an understanding of controlling parameters in softwood and hardwood pellets production. *Energ. Fuels*, *20*(6), 2686–2694.

Huang, Y., Finell, M., Larsson, S., Wang, X., Zhang, J., Wei, R., & Liu, L. (2017). Biofuel pellets made at low moisture content: Influence of water in the binding mechanism of densified biomass. *Biomass Bioenerg.*, *98*, 8–14.

Janssen, R., & Rutz, D. (Eds.). (2012). *Bioenergy for Sustainable Development in Africa*. New York, NY, USA: Springer-Verlag.

Kaliyan, N., & Morey, R. V. (2009). Factors affecting strength and durability of densified biomass products. *Biomass Bioenerg.*, *33*(3), 337–359.

Kaliyan, N., & Morey, R. V. (2010). Natural binders and solid bridge type binding mechanisms in briquettes and pellets made from corn stover and switchgrass. *Bioresour. Technol.*, *101*(3), 1082–1090.

Kenney, K. L., Smith, W. A., Gresham, G. L., & Westover, T. L. (2013). Understanding biomass feedstock variability. *Biofuels*, *4*(1), 111–127.

Križan, P. (2015). *The Densification Process of Wood Waste*. Berlin, Germany: Walter de Gruyter GmbH.

Larsson, S. H., Thyrel, M., Geladi, P., & Lestander, T. A. (2008). High quality biofuel pellet production from pre-compacted low density raw materials. *Bioresour. Technol.*, *99*(15), 7176–7182.

Lavergne, S., Larsson, S. H., Da Silva Perez, D., Marchand, M., Campargue, M., & Dupont, C. (2021). Effect of process parameters and biomass composition on flat-die pellet production from underexploited forest and agricultural biomass. *Fuel*, *302*, 121076.

Li, H., Jiang, L.-B., Li, C.-Z., Liang, J., Yuan, X.-Z., Xiao, Z.-H., Xiao, Z.-H., & Wang, H. (2015). Co-pelletization of sewage sludge and biomass: The energy input and properties of pellets. *Fuel Process. Technol.*, *132*, 55–61.

Mahadevan, R., Adhikari, S., Shakya, R., Wang, K., Dayton, D., Lehrich, M., & Taylor, S. E. (2016). Effect of alkali and alkaline earth metals on in-situ catalytic fast pyrolysis of lignocellulosic biomass: A microreactor study. *Energ. Fuel*, *30*, 3045–3056.

Mani, S., Tabil, L. G., & Sokhansanj, S. (2003). Compaction of biomass grinds-an overview of compaction of biomass grinds. *Powder Handl. Process.*, *15*(3), 160–168.

Mani, S., Tabil, L. G., & Sokhansanj, S. (2006a). Effects of compressive force, particle size and moisture content on mechanical properties of biomass pellets from grasses. *Biomass Bioenerg.*, *30*(7), 648–654.

Mani, S., Sokhansanj, S., Bi, X., & Turhollow, A. (2006b). Economics of producing fuel pellets from biomass. *Am. Soc. Agric. Biol. Eng.*, *22*(3), 421–426.

Matakala, P. W., Misael, K., & Jochen, S. (2015). *Zambia National Strategy to Reduce Emissions from Deforestation and Forest Degradation* (REDD+). Forestry Department, Ministry of Lands Natural Resources and Environmental Protection, FAO, UNDP, and UNEP, Government of the Republic of Zambia, Zambia.

Monteiro, E., Mantha, V., & Rouboa, A. (2012). Portuguese pellets market: Analysis of the production and utilization constrains. *Energy Policy*, *42*, 129–135.

Nguyen, Q. N., Cloutier, A., Achim, A., & Stevanovic, T. (2015). Effect of process parameters and raw material characteristics on physical and mechanical properties of wood pellets made from sugar maple particles. *Biomass Bioenerg.*, *80*, 338–349.

Nielsen, N. P. K., Gardner, D. J., Poulsen, T., & Felby, C. (2009). Importance of temperature, moisture content, and species for the conversion process of wood residues into fuel pellets. *Wood Fiber Sci.*, *41*, 414–425.

Nielsen, N. P. K. (2009). Importance of raw material properties in wood pellet production: Effects of differences in wood properties for the energy requirements of pelletizing and the pellet quality. Ph.D. Thesis, Faculty of Life Sciences, University of Copenhagen, Copenhagen, Denmark.

Obernberger, I., & Thek, G. (2004). Physical characterisation and chemical composition of densified biomass fuels with regard to their combustion behaviour. *Biomass Bioenerg.*, *27*(6), 653–669.

Pennise, D. M., Smith, K. R., Kithinji, J. P., Rezende, M. E., Raad, T. J., Zhang, J., & Fan, C. (2001). Emissions of greenhouse gases and other airborne pollutants from charcoal making in Kenya and Brazil. *J. Geophys. Res.*, *106*(24), D20, 143–155.

Peša, I. (2017). Sawdust pellets, micro gasifying cook stoves and charcoal in urban Zambia: Understanding the value chain dynamics of improved cook stove initiatives. *Sustain. Energy Technol. Assess.*, *22*, 171–176.

Pietsch, W. (1991). *Size Enlargement by Agglomeration.* Hoboken, NJ, USA: John Wiley & Sons.

Poddar, S., Kamruzzaman, M., Sujan, S. M. A., Hossain, M., Jamal, M. S., Gafur, M. A., & Khanam, M. (2014). Effect of compression pressure on lignocellulosic biomass pellet to improve fuel properties: Higher heating value. *Fuel*, *131*, 43–48.

Puig-Arnavat, M., Shang, L., Sárossy, Z., Ahrenfeldt, J., & Henriksen, U. B. (2016). From a single pellet press to a bench scale pellet mill — Pelletizing six different biomass feedstocks. *Fuel Process. Technol.*, *142*, 27–33.

Ramírez-Gómez, Á. (2016). Research needs on biomass characterization to prevent handling problems and hazards in industry. *Part. Sci. Technol.*, *34*(4), 432–441.

Ray, A. E., Li, C., Thompson, V. S., Daubaras, D. L., Nagle, N., & Hartley, D. S. (2017). "Biomass blending and densification: Impacts on feedstock supply and biochemical conversion performance. INL/MIS-16-38547, Rev. 1, Idaho National Laboratory, Idaho Falls, ID, USA.

Rhén, C., Gref, R., Sjöström, M., & Wästerlund, I. (2005). Effects of raw material moisture content, densification pressure and temperature on some properties of Norway spruce pellets. *Fuel Process. Technol.*, *87*(1), 11–16.

Rumpf, H. (1962). The strength of granules and agglomerates. In W. A. Knepper, (Ed.), *Agglomeration* (pp. 379–418). New York, NY, USA: Interscience Publishers.

Said, N., Abdel daiem, M. M., García-Maraver, A., & Zamorano, M. (2015). Influence of densification parameters on quality properties of rice straw pellets. *Fuel Process. Technol.*, *138*, 56–64.

Samuelsson, R., Larsson, S. H., Thyrel, M., & Lestander, T. A. (2012). Moisture content and storage time influence the binding mechanisms in biofuel wood pellets. *Appl. Energy*, *99*, 109–115.

Seki, M., Sugimoto, H., Miki, T., Kanayama, K., & Furuta, Y. (2013). Wood friction characteristics during exposure to high pressure: Influence of wood/metal tool surface finishing conditions. *J. Wood Sci.*, *59*(1), 10–16.

Serrano, C., Monedero, E., Lapuerta, M., & Portero, H. (2011). Effect of moisture content, particle size, and pine addition on quality parameters of barley straw pellets. *Fuel Process. Technol.*, *92*(3), 699–706.

Souza, G. M., Ballester, M. V. R., de Brito Cruz, C. H., Chum, H., Dale, B., Dale, V. H., Fernandes, E. C. M., Foust, T., Karp, A., Lynd, L., Filho, R. M., Milanez, A., Nigro, F., Osseweijer, P., Verdade, L. M., Victoria, R. L., & Van der Wielen, L. (2017). The role of bioenergy in a climate-changing world. *Environ. Dev.*, *23*, 57–64.

Stelte, W., Clemons, C., Holm, J. K., Ahrenfeldt, J., Henriksen, U. B., & Sanadi, A. R. (2011a). Thermal transitions of the amorphous polymers in wheat straw. *Ind. Crops Prod.*, *34*(1), 1053–1056.

Stelte, W., Holm, J. K., Sanadi, A. R., Barsberg, S., Ahrenfeldt, J., & Henriksen, U. B. (2011b). Fuel pellets from biomass: The importance of the pelletizing pressure and its dependency on the processing conditions. *Fuel*, *90*(11), 3285–3290.

Stelte, W., Holm, J. K., Sanadi, A. R., Barsberg, S., Ahrenfeldt, J., & Henriksen, U. B. (2011c). A study of bonding and failure mechanisms in fuel pellets from different biomass resources. *Biomass Bioenerg.*, *35*(2), 910–918.

Stelte, W., Sanadi, A. R., Shang, L., Holm, J. K., Ahrenfeldt, J., & Henriksen, U. B. (2012a). Recent developments in biomass pelletization — A review. *Bioresources*, *7*(3), 4451–4490.

Stelte, W., Clemons, C., Holm, J. K., Ahrenfeldt, J., Henriksen, U. B., & Sanadi, A. R. (2012b). Fuel pellets from wheat straw: The effect of lignin glass transition and surface waxes on pelletizing properties. *Bioenergy Res.*, *5*(2), 450–458.

Thek, G. & Obernberger, I. (2010). *The Pellet Handbook.* Earthscan Ltd. https://doi.org/10.4324/9781849775328.

Thompson, D. N., Campbell, T., Bals, B., Runge, T., Teymouri, F., & Ovard, L. P. (2013). Chemical preconversion: Application of low-severity pretreatment chemistries for commoditization of lignocellulosic feedstock. *Biofuels*, *4*(3), 323–340.

Tumuluru, J. S. & Fillerup, E. (2020). Briquetting characteristics of woody and herbaceous biomass blends: Impact on the physical properties, chemical composition, and calorific value, *Biofuel. Bioprod. Biorefin.*, *14*(5), 1105–1124.

Tumuluru, J. S., Rajan, K., Hamilton, C., Pope, C., Rials, T. G., Labbe, N., & Andre, N. O. (2022). Pilot-scale pelleting tests on high-mositure pine, switchgrass, and their blends: Impact on pellet physical properties, chemical composition, and heating values. *Front. Energy Res.*, *9*, 788284.

Tumuluru, J. S. (2019). Pelleting of pine and switchgrass blends: Effect of process variables and blend ratio on the pellet quality and energy consumption. *Energies*, *12*(7), 1198.

Tumuluru, J. S. (2019a). Effect of moisture content and hammer mill screen size on the briquetting characteristics of Woody and Herbaceous biomass. *KONA Powder Part. J.*, *36*, 241–251.

Tumuluru, J. S., Wright, C. T., Hess, J. R., & Kenney, K. L. (2011). A review of biomass densification systems to develop uniform feedstock commodities for bioenergy application. *Biofuel. Bioprod. Biorefin.*, *5*(6), 683–707.

Tumuluru, J. S. (2021). Explore the potential to formulate feedstock blends from diverse biomass inputs for improved processing performance at lower costs. INL/EXT-21-62176, Idaho National Laboratory, Idaho Falls, ID, USA.

Tumuluru, J. S. (2016). Specific energy consumption and quality of wood pellets produced using high-moisture lodgepole pine grind in a flat die pellet mill. *Chem. Eng. Res. Des.*, *110*, 82–97.

Uasuf, A., & Becker, G. (2011). Wood pellets production costs and energy consumption under different framework conditions in Northeast Argentina. *Biomass Bioenergy*, *35*(3), 1357–1366.

Whittaker, C., & Shield, I. (2017). Factors affecting wood, energy grass and straw pellet durability — A review. *Renew. Sust. Energ. Rev.*, *71*, 1–11.

Wilson, T. O. (2010). Factors affecting wood pellet durability. Masters Thesis, Department of Agricultural and Biological Engineering, The Pennsylvania State University, State College, PA, USA.

World Health Organization (WHO) (2016). *Burning Opportunity: Clean Household Energy for Health, Sustainable Development, and Wellbeing of Women and Children.* Geneva, Switzerland: World Health Organization.

Yancey, N. A., Tumuluru, J. S., & Wright, C. T. (2013). Drying, grinding, and pelletization studies on raw and formulated biomass feedstock's for bioenergy applications. *J. Biobased Mater. Bio.*, *7*(5), 549–558.

https://doi.org/10.1142/9781800613799_0005

Chapter 5

The Role of Bulk Solids Flow Behavior in Biomass Densification

Jayant Khambekar[*,§], Carrie E. Hartford[†,¶], and John W. Carson[‡,||]

[*]*Jenike & Johanson, Houston, Texas, USA*
[†]*Jenike & Johanson, San Luis Obispo, California, USA*
[‡]*Jenike & Johanson, Tyngsboro, Massachusetts, USA*
[§]*jkhambekar12@jenike.com*
[¶]*chartford@jenike.com*
[||]*jwcarson@jenike.com*

Abstract

Applications involving processing of biomass feedstocks have increased significantly in recent years. Reliable storage and handling of biomass are critical for the success of these applications. Given the potential flow issues associated with inadequately designed biomass handling and processing equipment, severe consequences could occur to the process reliability and efficiency. Thus, a thorough understanding of the flow behavior of the given biomass feedstock and the use of this information in the design of storage, handling, and processing systems are essential. Flow behavior of bulk solids is a complex phenomenon. Multiple parameters associated with bulk solids flow need to be characterized and studied to obtain a complete understanding. In addition, effects of environmental conditions, process, and operation-related parameters on the flow behavior also need to be studied. This chapter introduces the basic principles and methods used for characterizing the flow behavior of bulk solids, with an emphasis on herbaceous biomass feedstocks. Then, illustration is provided about how to use this information obtained from characterizing flow behavior in design for handling equipment for pelletization of biomass feedstocks. Finally, principles associated with good project management are also discussed.

Keywords: Bulk solids, flow behavior, flow patterns, hoppers, transfer chutes, feeders, particle size and shape factors, densification

5.1. Introduction

The process of densification involves many unit operations where handling and storage of biomass must be performed. Even after densification, the pelletized or briquetted biomass needs to be stored and handled. Thus, storage and handling of biomass are substantial and critical activities for the densification process. As such, any successful densification process must have effective systems for biomass storage and handling.

In spite of the important role of storage and handling operations, industrial practice often relies on using similar if not the same standard equipment at different plants. If one particular design of storage or handling system has worked for one application, there is a tendency to use the same design for other applications. As can be expected, this frequently leads to many *war stories* of how various flow problems occurred during commissioning of plants and how monumental efforts were required to address such issues.

A major reason for these operational failures is that designers of bulk solids storage and handling equipment are either unaware of the science and the details involved in bulk solids storage and handling aspects or they are only partly familiar. As a result, they end up using an inappropriate approach. Proven scientific methods are available for designing storage and handling equipment which take into consideration the flow behavior of the particular bulk solid(s) being handled.

This chapter starts with a brief history of the development of the science for the storage and flow of bulk solids. Then, various characteristics of bulk solids that are associated with their flow behavior are described. After that, connections are shown between these characteristics and typical operations involved during densification of biomass materials so that a systematic and scientific approach for designing storage and handling equipment can be used.

Finally, some project management aspects, which establish which tasks get completed and when, are touched upon so as to ensure that the systematic and scientific design approach outlined here can be practically executed in today's *project stage focused* approach.

5.2. Brief History of Development of Science of Bulk Solids Flow

When one considers that scientific discovery has existed for several millennia, the science of bulk solids flow is in its infancy. Prior to the 1950s, the understanding of this field was primarily phenomenological. Unlike liquids, bulk solids form piles, so measurements were made of their angle of repose. Bulk solids are frictional — both between particles and between particles and a surface on which they move. Their bulk densities vary over a wide range, even for solids that have the same generic name. Most bulk solids increase in bulk density the greater the pressure that is applied to them. Finally, bulk solids are cohesive as evidenced by the ability to form snowball-like structures. These structures often become stronger with increased pressure and sometimes with the length of time that pressure is applied.

Other interesting observations included the discovery that a bulk solid (often sand) is an excellent material to use in an hourglass since, unlike liquids, its rate of discharge does not vary with level. Some noted that the discharge pattern in a bin or silo varies depending on the shape and smoothness of the hopper section. A German engineer, HA Janssen, published a theory in 1895 that explained why the pressure that grain exerts on the walls of a silo does not increase linearly from top to bottom as is the case with a liquid.

Unfortunately, there was no theory describing how bulk solids behave in storage vessels, so there was no way to use these measurements and observations to design vessels for a particular flow pattern or to avoid flow problems. Design of storage and handling equipment was essentially a black art. Most hoppers were either 45° or 60° because those were the common triangles that all engineers carried around with them. Pyramidal hoppers were common since they are easier to fabricate than cones and more unusual shapes. The outlet size was set by matching the dimensions of the feeder inlet with no thought given as to whether this would be large enough to avoid arching or ratholing or to provide the required discharge rate. In short, no one gave much thought to the properties of the material being stored or to the design of the storage vessel. After all, it's "just a bin".

It took a Polish-born, American engineer to bring science and engineering to this field. Andrew Jenike was born in Poland in 1914 and graduated from Warsaw Polytechnic Institute with a B.S. in Mechanical Engineering in 1939. He received his Ph.D. in Structural Engineering from the University of London in 1949 and soon thereafter emigrated to Canada and then to the United States, eventually settling in Salt Lake City.

While working as a structural engineer during the day, he spent his evenings and weekends researching possible topics that would allow him to do something unique, something that would set himself apart, something that would be worthwhile. Finally, on his 39th birthday, he made his decision. The topic he chose was the design of bins and hoppers for storage and flow of bulk solids.

He spent seven years as the staff of the University of Utah School of Engineering where, with funding from the National Science Foundation and other organizations, he developed basic theories and test equipment. He set forth design procedures in two famous University of Utah Engineering Experiment Station Bulletins (No. 108 (Jenike, 1961) and 123 (Jenike, 1964)), the latter being the most cited work in the field today.

One key contributor to Jenike's success in developing the basic theories of solids flow came from an unlikely source. While on a trip back to his native Poland, he came across a Polish translation of Sokolovskii's now famous book, *Statics of Soil Media.* At the time, this book had just been published in Russian and had not been translated into English. Jenike immediately recognized that Sokolovskii's concept that soil stress could vary directly proportional to the distance from a point fit nicely with some crude bin pressure measurements that he had conducted using water-filled diaphragms. This gave rise to his Radial Stress Field concept which he used to describe the stresses in silos, especially in the hopper section. He went on to determine the conditions under which bulk solids flow along hopper walls — which he named *mass flow* — and when they do not — which he named *funnel flow.*

On the basis of his theoretical considerations, Jenike derived a procedure for designing silos for unobstructed flow. This included calculation of the slope and smoothness of hopper walls needed to ensure mass flow and prediction of minimum outlet dimensions required to avoid stable arches or ratholes. Jenike also worked on the practical design of silos, such as the influence of the feeder on flow in a hopper, discharge rates through orifices, and material-induced pressures on silo walls.

To measure the flow properties of bulk solids that are required to use his design procedure, Jenike developed a tester that is widely known today as the Jenike shear tester. Other testers have been developed since, but the Jenike design procedure is still the most widely used method in use today.

Some other key contributors to this field include the following:

- **Jerry Johanson:** He was one of Andrew Jenike's key graduate students, obtaining his Ph.D. in 1962. He went on to work for four years at US Steel Research and then he joined Jenike to form Jenike & Johanson, Inc. Now, some 55 years later, it is the largest and most experienced consulting engineering firm in the world dedicated to the science and engineering of bulk solids flow, storage, and processing. Johanson's unique contributions to this field include numerical analysis of stress and velocity fields in converging hoppers, settlement of fine powders and their limiting discharge rates, and application of radial stress field to roll compaction.
- **Jörg Schwedes:** He received his Ph.D. working under the direction of Prof. Hans Rumpf at the Technical University of Karlsruhe. His thesis topic was the shear behavior of slightly compacted cohesive bulk materials. After a short stint in industry at Bayer AG, he spent the rest of his career at the Institute of Mechanical Process Engineering at the Technical University of Braunschweig

where he was known for his research and writing as well as for the many excellent graduate students that he advised.

- **Dietmar Schulze:** He obtained his Ph.D. at the Institute of Mechanical Process Engineering at the Technical University of Braunschweig under the direction of Prof. Jörg Schwedes. Soon after completing his Ph.D., he and Schwedes formed the consultancy "Schwedes + Schulze Schüttguttechnik". Schulze has for many years been a Professor at the Ostfalia University of Applied Sciences where he lectures and performs research in this field. He is best known for his development of the Schulze Ring Shear Tester and for publication of his comprehensive book, *Powders and Bulk Solids: Behavior, Characterization, Storage and Flow* (Schulze, 2008).
- **Gisle Enstad:** He obtained his Ph.D. at Chr. Michelsen Institute under the direction of Prof. Rolf Eckhoff. Enstad worked at the Institute from 1972 through 1988, at which time the activity was moved to Tel-Tek in Porsgrunn, Norway, where he relocated. One of his main interests for a number of years was characterizing the flow properties of powders. A major research project involved developing new tools for design of silos based on computer simulation of flow of powders in silos and new ways of characterizing constitutive properties of powders. He developed an alternate theory for arching in hoppers and perfected a uniaxial tester. He retired as Chief Scientist at Telemark University College's Powder Science and Technology Department in Norway (Tel-Tek Dept. POSTEC).
- **Alan Roberts:** He was first introduced to Andrew Jenike while on a sabbatical at Cornell University in the mid-1970s. Thus, began a life-long friendship and dedication to this field. Roberts led the Bulk Solids Research Group at the University of Newcastle (Australia) and also the University of Newcastle Research Associates (TUNRA Bulk Solids). He is known for his teaching, writing, research, and consulting.
- **Peter Arnold:** He was a close associate of Prof. Alan Roberts. On the staff of the University of Wollongong, his research interests included the storage, flow, and conveying of bulk solid materials.

He had special interest in the measurement of flow properties of bulk solids, design of gravity storage bins and bottom reclaim systems, feeder design (especially screw and belt feeders), chute design, storage and handling of fine powders, and pneumatic conveying. He is perhaps best known as the lead author (along with A. G. McLean and Alan Roberts) of *The Design of Storage Bins for Bulk Solids Handling* (Arnold *et al.*, 1980).

After these initial and notable contributions, many others have joined the field of bulk solids storage and flow and made contributions in one way or another. Today, the science for the storage and flow of bulk solids keeps growing, allowing us to predict more phenomena and more accurately.

5.3. The Science of Bulk Solids Flow

5.3.1. *What Is a Bulk Solid?*

Traditionally, a bulk solid is defined as a collection of discrete, solid particles. As long as there is a sufficiently large number of discrete particles present together, we can call the collection a *bulk solid.* As such, there is a wide variety of materials that come under the broad category of bulk solids (see Fig. 5.1). For example, this includes industrial materials such as coal, limestone, cement, and gypsum, food products such as pasta and cereals, pharmaceutical powders such as active ingredients and fillers/excipients, plastic pellets and polymer resins, mining raw materials such as ores of iron, copper, gold, and nickel, and of course, various types of biomass materials.

Figure 5.1. Examples of bulk solids.

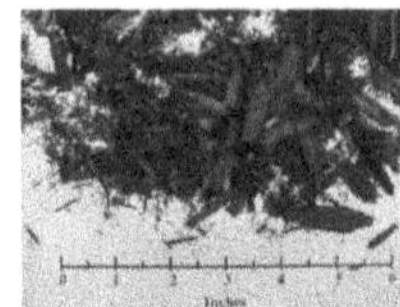
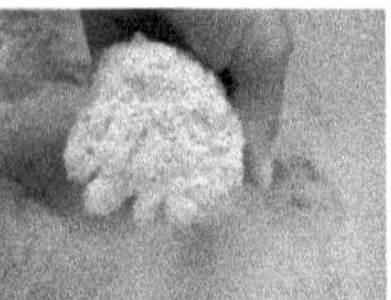

Figure 5.2. Examples of various biomass materials.

Biomass materials derived from dry plant matter are often called *lignocellulosic biomass*. Lignocellulosic biomass can be broadly classified into two categories: woody biomass and herbaceous biomass. Woody biomass is a term typically used for biomass materials derived from tree trunks. These include woodchips, waste wood (also called construction waste), as well as wood shavings and sawdust. Herbaceous biomass is a term typically used for biomass derived from grass-based feedstock. Examples include corn stover, switchgrass, miscanthus, juniper, etc. Figure 5.2 shows some examples of woody biomass as well as herbaceous biomass.

In general, herbaceous biomass has higher nutrient content and lower lignin content than woody biomass. Herbaceous biomass varies in composition depending on the type of the plant tissue. Also, its composition can vary depending on the time of the year as well as the availability of minerals and nutrients in the soil. These variations can also influence the flow behavior of such biomass.

In what follows, first, we will describe the science of flow behavior in general terms, which applies to any bulk solid. Then, in the next section, we will apply these principles to the flow behavior of biomass materials as they are commonly encountered in densification processes. Given the context of this book, our focus will be primarily on herbaceous biomass.

5.3.2. *Why Do We Need to Describe Flow Behavior?*

In most industrial applications, bulk solids (including biomass materials) are used as raw materials or feedstocks. These bulk solids need to be received, stored, conveyed, metered, and processed.

All these steps rely on the ability of the bulk solid to discharge smoothly from a given piece of equipment or storage system. Without a good understanding of flow behavior, storage and handling equipment often experience significant issues. Stoppage of discharge from silos and hoppers, arching, ratholing, erratic flow behavior, and discharge rate limitations are unfortunately still common problems due to inadequate design of storage and handling equipment. Such problems have significant implications. They can affect the operation of the downstream process/reactor/furnace/boiler, they require manual intervention to restart flow, they result in lost production time, and they can also result in safety issues. Such poorly operating systems reduce the reliability of the entire process.

Thus, an understanding of flow behavior of bulk solids is critical for reliable operations. This need is utmost when storing bulk solids in silos, bins, and hoppers as well as when metering discharge from the outlets of these vessels.

5.3.3. *Flow of Fluids Versus Flow of Bulk Solids*

When one thinks about flow behavior of a bulk solid, flow of fluids naturally comes to mind. Fluids, that is liquids and gases, are all around us. We experience their flow and motion every day, in every breath. Bulk solids also are omnipresent as is evident from the various examples of bulk solids described previously. So, why do we think of fluids when we think about flow?

It seems that the reason is likely related to the fact that many of us have been exposed to at least some theory associated with the flow of fluids, such as in fluid mechanics, gas dynamics, or other similar courses during typical undergraduate engineering curricula. In comparison, the theory for flow of bulk solids was developed relatively more recently as described in the earlier section. As such, many of us have not received formal training in it.

While it is understandable to compare and contrast bulk solids flow with fluids flow, the principles of fluid mechanics cannot be used to describe the flow behavior of bulk solids. To illustrate this point, consider a few key characteristics of fluids and bulk solids and compare them against each other.

Let's begin with friction. Many fluids can be considered low-friction mediums. Flow of water out of reservoirs or storage tanks usually does not lead to significant wear of the tank, as a result of flow only (ignoring impact or corrosion forces). Bulk solids, on the other hand, are quite frictional in nature, comparatively speaking. For example, a silo storing aggregate material can wear fast due to motion of particles against the silo walls.

Another key characteristic is the ability to support/sustain shear force. Fluids typically cannot sustain shear forces, so they flow or expand to the surfaces confining them. When poured on a flat surface, liquids just spread out and cannot form a pile. In other words, liquids have zero angle of repose. Bulk solids, however, can support shear forces. If you pour a bulk solid on a flat surface, it forms a pile. That is why bulk solids have a non-zero angle of repose. For example, consider a pile of woodchips, as shown in Fig. 5.3.

In addition to these key differences, most liquids are nearly incompressible when common applications are considered. In contrast, typical bulk solids are compressible. When a bulk solid is stored in a silo, its bulk density changes along the silo height due to variation in consolidation pressure.

Also, such consolidation can increase the cohesive strength of bulk solids, whereas liquids do not show such behavior in commonly applied consolidation pressure ranges. For example, if the outlet of a silo is not large enough, a silo full of a bulk solid may experience the formation of an arch above the outlet. This can completely stop the

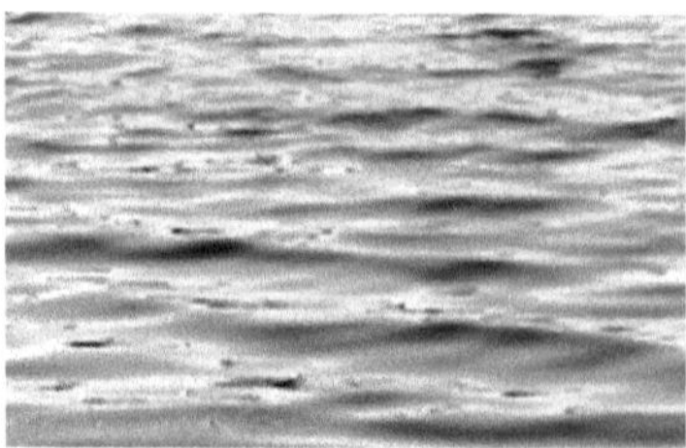

Figure 5.3. Liquids cannot form piles like bulk solids.

discharge from the silo. On the other hand, a tank full of liquid will be able to discharge at any typical outlet size.

The above points illustrate why principles of fluid mechanics cannot be used to model flow behavior of bulk solids. Andrew Jenike recognized this early on and set forth to develop a theory for the flow of bulk solids.

The following sections describe key aspects associated with this theory. Given the scope of this chapter, a mathematical description of the theory is not presented here. The listed references at the end of this chapter provide the mathematical formulation and associated details to any interested reader (Jenike, 1961, 1964; Schulze, 2008).

5.3.4. *Fundamentals of Bulk Solids Flow*

5.3.4.1. *Defining flowability*

People who have worked with bulk solids recognize how complex their flow behavior can be. As such, no single test or index can quantify entire bulk solids flow behavior. While this is true, to talk about a bulk solid's flow behavior at least from a qualitative standpoint, often the term "flowability" is used. A simple way to look at flowability is to say it is the ability of a given bulk solid to flow. Sometimes, this leads to thinking of flowability as a one-dimensional characteristic of a bulk solid, whereby bulk solids can be ranked on a scale from "free-flowing" to "non-flowing." Unfortunately, this simplistic view lacks science and understanding sufficient to address common problems encountered in industrial applications.

Flowability can never be expressed as a single value or index. In fact, flowability is not an inherent material property at all. It is the result of the interaction of several of a bulk solid's characteristics as well as characteristics of the equipment used for handling, storing, or processing the bulk solid. Thus, *easy flowing* bulk solids (e.g., sand and plastic pellets) may exhibit poor flowability when put into or handled through inadequately designed equipment. On the other hand, *difficult-to-flow* bulk solids can exhibit good flowability when handled through robustly designed equipment. Therefore, a more accurate definition of flowability is the ability of a given bulk solid to flow in a desired manner in a specific piece of equipment. Thus, to talk about good flow behavior, one needs to look not only at the flow properties of the bulk solid but also at the design of the handling equipment. In what follows, we connect these two aspects of flowability: characteristics of the bulk solid and equipment design features.

In the following, we discuss principles of flow behavior in a bulk solids handling/storage vessel. These vessels may be called silos, bins, hoppers, columns, or tanks, depending on the industry, company culture, or size of the equipment. From a bulk solids flow perspective, they are all fundamentally storage vessels. All such vessels will be hereafter referred to as *silos* in this chapter.

Typically, silos have a portion where the geometry of the cross-section is constant. This cross-section can be circular, square, or rectangular in shape. It is commonly called the *cylinder* section. The portion of the silo below the cylinder is commonly referred to as the *hopper* section. Its cross-sectional area may increase or decrease over the height, and it can be conical, wedge-shaped, pyramidal, or some unique geometry. The hopper section itself can also consist of sub-sections, each having different shapes and/or different slopes/steepness.

5.3.4.2. *Flow patterns*

There are two primary flow patterns that can develop in a silo during discharge: *funnel flow* and *mass flow*. Both patterns are shown in Fig. 5.4.

In *funnel flow*, an active flow channel forms above the silo outlet, with stagnant material at the periphery. As the level of material in the silo decreases, material from stagnant regions may or may not slide into the flowing channel, depending on the bulk solid's cohesive strength. When the bulk solid has sufficient cohesive strength, the stagnant material does not slide into the flow channel, which results in the formation of a stable rathole. In addition to flow stoppages that occur as a consequence of ratholing, funnel flow can cause material degradation, results in a first-in-last-out flow sequence, and increases

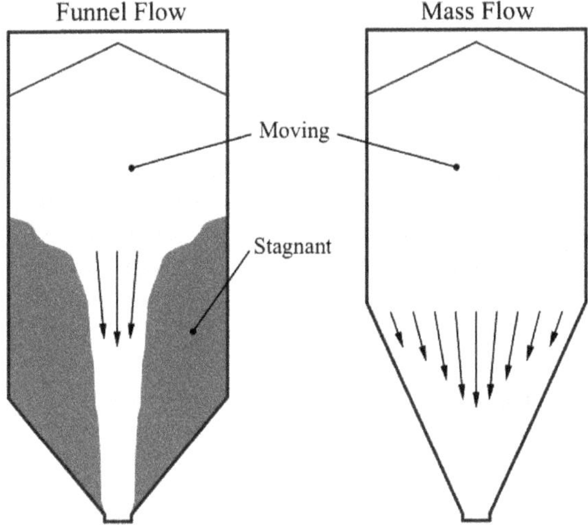

Figure 5.4. Funnel and mass flow patterns.

the extent to which sifting segregation impacts the uniformity of the discharging material.

In *mass flow*, all of the material is in motion whenever any is withdrawn from the silo. Material from the center as well as the periphery moves toward the outlet. Mass flow silos provide a first-in-first-out flow sequence, eliminate stagnant material, reduce sifting segregation, and provide a steady discharge with a consistent bulk density and a flow that is uniform and well controlled. Requirements for achieving mass flow include sizing the outlet large enough to prevent arching and achieve the required discharge rate plus ensuring the hopper walls have sufficiently low wall (material/surface boundary) friction and are steep enough to achieve flow at the walls.

The third type of flow pattern, called *expanded flow*, can develop when a mass flow hopper (or hoppers) is placed beneath a funnel flow hopper, as shown in Fig. 5.5. The mass flow hopper is designed to activate a flow channel in the funnel flow hopper that is sized to prevent the formation of a stable rathole. Particularly for large-diameter silos, the major advantage of an expanded flow discharge pattern is the savings in headroom, as compared to a complete mass flow design. This approach not only reduces the capital cost

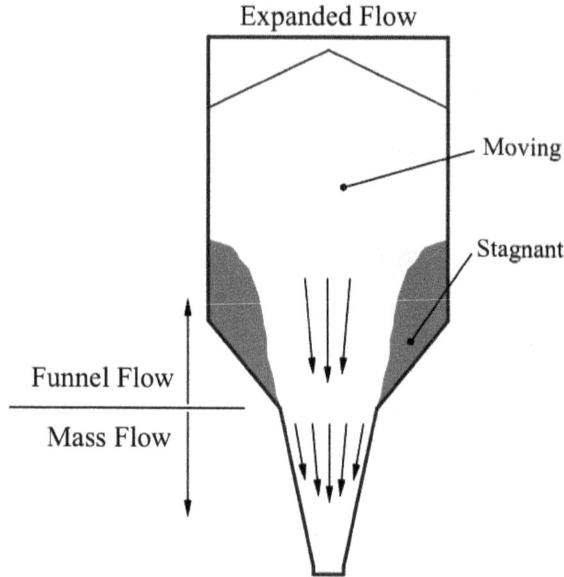

Figure 5.5. Example of expanded flow pattern.

but also facilitates retrofitting existing silos by minimizing the additional headroom requirement. The mass flow hopper beneath the funnel flow hopper still has the benefit of discharging material reliably with a consistent bulk density. Note that segregation and material degradation problems are not necessarily minimized with an expanded flow pattern.

5.3.4.3. *Axisymmetric and plane-flow geometries*

When analyzing the flow behavior of a bulk solid in a hopper section, it is important to recognize the symmetries of key variables with respect to the flow region within the hopper. In his work, Jenike focused on two main types of hopper geometries, namely, *axisymmetric* and *plane-flow* (Jenike, 1961, 1964).

An axisymmetric hopper is one where flow behavior can be assumed to be identical for any plane that passes through the axis of symmetry of the hopper, within the flow region confined by the hopper. A common example is a conical hopper (Fig. 5.6).

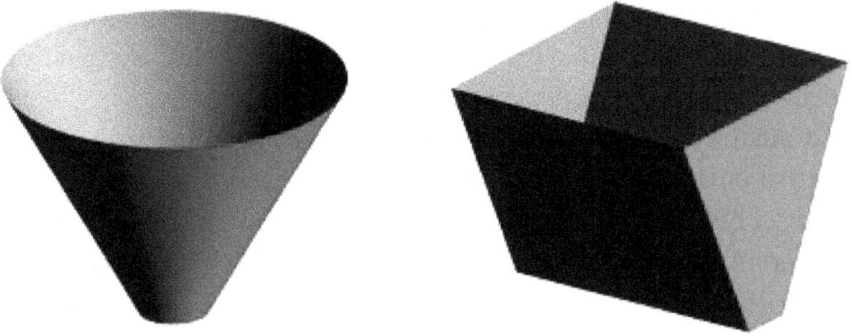

Figure 5.6. Examples of axisymmetric and plane-flow hoppers.

A plane-flow hopper is one where flow behavior can be assumed to be identical for any plane perpendicular to the longest dimension of the plane-flow geometry, within the flow region confined by the hopper. A common example of a plane-flow hopper is a wedge hopper, where the outlet of the wedge hopper has a length of at least three times longer than the outlet width.

Understanding whether a given hopper geometry is axisymmetric or plane-flow is important in order to apply appropriate analysis, as we will see in the following sections.

5.3.4.4. *Importance of proper feeder design*

Most silos have a feeder connected to the hopper outlet to control the rate of discharge of the stored bulk solid. To achieve reliable flow, it is not sufficient to ensure only that the silo design is done correctly; the feeder below the silo is also important. This is because the flow pattern in a silo is strongly influenced by the manner in which the feeder withdraws material from the outlet.

A proper feeder design must accomplish the following:

1. Provide reliable and uninterrupted flow of bulk solids from the silo.
2. Withdraw material from the entire cross-section of the silo outlet.
3. Control/meter the discharge rate from the silo.

The feeder inlet geometry must avoid formation of a lip/ledge where material can build up. Otherwise, it will create an obstruction to material flow and thus eliminate mass flow in the silo.

To ensure that a feeder is capable of withdrawing material from the entire outlet, it must provide increasing capacity in the direction of feed. This can result in different requirements/design features for different types of feeders. This topic will be discussed in more detail in the next section while considering typical biomass handling operations.

5.3.4.5. *Flow through a transfer chute*

When transferring a bulk solid from one mechanical conveyor onto another conveyor or when filling a silo, a *transfer chute* (sometimes called a *transfer point* or simply a *chute*) is often used. It is important to follow the principles outlined in the following in order to achieve good performance from chutes.

The primary function of a chute is to transfer bulk solid. If bulk solid accumulates in the chute, it can fill up to the point that it quickly turns into a hopper; hopper design principles then need to be applied. This typically results in requiring significantly steeper chute walls and larger opening sizes; otherwise arching and ratholing problems could occur. To avoid such a scenario, a good chute design allows sufficient cross-sectional area for the passing bulk solid stream. A general rule of thumb is to keep the chute cross-sectional area less than one-third full (Stuart-Dick and Royal, 1992).

Free fall of bulk solid inside a chute must be avoided or minimized. Otherwise, the falling bulk solid may experience significant impact pressure at the drop location. High impact pressure increases the chances of formation of material buildup. If buildup starts occurring in a chute, it can grow quickly over time and plug the chute leading to stoppage of flow and significant material spillage.

A good chute design principle is to minimize the impact angle between the trajectory of the bulk solid and the plane of the chute surface at the point of impact. High impact angles increase impact pressure and result in loss of momentum of the bulk solid stream.

This causes the stream to significantly slow down after impact and can lead to the bulk solid coming to a stop on the chute surface.

5.3.5. *Bulk Solids Flow Properties and Their Measurement*

In the last section, we discussed the fundamentals of bulk solids flow behavior through various equipment and talked about good design principles. However, we have not yet discussed what data can be used as the basis for conducting such a design. This is the focus of this section. Here, we discuss the properties that together describe the flow behavior of a given bulk solid. We also go over the test methods used to measure these properties as well as the factors that influence these properties.

5.3.5.1. *Typical flow properties that describe bulk solids flow behavior*

There are several properties that are relevant to describing the flow behavior of a bulk solid. These properties are based on the hypothesis that bulk solids can be represented as a continuum. Thus, it assumes that bulk solids flow behavior can be described as a gross phenomenon, neglecting the interaction between individual particles. These properties are collectively referred to as *flow properties.*

Flow properties can be defined as the specific characteristics and properties of a bulk solid that affect its flow behavior at a fundamental level and can be measured. These arise from the collective forces acting on the individual particles, such as van der Waals, electrostatic, surface tension, interlocking, and friction (Cheng *et al.*, 2021).

In the following, we describe some of the most common flow properties of interest in industrial applications of bulk solids handling.

Cohesive strength: As the name suggests, cohesive strength is a quantitative measure of the cohesiveness of a bulk solid. As a common example of this property, consider squeezing wet sand or snow in your hand. If it gains sufficient cohesive strength, it will retain its shape once you open your hand.

In the lab, cohesive strength of a bulk solid is measured under controlled conditions of consolidation and shear (Jenike, 1964). A yield locus, which is a plot of the measured shear force as a function of the compressive load at which shear occurs, is determined at various levels of consolidation pressure. Using this, a relationship describing the cohesive strength of the bulk solid as a function of the consolidating pressure can be developed. This relationship, known as a *flow function*, is important in determining the minimum outlet sizes for silos to avoid flow issues such as arching and ratholing.

Internal friction: Internal friction, which is caused by solid particles moving against each other, is expressed as an angle of internal friction. This angle can be determined using the data obtained from the cohesive strength test (Jenike, 1964). Internal friction value is an important parameter that in conjunction with other properties helps determine hopper design parameters.

Interlocking strength: When a bulk solid contains particles that are significantly non-uniform with respect to the aspect ratio of the particles and/or that are large relative to the size of the outlet, the particles may interlock during discharge. A common day-to-day life example is interlocking of vitamin capsules over the opening when the containing bottle is suddenly inverted. Many biomass materials are sensitive to interlocking. Herbaceous biomass materials are also sensitive to "nesting" phenomenon which can exacerbate the interlocking behavior.

Interlocking tendency is often measured using a scaled-down laboratory setup. In a such test, a representative sample of bulk solid is filled into a specialized hopper that allows for change in hopper outlet size as well as change in consolidation pressure acting on the bulk solid. An appropriate test program is developed based on the underlying application of interest. Accordingly, tests are run at several different levels of consolidation pressures and outlet sizes to experimentally understand the interlocking tendency. From this testing, an outlet size required to avoid interlocking can be determined.

Wall friction: This is a quantitative measure of how frictional a given bulk solid is against a specific surface. Used in a continuum model, it is expressed as the wall friction angle or coefficient of friction. It is important to quantify this value in order to determine the hopper slope at which the bulk solid will slide along the hopper walls. The lower the coefficient of friction, the less steep hopper or chute walls need to be for a bulk solid to flow along them.

In a lab, wall friction is measured by determining wall yield loci (Jenike, 1964). Each wall yield locus is a plot of the measured shear force as a function of the applied compressive force. Variations in the bulk solid and the wall surface finish can have a dramatic effect on the resulting values of wall friction and hence on the hopper angles required to achieve mass flow.

Bulk density: Most readers are probably familiar with the concept of bulk density as the density of a given material in its bulk form. However, it is important to note that bulk density is not a single value, but it varies as a function of consolidating pressure.

There are various methods used in industry to measure bulk density (Schulze, 2008; Carson *et al.*, 2020; ASTM D6683; ASTM D7481). In a typical test, a container is filled with a known mass of the bulk solid and then subjected to progressive consolidation while measuring the change in volume. The test results are often expressed as a straight line on a log–log plot. In bulk solids literature, the slope of this line is typically called *compressibility*. The resulting data can be used to accurately determine capacities for storage and transfer equipment, as well as to provide information to evaluate wall friction, minimum outlet size to avoid arching, and feeder operation requirements.

Permeability: Permeability can be described as the tendency of a bulk solid to resist the flow of a gas through a packed bed of the bulk solid. It is measured as a function of bulk density. During testing, the flow rate of air/gas at a specific pressure drop through a sample of known bulk density and height is measured. Measurements are done at different values of bulk density to plot the relationship between permeability and bulk density.

Flow rate limitations may occur when handling fine-powdered bulk solids. Permeability test data can be used to calculate critical bulk solid discharge rates that will be achieved for steady flow conditions through various outlet sizes. Based on this, outlet size to achieve the desired discharge rate can be determined.

Chute angles: In order to maintain bulk solids flow in a chute, its inside surface walls must be steep enough and have sufficiently low friction to allow the bulk solid to flow along them, especially after an impact. A chute angle test measures the critical chute angles required for clean-off as a function of impact pressure.

The test consists of subjecting a given bulk solid to a range of loads to represent different impact pressures while in contact with a given chute surface. For each impact pressure value, the angle from horizontal at which the bulk solid starts moving against the chute surface coupon is measured by inclining the coupon about a distant pivot point. The angle at which the bulk solid slides is plotted as a function of impact pressure. These angles are used to determine the minimum chute angle required at an impact point to overcome adhesion and ensure flow.

Segregation potential: When two different bulk solids or ingredients are mixed together or when a bulk solid having a wide particle size distribution (e.g., presence of fines as well as coarse particles) is handled, the fine and coarse fractions have a natural tendency to separate from each other during handling. Segregation can also occur due to differences in particle density, particle shape, etc. There are three primary mechanisms of segregation: sifting, fluidization, and dusting (Prescott and Barnum, 2000).

In a typical test, a representative sample of bulk solid is subjected to different mechanisms and the resulting separation/segregation is measured using particle size analysis or chemical assay (ASTM D6940; ASTM D6941).

Understanding of segregation potential is important as segregation can result in discharging or handling a flux of fines or coarse particles at different intervals. If the downstream process requires a uniform mix or uniform particle size distribution, segregation is detrimental. The information obtained from segregation testing is useful in designing equipment reliably to handle a mix of ingredients or bulk solids with wide particle size distribution.

Angle of repose: As described earlier, bulk solids are distinct from liquids in that they can form a pile and thus exhibit a non-zero angle of repose. Because of the ease of such measurement, unfortunately, the angle of repose information is used by uninformed designers anywhere and everywhere. It is certainly an important piece of information and can aid in estimating storage capacity of silos or stockpiles. However, recognizing that an angle of repose measurement is *independent* of the hopper internal surface, it should never be used to calculate hopper angles to achieve reliable flow.

5.3.5.2. *Effect of various parameters on flow behavior*

There are several parameters which influence the flow properties described in the earlier section. Through such influence on flow properties, these parameters ultimately influence the flow behavior of a bulk solid (Carson *et al.*, 2020). To give a general feel to the reader, we present a *simplified* view of the effect of these parameters. The intention here is to help the reader develop a *general* understanding of how changing these parameters can change flow behavior, mainly from the perspective of propensity of flow problems such as arching and ratholing. In reality, the flow behavior of bulk solids is quite complex. The best way to understand the effect of any parameter on flow properties is to conduct actual tests.

Effect of particle size: Particle size has a strong influence on the cohesive strength of a bulk solid. In general, particles with smaller sizes exhibit poorer flow behavior compared to larger particles. Naturally, smaller particles provide greater specific surface area for cohesive forces to interact. As a result, compared to bulk solids consisting of larger particles, a bulk solid consisting of smaller particles typically requires larger outlet size to avoid flow obstructions.

The effect of particle size on wall friction is not always the same as for cohesive strength (Cheng *et al.*, 2021). When considering wall friction, it is not solely dependent on the bulk solid but also on the wall surface. So, depending on the wall surface profile and the competitive mechanisms of sliding and rolling friction, the influence of particle size on wall friction is not always the same.

Effect of particle shape: Typically, particles with irregular shapes and large aspect ratios (length-to-width) exhibit poorer flow behavior than spherical particles. Irregularly shaped or elongated particles such as wood chips, grass strands, and straw exhibit a higher degree of mechanical interlocking. Angular particles tend to be more frictional than smooth particles and cause higher abrasive wear.

Effect of consolidating pressure: As described earlier, most flow properties are a function of the level of consolidation of bulk solid. The larger the consolidating pressure, the more tightly the particles are packed together and the higher the bulk density. Greater cohesive strength makes a bulk solid more prone to flow issues.

Effect of moisture content: The moisture of a bulk solid can change due to rain, heat, change in relative humidity, hygroscopic tendency of the bulk solid, or due to processing. For most bulk solids, cohesiveness increases with increases in moisture content. This is caused by the formation of liquid bridges between particles which increase interparticle cohesive forces due to their surface tension. After a point, especially when the moisture content of a bulk solid reaches near its saturation moisture content, the presence of water can have a lubricating effect on the bulk solid, and cohesiveness starts to decrease.

Effect of temperature: Flow behavior can change substantially when a bulk solid is subjected to varying temperatures. On the one hand, bulk solids such as plastics become sticky and cohesive at elevated temperatures. On the other hand, particles of minerals such as coal and mined ores can turn into huge boulders in freezing conditions. Thus, changes in handling temperature influence flow behavior of many bulk solids; however, the specifics may vary depending on the nature of the material. For woodchips and herbaceous biomass, low temperatures typically result in elevated flow concerns due to freezing of any liquid bridges present between the particles, further increasing the cohesive strength.

Effect of storage time at rest: When a bulk solid remains at rest (i.e., without motion) in a silo for a period of time, its cohesiveness may increase. Typically, such an increase in strength is the result of settling and compaction, crystallization, or chemical reaction.

If allowed to remain stationary on a wall surface, many bulk solids experience an increase in wall friction as well due to increased adhesion. Thus, storage at rest often results in requiring larger outlet dimensions and steeper hopper walls to avoid the potential for flow issues.

Effect of source of bulk solid: Generally speaking, the flow behavior of a bulk solid shows dependance on the source or supplier of the material. For example, depending on the excavation site, ore geology, mining seam, nutrients in the soil, supplier manufacturing process/technology, supplier process control, etc. variations can occur in the same (at least chemically speaking) bulk solid that can lead to differences in flow behavior. This is why, in spite of meeting contract specifications (based on particle size or moisture content or other parameters), it is not uncommon to hear about differences in the exhibited flow behavior from the same bulk solid but coming from different sources. A prudent approach is to test representative samples from such different sources to ensure that equipment is designed to handle such source-based variations.

5.3.5.3. *Jenike shear tester*

In the 1950s, Andrew Jenike developed a translational shear tester to measure the flow behavior of bulk solids under low levels of stress commonly encountered in bulk solids handling applications. This tester is similar to but quite distinct from testers used in soil mechanics applications where the level of stress is much higher. Known as the Jenike Shear Tester, both European and American (ASTM) standard test methods have been developed for its use (Jenike, 1964; Schulze, 2008; Cheng *et al.*, 2021; Carson *et al.*, 2020; ASTM D6128).

While the Jenike Shear Tester is considered the gold standard for measuring bulk solids flow behavior, following Jenike's contributions and publications, a few other types of testers have also been developed. A detailed discussion about these testers is beyond the scope of this chapter. However, their pros and cons are described in other publications referenced at the end of this chapter (Schulze, 2008; Cheng *et al.*, 2021; Carson *et al.*, 2020; ASTM D8081).

The Jenike Shear Tester characterizes flow behavior of bulk solids by shearing representative samples at various levels of consolidation. Using this information, yield loci and wall yield loci are determined at different levels of consolidation. The shearing occurs within a shear cell, which consists of a stationary base, a ring of the same diameter positioned above the base, and a cover on top of the ring (Fig. 5.7).

For the measurement of a point on a yield locus of a bulk solid, the shear cell is filled with a representative sample of the bulk solid (see Fig. 5.8). Then, the cover is loaded centrally with a vertical

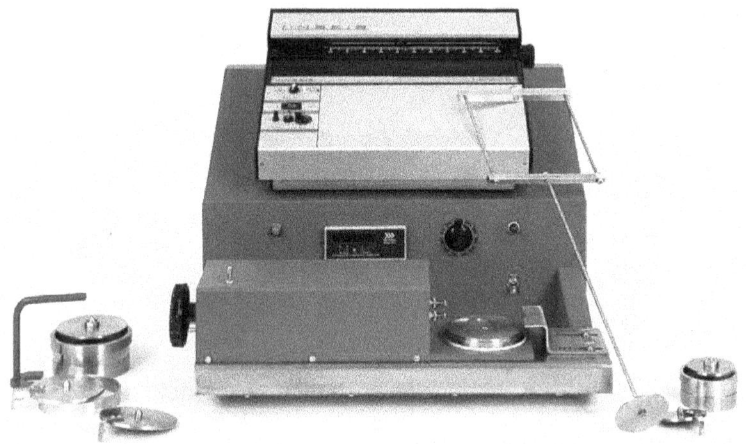

Figure 5.7. Jenike Shear Tester, shear cells, and accessories.

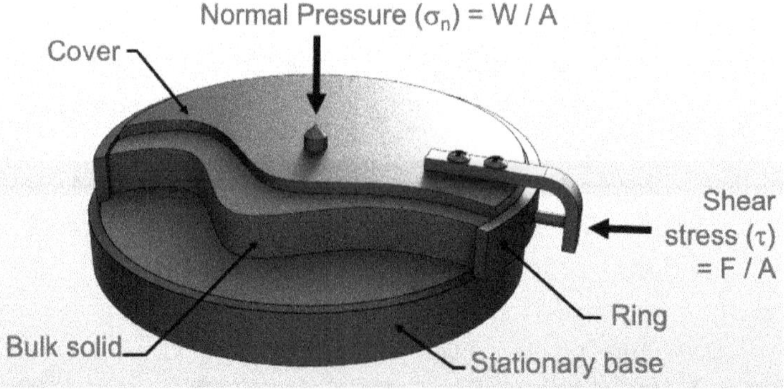

Figure 5.8. Schematic of shear cell loading for yield locus measurement.

force. The upper part of the shear cell, including the ring, the cover, and the bulk solid within, is forced to move horizontally against the fixed stationary base by a motor-driven stem which pushes against a bracket attached to the cover. The shear force exerted by the stem is continuously measured. Due to the displacement of the ring and the cover against the stationary base, the bulk solid undergoes shear deformation. The test is carried out until the sample is sheared to failure. The vertical force and shear force are converted to normal (i.e., perpendicular to the shear plane) and shear stresses based on the cross-sectional area of the shear cell.

When measuring data for the next point on the yield locus, the shear cell is filled with a new sample of the bulk solid, and the procedure is repeated. Often, manual pre-consolidation and pre-shearing are required to facilitate shearing to failure within the limited available travel of the ring (ASTM D6128).

A similar procedure is used when determining wall yield loci, except the stationary base is replaced with a representative coupon of the internal surface of the hopper against which the bulk solid slides (see Fig. 5.9).

While yield loci and wall yield loci represent the primary measurements associated with characterizing flow behavior, there are other measurements as well which require additional test equipment.

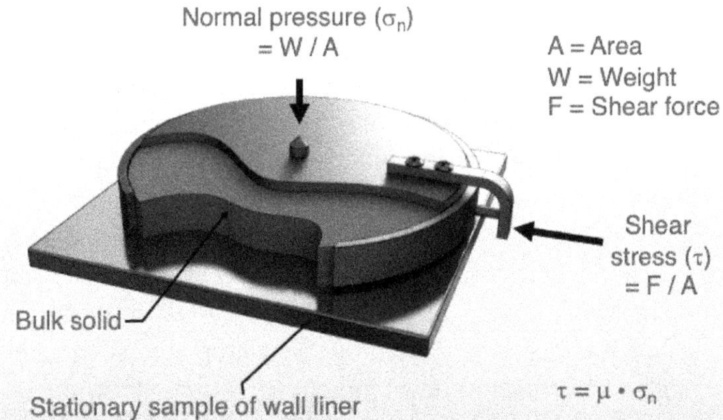

Figure 5.9. Schematic of loading for wall yield locus measurement.

Details about such equipment can be found in the referenced publications (Jenike, 1964; Schulze, 2008).

5.3.6. *Applications of the Measured Test Data*

Once testing has been conducted on a representative sample(s) of the bulk solid at the representative handling conditions of moisture content, temperature, storage time, etc., the resulting set of data is a very powerful tool that allows robust handling equipment design to avoid potential flow problems. From a reliable flow perspective, achieving mass flow is critical for most applications. In this section, we discuss how one can calculate hopper design parameters for achieving mass flow using measured test data.

5.3.6.1. *Determination of hopper angles to achieve mass flow*

In order to achieve mass flow, two conditions must be met. First, the hopper walls must be steep enough and low enough in wall friction to allow flow along them, and second, the hopper outlet must be large enough to prevent arching. How steep such a hopper must be to achieve flow and how large a hopper outlet must be to prevent arching can be calculated based on the flow properties of the bulk solid to be handled.

For a bulk solid to slide on the internal surface of a hopper, friction between the two must be overcome. As explained earlier, a wall friction test, in conjunction with cohesive strength and bulk density tests, allows measurement of shear stress as a function of applied normal pressure. Figure 5.10 shows the results of a typical wall friction test. Wall friction angle, designated as ϕ', is defined as the angle formed by a line drawn from the origin to a point on the curve. The coefficient of friction μ is defined as the ratio of friction force to the force perpendicular to the surface; hence, $\mu = \tan \phi'$.

Once the wall friction angle for a given bulk solid and hopper inside surface has been measured, the next step is to determine what hopper angles will result in development of mass flow. Andrew Jenike developed design charts that show the relationship between allowable hopper angles for mass flow and wall friction angle (Jenike, 1964).

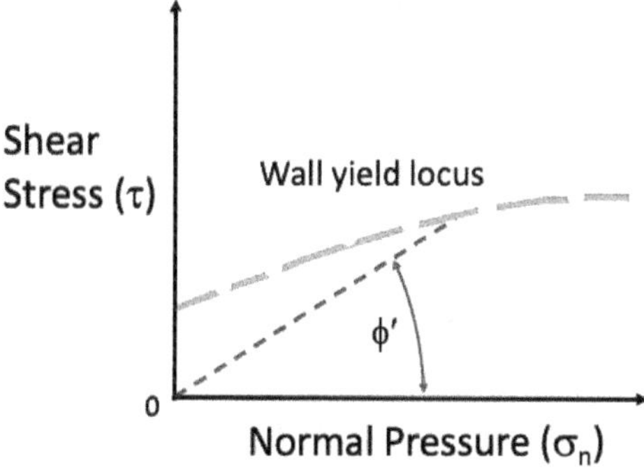

Figure 5.10. Typical wall yield locus and wall friction angle.

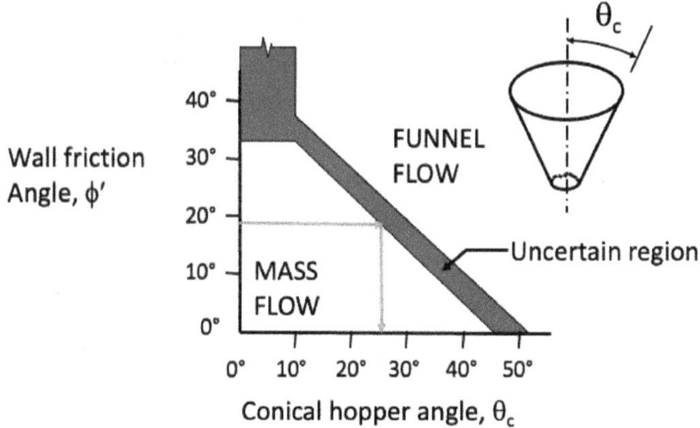

Figure 5.11. Typical mass flow design chart for conical hoppers.

Figure 5.11 shows a typical chart for a conical hopper geometry. On the horizontal axis are values of hopper angles (θ_c) measured in degrees from vertical while the vertical axis has wall friction angles, ϕ'. Using such a chart, one can determine the maximum (i.e., shallowest) hopper angle that will allow mass flow for a certain wall friction angle. When a measured value of ϕ' is selected, and a horizontal line is drawn from the ϕ' value to the edge of the mass flow region, the intersection point is projected down to the x-axis to calculate the maximum hopper angle, which is the shallowest angle to achieve mass flow.

Note that there is a region labeled *uncertain* which lies between funnel flow and mass flow. This represents a margin of safety to cover slight variations in bulk solids properties.

Jenike also developed similar charts for plane-flow (i.e., wedge) hoppers. For the same value of wall friction angle, mass flow can be achieved at shallower angles (typically 10°–12° less steep) in a plane-flow hopper compared to an axisymmetric hopper.

5.3.6.2. *Determination of hopper outlet size to avoid arching*

While a rathole cannot form in a mass flow hopper, it is still possible for an arch to form over the hopper outlet. Two types of arches can form: cohesive arches and interlocking arches. Typically, cohesive arching is associated with the presence of a significant quantity of fines or moisture in the bulk solid, whereas interlocking arching is associated with the presence of particles that are relatively large and/or have a large aspect ratio.

As explained earlier, a cohesive strength test, in conjunction with the bulk density test, allows measurement of a bulk solid's flow function (the strength/consolidating pressure relationship). This strength, called unconfined yield strength, is directly related to the ability of the bulk solid to form cohesive arches and ratholes.

Once a flow function has been determined, the minimum opening size to prevent arching can be calculated by the use of the hopper's *flow factor*. Similar to the mass flow angle design charts, Jenike also developed charts for calculating flow factors for a given hopper geometry and wall friction angle. The intersection of the flow factor and flow function corresponds to the minimum hopper outlet size required to prevent a cohesive arch from forming.

There are separate flow factor charts for axisymmetric and plane-flow hoppers. In addition to the benefits offered by plane-flow over axisymmetric flow in terms of less steep hopper walls required to achieve mass flow, the required width of a slotted opening at the outlet of a wedge hopper to prevent arching is about half that required for the diameter of a conical hopper outlet. Note that the length of a slot must be at least three times its width for the plane-flow condition to be satisfied.

Interlocking arching can occur when handling bulk solids with particles that are relatively large compared to the hopper outlet size. One rule of thumb is that a conical hopper outlet must have a diameter of at least six to eight times the longest dimension of the bulk solid particles to prevent interlocking arching. For wedge hoppers, the hopper outlet width must be at least three to four times the longest dimension of the particles provided the outlet length is at least three times its width. Unfortunately, for biomass feedstocks, these rules of thumb often fall short in providing reliable flow without interlocking. As a result, the interlocking strength test described earlier is quite helpful in getting a good estimate of the outlet dimension required to prevent interlocking arching.

When dealing with biomass feedstocks, especially when the feedstock particles are fibrous or have jagged edges, some additional factors need to be taken into account. For such feedstocks, due to the flexible nature of particles as well as the presence of intertwining fibers, *nesting* of particles can occur. This results in much higher propensity for interlocking arching. While the interlocking strength test is still a good method to estimate the outlet size requirement for such fibrous feedstocks, there are practical limitations as to how many different particle alignments can meaningfully be created during the test. As a result, prediction of outlet dimension to avoid interlocking arching for biomass feedstocks involves some uncertainty.

5.3.6.3. *Determination of hopper outlet size to achieve desired flow rate*

In addition to avoiding arching, a hopper outlet must also be sized to achieve the desired discharge rate. All bulk solids have a maximum steady-state discharge rate corresponding to a given outlet/opening size. For bulk solids consisting primarily of coarse particles, achieving the desired discharge rate is seldom an issue.

For fine bulk solids, the maximum obtainable discharge rate can be much lower than for a coarse bulk solid with the same bulk density. This is due to the interaction between air (or gas) and fine particles as described by the permeability of the bulk solid. When handling bulk solids with a significant portion of fines, a detailed analysis needs to be conducted by considering the two-phase solids/gas interactions, based on the measured flow properties of the bulk solid as well as the specific hopper design features (Purutyan *et al.*, 2008).

5.4. Application to Densification of Herbaceous Biomass

Densification of biomass is an area of significant interest to the industry. Often, low bulk density of ground/milled herbaceous biomass is a major barrier to its widespread use. The low bulk density, coupled with low energy content compared to traditional fuels, implies that often a large volume of biomass feedstock is required for use in the application, whether it is related to power generation or biofuels production.

One of the ways to address this issue is to densify herbaceous biomass. Pelletization is a very widely used method of densification. Pelletization can increase the bulk density up to 4–5 times the bulk density of herbaceous biomass. This can significantly lower the feedstock volume required to be transported, stored, as well as handled. It can provide some operational simplicity and may also save costs.

In addition to this, pelletization of herbaceous biomass can also result in the improvement of its flow behavior. When handling milled herbaceous biomass, the low bulk density coupled with the nesting tendency of milled fibers can often result in poor flow characteristics. The presence of fines and moisture further increases challenges to flow behavior. In comparison, handling pelletized biomass may provide several advantages. Pellets are typically dry and have low fines content. Often, pellet surface is smooth and nesting of fibers is not likely to occur. As a result, generally speaking, pelletization can improve flow behavior of milled herbaceous biomass. However, when handling pelletized feedstock, interlocking of pellets must be avoided. Furthermore, if conditions during transport, storage, and handling of pellets result in breakage, fines generation, and exposure to moisture, significant flow issues can occur for pelletized biomass also. Specifics of pellet flow behavior will depend on the feedstock, pellet size, shape, moisture content, presence of binders, starch, protein, fiber, inorganic matter, etc.

So far, we have discussed what is a bulk solid, how flow behavior of a bulk solid is different from flow behavior of a fluid, what general principles guide the design of bulk solids storage and handling equipment to achieve reliable flow behavior, what properties determine the flow behavior of a bulk solid, how to measure these properties, and how to use this information to determine key hopper design parameters. Now, we are at a point where we can apply this information to bulk solids storage/handling operations involved in a typical densification process for herbaceous biomass. Various biomass densification technologies have been reviewed in detail in the literature (Tumuluru *et al.*, 2010). Also, we recognize that other chapters in this book go over the details of such operations from a process standpoint; we cover these operations from a reliable bulk solids storage/handling perspective.

5.4.1. *Receiving of Feedstock in the Yard*

Typically, most biomass feedstocks are delivered to densification plants by trucks. These feedstocks may be in the form of bales of herbaceous biomass such as corn stover, switch grass, or miscanthus. In some cases, loose feedstock such as sugarcane bagasse may also be delivered by truck. These trucks often unload the feedstock in the receiving yard by either a self-reclaiming truck unloader such as a Walking Floor® or using platforms that elevate the trailer on an adjustable sloping ramp (Fig. 5.12).

Depending on the size of the plant, its throughput, and the nature of the feedstock, the storage and handling system may be simple or sophisticated (Khambekar and Barnum, 2012). Simple systems often use front-end loaders to move the feedstock unloaded from trucks to a storage location. More complex systems may employ belt conveyors to move the loose feedstock to form a pile with an automatic stacker and reclaimer mechanism.

Ideally, the feedstock storage location should be covered so as to keep the feedstock away from impurities and offer adequate

Figure 5.12. Typical feedstock receiving yard.

protection from rain and other sources of moisture. As discussed earlier, the presence of moisture typically worsens flow behavior. As such, additional capital required for building covered storage is well justified given improvement in flow behavior as well as reduced processing such as drying and impurity removal operations downstream.

For achieving densification, biomass feedstock needs to be pelletized or briquetted. While we recognize that the requirements for feedstock particle size and moisture content may change depending on the choice of pelleting or briquetting, from a bulk solids handling standpoint, the handling operations and general design principles are quite similar. So, while the following discussion is presented for pellet mills, it also applies to briquette presses.

5.4.2. *Size Reduction and Drying*

For successful pelletization operation, the feedstock needs to be reduced to a small enough particle size and adjusted to appropriate moisture content. Most pellet mills rely on receiving –3 mm material as input (Garcia-Maraver and Carpio, 2015). The feedstock should not be too dry or too wet. Too dry feedstock can make pellets brittle and prone to cracking, while too wet feedstock will result in pellets that crumble apart easily. Thus, size reduction and drying are critical parts of the pelletization process (Tabil *et al.*, 2011).

Most pellet plants use a two-step size reduction process (Garcia-Maraver and Carpio, 2015). First, depending on the nature of the biomass feedstock, a shredder or coarse hammermill is used to break straws to a minus 1 inch length. Depending on the moisture content of the feedstock, this coarse feedstock is then sent for drying or directly to a second, final-stage hammermill, which grinds the feedstock to −3 mm.

During these size-reduction and drying operations, typically the feedstock is transferred using belt conveyors and transfer chutes. This is where the chute design principles apply, in conjunction with the measured flow properties at the representative handling conditions. It is important to ensure that the mill-feed-chute and/or the dryer-feed-chute don't operate with full cross-section. A chute with a full cross-section acts as a hopper and thus becomes susceptible to the occurrence of arching and ratholing. Under this scenario, discharge from the chute will stop completely or become highly limited.

Good chute design principles must be followed in order to ensure that hammermills and dryers operate reliably without starving. Discrete Element Modeling (DEM) can be used to visualize expected flow behavior through chutes and adjust the chute design as necessary prior to fabrication (see Fig. 5.13).

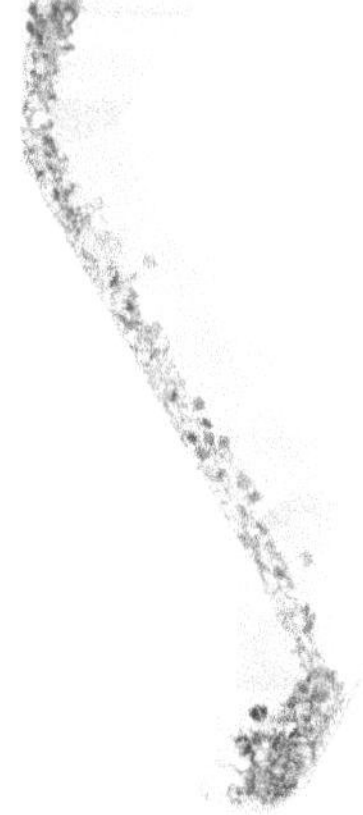

Figure 5.13. Examples of problematic (left) and improved (right) chute design for woodchips.

5.4.3. *Buffer Storage*

Once the biomass feedstock is sized and has the correct range of moisture content [typically, between 10 and 20% (Garcia-Maraver and Carpio, 2015)], it is useful to have a buffer storage of such feedstock. Such storage allows for pelletization operation even when the hammermill(s) and/or dryer are undergoing maintenance.

As discussed earlier, it is important to ensure that the buffer storage silo is designed to achieve mass flow. This requires sizing the outlet large enough to avoid arching and ensuring that hopper walls are smooth/steep enough to achieve flow along them. If the silo has smooth/steep enough hopper walls but the outlet is not appropriately sized, arching can still occur.

When arching occurs, an obstruction forms over the hopper outlet (see Fig. 5.14). Arching can occur due to the cohesive strength of the feedstock or due to mechanical interlocking/nesting of large particles. The presence of fines and moisture in the feedstock increases the possibility for cohesive arching (for example, hammermilled herbaceous feedstock), whereas the presence of particles with large aspect ratios increases the possibility for interlocking arching (e.g., woodchips). As seen in Fig. 5.14, the presence of straws, fibers, and splinters often further increases the arching potential. When the feedstock forms a

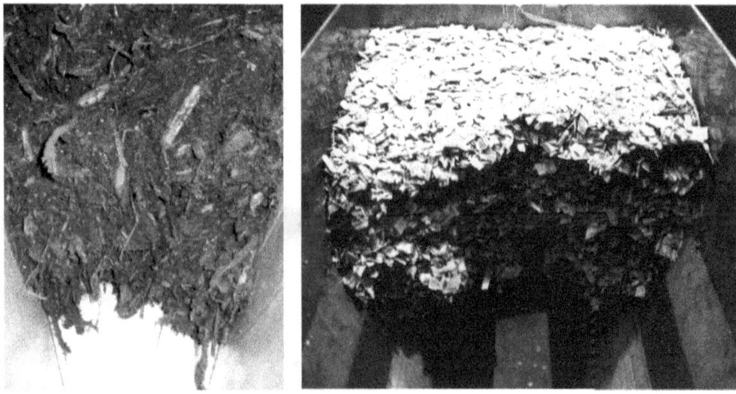

Figure 5.14. Examples of cohesive and interlocking arching for biomass materials.

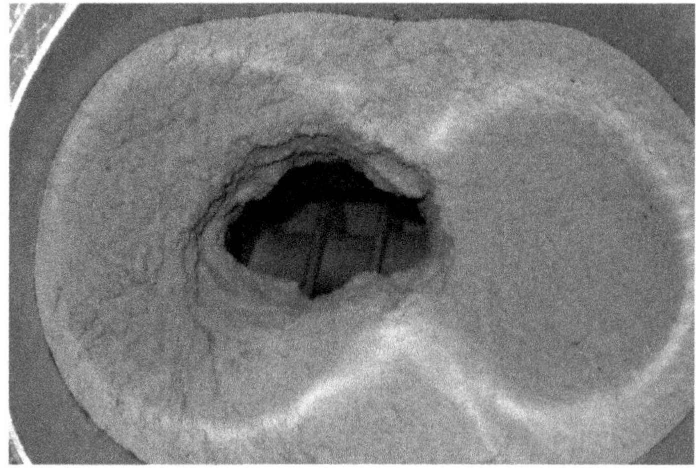

Figure 5.15. Example of ratholing.

stable arch above an outlet, material discharge is prevented resulting in a no-flow condition which starves the downstream pellet mill.

If the buffer storage silo is not designed to achieve mass flow, funnel flow results, often with accompanying rathole formation (see Fig. 5.15).

Initiation of a rathole occurs as the material moves towards the hopper outlet through a steep, funnel-shaped flow channel surrounded by stagnant material near the walls of the shallow hopper geometry. As the level of material in the flow channel drops, layers of material from the top surface of the stagnant region may continue sliding into this active flow channel. If this fails to occur, the flow channel empties and a stable rathole forms. Under this condition, discharge from the silo stops until the material is filled again into the empty flow channel.

Whenever stagnant material is present inside a silo, it will result in *limited live storage capacity.* Such stagnant material occupies valuable storage space in the silo and will not discharge on its own. Funnel flow also leads to a first-in-last-out flow sequence which can affect the quality of pellets produced. Thus, it is important to ensure that the buffer storage silo is designed to achieve mass flow.

5.4.4. *Feeding a Pellet Mill*

A pellet mill requires a consistent, steady supply of feedstock. Thus, a properly designed feeder is essential to meter discharge from the buffer silo. Screw feeders and belt feeders are commonly used in the biomass industry. No matter which feeder is used, as described in the earlier section, it is important to ensure that the feeder withdraws material from the entire cross-section of the silo outlet.

If a screw feeder has a constant pitch, the material will flow through a channel near the back wall of the hopper creating the potential for ratholing as well as decreasing the achievable discharge rate (see Fig. 5.16). If the screw feeder is designed with an appropriate increase in capacity along its length, it will remove material from the entire cross-section of the silo outlet. This type of screw feeder is called a *mass flow screw feeder* (Marinelli and Carson, 1992).

If a belt feeder is used, appropriate increasing capacity must be provided in the direction of feed (similar to a screw feeder) to remove material from the entire outlet cross-section.

If an improperly designed belt feeder interface is installed, preferential flow from the back or front of the hopper will occur, leading to the occurrence of flow problems (see Fig. 5.17). This can be avoided by using a tapered interface that provides increasing capacity. In the interface section, the increasing capacity along the

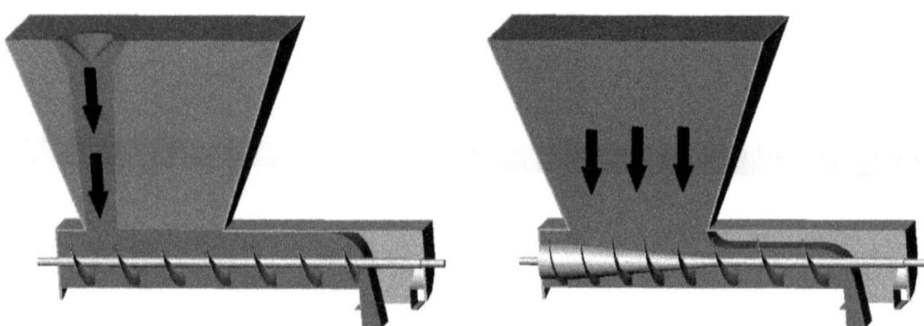

Figure 5.16. Constant pitch screw feeder (left) and mass flow screw feeder (right).

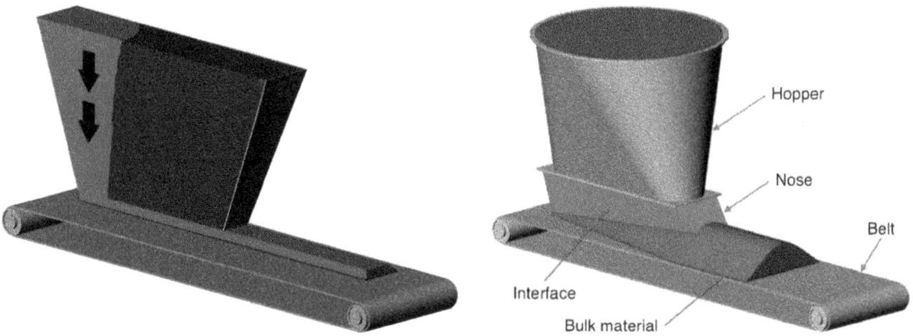

Figure 5.17. Lack of increasing capacity (left) and properly designed interface with increasing capacity (right).

length is achieved by the increase of the height of the interface above the belt as well as the increase in width (Carson *et al.*, 2008).

5.4.5. *Handling of Pellets*

Once the herbaceous feedstock is converted into pellets or briquettes, its nature changes but it is still a bulk solid. This bulk solid consists of relatively large particles that often have a large aspect ratio. As such, after pelletization/briquetting, one is more concerned about the occurrence of interlocking rather than the occurrence of cohesive arching.

Pellet storage bins and feeders also need to take into consideration that the pellets can break during handling. Breakage not only results in pellet quality issues but also generates dust, which can be an explosion or health hazard. Hence, material stream impacts should be avoided in post-pelletization handling, with the focus on gentle transfers.

5.4.6. *Final Word on Bulk Solids Flow Behavior*

Flow behavior is an important consideration which can make or break the densification process. A successful plant operation requires knowledge of the characteristics of the feedstock being handled and equipment designs that follow established design principles using measured test data on representative samples. Not doing so often leads to the occurrence of flow issues. When flow problems occur, valuable production capacity and time are lost, excessive maintenance and housekeeping costs are incurred, and health and safety issues arise.

Next, we discuss some project management aspects. In today's project-stage-focused approach, it is important to understand which types of design activities should be conducted at which project stage. This will ensure successful incorporation of proper bulk solids handling system design principles into your next project.

5.5. Project Management of Biomass Projects

Due to the variability of biomass feedstocks, handling systems need to be designed robust enough to handle the range of potential biomass characteristics.

Analysis of a mining and metals study gives insight into how to move through projects involving highly variable raw material feedstocks (such as biomass) with increased success. Twigge-Molecey examined how poor utilization of knowledge in the mining and metals industry resulted in over US$20 billion in non-productive capital investments over the last decade in 43 specific projects (Twigge-Molecey, 2003). From the analysis of the failures, he identified key issues relating to risk management in capital projects. The results from this study can be correlated to practices related to scale-up and production in biomass densification processes.

Analysis of these projects that failed to meet industry criteria for success indicated that over two-thirds of the projects did not have proper phasing, over two-thirds had no continuity in the execution teams from concept to start-up, and over 40% had serious "front-end development" issues. In most cases, at least two of the above factors were present.

According to Twigge-Molecey, the key reasons that specific technology development programs fail are as follows:

1. lack of strategic alignment with the business, resulting in lack of adequate support and resources,

2. lack of a disciplined, multi-function "stage gating" process (or phasing) during development allowing projects to progress that should not,
3. lack of a corporate champion who will maintain the momentum over the many years required to bring a new technology to successful commercial application, and
4. not bringing the best possible minds and experience to the program.

Densification projects typically involve innovative technology. As such, the project must be set up correctly at the beginning by bringing the best minds and experience to the program. Projects require a multi-disciplinary team as no single individual or organization has the expertise to devise a complete solution. While it can be tempting to jump into action, it is critical to first identify all the competencies and skills that may be needed for the project. Even though the exact pathway of the project is sometimes a bit of an unknown, it is usually possible to identify the competencies required. Once identified, the competencies can then be used as a framework onto which each of the proposed participants (individuals and/or organizations) can be mapped. Each participant needs to be a master of a primary competency listed. In most cases, they will have some experience with the other competencies on the list as well.

Each participant should be rated (0 = No knowledge to 10 = Mastery) against each competency (Wellwood, 2016a). The aim is to cover all competencies with an 8–10 rating participant, without too much overlap above 5. Where there are two or more participants with high ratings in the same competency, unwelcome demarcation issues can arise, especially if they are geographically dispersed.

As indicated in the example map in Fig. 5.18, it is often not until you map it out like this that the gaps become evident. Gaps can also occur with regard to computer modeling and simulation such as DEM and CFD to select the correct process equipment for biomass densification.

Direct Funding Opportunity (DFO) projects, through the Feedstock-Conversion Interface Consortium (FCIC) consisting of

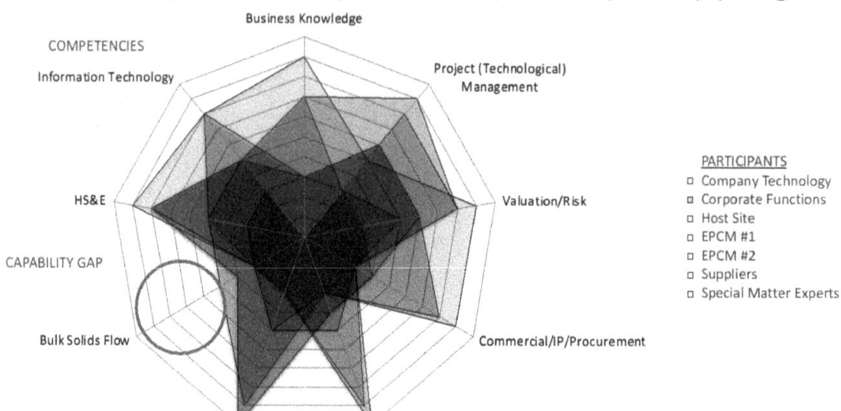

Figure 5.18. Capability mapping.

nine US Department of Energy (DOE) National Laboratories, are helping industry address challenging problems related to the handling of biomass to improve operational reliability. Through continued efforts, these national laboratories have developed significant capabilities in process research that can be used by industry to fill competency gaps and reduce project risk.

Once the multi-disciplinary team is formed, it should evaluate the project and list the adverse events that may occur, assess the likelihood that they will occur, and estimate the severity of each event (Bell, 2005). The key to this type of study is to be realistic concerning the likelihood and severity of each event.

When taking on a project, it is important to consider four key dimensions (Wellwood, 2017):

- basis of design
- options analysis and flowsheet development
- equipment selection
- project governance and best practice

Within these categories, there are critical questions that need to be asked of the biomass project team to avoid problems involving bulk solids handling (Wellwood, 2016b). Questions circle around the following:

1. Have you included flow-related impacts in the sensitivity analysis?
2. Is the basis of design informed by physical testing of the feedstock and all its variation?
3. Have you tested the feedstocks, mixtures, additives, and waste throughout the process?
4. Have you measured all the characteristics necessary to inform the design stage?
5. Has the testing considered variations between suppliers to explore any sensitivity?
6. How is the natural behavior of biomass being dealt with in the flowsheet?
7. What steps have been taken to identify and explore all the "What If's"?
8. Have all the material flow risks been captured in the risk assessment and assigned realistic likelihood and consequence scores?
9. Have equipment items been matched to the characteristics of the materials being handled?
10. Have material handling equipment items been properly specified?
11. How were the suppliers selected, and what factory and site acceptance tests have been agreed to?
12. Do you have the right competencies at the design table? How do you know?
13. Has the process been independently reviewed by a competent expert in the area?
14. Have you defined what constitutes a successful outcome in terms of flow and its impacts?
15. Who is accountable for achieving controlled flow?

5.5.1. *Closing Thoughts on Project Management*

Sometimes, it is easy to lose sight of bulk solids handling operations while focusing on high-profile process equipment such as hammermill, dryer, and pellet mill. However, it is often the bulk solids handling that becomes the weakest link and can prevent a plant from reaching commercial success.

To improve project success, great care must be taken in forming multi-disciplinary teams to cover all aspects of the project from the process side to the bulk solids handling side and everything in between. Risk assessments must be performed with realistic expectations of likelihood and severity to develop a plan to mitigate those risks, which may include physical modeling and/or computational modeling. Any type of modeling should be performed with care as any model is only as good as the inputs and model developed. Using vendor-agnostic specialists in different areas such as bulk solids handling, preprocessing, and processing at the beginning of the design ensures that the feed is well characterized and variability understood. It also helps the project move through the iterative design cycles quicker and navigate suppliers and equipment selection by virtue of their independence.

Setting aside sufficient amount of money and time to measure the flow characteristics of the bulk solids and run pilot plants and modeling is critical, and assembling the correct team reduces project risks. Taking this proactive approach to project planning ensures success in biomass handling operations.

5.6. Conclusions

Given the increasing use of biomass feedstocks in various applications, aspects associated with reliable flow of biomass are receiving greater attention. The ability of feedstock to discharge smoothly from a given piece of equipment or storage system is critical in many processes. Without a good understanding of flow behavior, storage and handling equipment often experience significant issues. Such problems can be avoided by conducting a thorough analysis of flow behavior and using a systematic design approach for storage and handling equipment.

Stoppage of discharge from silos and hoppers, arching, ratholing, erratic flow behavior, and discharge rate limitations are the main problems due to inadequate design of storage and handling equipment. These problems can affect the operation of the downstream process/reactor/furnace/boiler, they require manual intervention to restart flow, they result in lost production time, and they can also result in safety issues. Such poorly operating systems reduce the reliability of the entire process.

There are two primary flow patterns that can develop in a silo during discharge: *funnel flow* and *mass flow*. In *funnel flow*, an active flow channel forms above the silo outlet, with stagnant material at the periphery. In *mass flow*, all of the material is in motion whenever any is withdrawn from the silo. Many of the flow issues are associated with the funnel flow discharge pattern and can be solved by designing the equipment to achieve reliable mass flow, based on the flow characteristics of the representative sample of bulk solid. Key characteristics include cohesive strength, internal friction, interlocking strength, wall friction, bulk density, permeability, chute angles, and segregation potential. Parameters such as consolidating pressure acting on the bulk solid, moisture content, temperature, and storage time at rest can also noticeably influence flow behavior. The measured data from testing are analyzed to determine silo design parameters such as hopper angles and outlet sizes. Such a systematic design process can be applied to hoppers and silos associated with receiving feedstock in the yard, size reduction, drying, buffer storage, feeding pellet mill, and handling of pellets.

To improve project success, it is helpful to form multi-disciplinary teams to cover all aspects of the project from the process side to the bulk solids handling side and everything in between. Risk assessments must be performed with realistic expectations of likelihood and severity to develop a plan to mitigate those risks. Taking a proactive approach to project planning can go a long way in ensuring success in biomass handling operations.

References

Arnold, P. C., McLean, A. G., & Roberts, A. W. (1980). *Bulk Solids: Storage, Flow and Handling.* Australia: TUNRA Bulk Solids Handling Research Associates.

ASTM D6683. Standard test method for measuring bulk density values of powders and other bulk solids.

ASTM D7481. Standard test methods for determining loose and tapped bulk densities of powders using a graduated cylinder.

ASTM D6940. Standard practice for measuring sifting segregation tendencies of bulk solids.

ASTM D6941. Standard practice for measuring fluidization segregation tendencies of powders.

ASTM D6128. Standard test method for shear testing of bulk solids using the Jenike Shear cell.

ASTM D8081. Standard guide for theory and principles for obtaining reliable and accurate bulk solids flow data using a direct shear cell.

Bell, T. (2005). Challenges in the scale-up of particulate processes — An industrial perspective. *Powder Technol.*, *150*, 60–71.

Carson, J. W., Del Cid, L., & McInerney, J. (2020). Flow measurement of powders and bulk solids. In *Kirk-Othmer Encyclopedia of Chemical Technology* (pp. 1–23). John Wiley & Sons, Inc.

Carson, J. W., Cabrejos, F., & Rulff, M. (2008). Effective design of belt feeder interfaces to achieve reliable operation. Proceeding of International Symposium, Reliable Flow of Particulate Solids IV, Tromso Norway, June 10–12.

Cheng, Z., Leal, J. H., Hartford, C. E., Donohoe, B. S., Craig, D. A., Xia, Y., Daniel, R. C., Ajayi, O. O., & Semelsberger, T. A. (2021). Flow behavior characterization of biomass feedstocks. *Powder Technol.*, *387*.

Garcia-Maraver, A. & Carpio, M. (2015). Biomass pelletization process. In *Biomass Pelletization Standards and Production*, Chapter 4, WIT Transactions on State-of-the-art in Science and Engineering.

Jenike, A. W. (1961). *Gravity Flow of Bulk Solids.* Bulletin No. 108, Utah State University.

Jenike, A. W. (1964). *Storage and Flow of Solids.* Bulletin No. 123, Utah State University.

Khambekar, J. V. & Barnum, R. A. (2012 September). Cofiring biomass: Material handling challenges and solution approach. *Biomass Mag.*, *6*(9), 42–44.

Marinelli, J. & Carson, J. W. (1992 December). Use screw feeders effectively. *Chem. Eng. Prog.*, *88*(12), 47–51.

Prescott, J. K. & Barnum, R. A. (2000 October). On powder flowability. *Pharm. Technol.*, *24*(10), 60–84.

Purutyan, H., Del Campo, A., & Barnum, R. A. (2008). Using flow properties to solve flow problems with hard-to-handle powders in the ceramics industry. Presented at the 6th International Latin-American Conference on Powder Technology, November 7–10, 2007, Rio de Janeiro, Brazil. And Materials Sciences Forum Vols. 591–593, pp. 620–627.

Schulze, D. (2008). *Powders and Bulk Solids: Behavior, Characterization, Storage and Flow.* Springer.

Stuart-Dick, D. & Royal, T. A. (1992 September). Design principles for chutes to handle bulk solids. *Bulk Solids Handling*, *12*(3), 447–450.

Tabil, L., Adapa, P., & Kashaninejad, M. (2011). Biomass feedstock preprocessing — Part 1: Pre-treatment. In *Biofuel's Engineering Process Technology.* IntechOpen.

Tumuluru, J. S., Wright, C. T., Kenney, K. L., & Hess, J. R. (2010). A review on biomass densification technologies for energy applications. Tech. Report INL/EXT-10-18420, Idaho National Laboratory.

Twigge-Molecey, C. (2003 December). Knowledge, technology and profit 2003. Cobre 2003; Fifth International Conference, Santiago, Chile, pp. 41–57.

Wellwood, G. (2016a). Fail to plan? Plan to fail! — The case for capability mapping. *LinkedIn Post*, March.

Wellwood, G. (2016b July/August). 15 good questions for mining and minerals project teams [online]. Australian Bulk Handling Review, pp. 26–30.

Wellwood, G. (2017 July). One perfect (production) day; a bulk solids handling perspective. Iron Ore 2017 Conference. Paper Number 46.00.

https://doi.org/10.1142/9781800613799_0006

Chapter 6

Thermal Pretreatment Impact on Densified Biomass Physical Properties

Bahman Ghiasi[*,‡], Jaya Shankar Tumuluru[†,§], and Shahab Sokhansanj[*,¶]

[*]*Chemical and Biological Engineering Department*
The University of British Columbia, Vancouver
British Columbia, Canada
[†]*Southwestern Cotton Ginning Laboratory*
United States Department of Agriculture
Agriculture Research Service, Las Cruces, New Mexico, USA
[‡]*bahman.gs@gmail.com*
[§]*jayashankar.tumuluru@usda.gov*
[¶]*shahab.sokhansanj@ubc.ca*

Abstract

Lignocellulosic biomass has low energy content and is high in oxygen content. The proximate and ultimate composition of lignocellulosic biomass is inferior compared to coal. The grinding properties of biomass are completely different compared to coal. Biomass is fibrous, whereas coal is brittle. One way to make biomass look like coal is through torrefaction. The biomass is roasted in an oxygen-free environment at temperatures of 200–300°C for different residence times. During torrefaction, the biomass loses the low energy content of volatiles and produces a solid product that is high in energy content. The solid fraction rich in carbon is torrefied biomass or 'biocoal'. Biocoal represents a renewable energy commodity that can substitute coal. The torrefied biomass has superior biomass in terms of proximate and ultimate composition and physical properties such as grinding and particle size, but the challenge is low in bulk density. The low bulk density is primarily due to the loss of low energy content volatiles during the torrefaction process. One way to increase the density of the torrefied biomass is through densification. The densification systems commonly used today are pellet mills and briquette presses. Densification helps improve the transportation and handling of low-density torrefied biomass. The challenge is in making a durable pellet using torrefied biomass as the biomass loses its binding ability. Typically, binders are used for making torrefied and densified biomass. The challenge of adding binders is introducing foreign material to the biomass and changing the composition. In addition, adding binders can change the torrefied material properties, such as hydrophobicity. The other option to overcome this limitation is to torrefy the densified biomass. This chapter looks at the production of torrefied and densified biomass and its physical properties.

Keywords: Torrefaction, steam explosion, pelleting, pellet quality, unit density, durability, ash content

6.1. Introduction

Based on Energy Information Administration (EIA) (2018), energy from biomass is called bioenergy and is considered renewable energy. Biomass is one of the significant energy sources for humanity and is presently estimated to contribute 10–14% of the world's total energy supply. A joint US Department of Energy (DOE) and Oak Ridge National Laboratory (ORNL) published report suggests there are more than 1.2 billion tons of biomass currently produced in the US annually, which could sustainably be used for energy production (DOE, 2016). Various types of biomasses are typically used for bioenergy production, including woody, herbaceous, fruit, and aquatic biomass (Tumuluru, 2020). The woody biomass includes forest, plantation, shrubs, and other virgin wood; herbaceous biomass comes from plants and non-woody material; fruit biomass is from orchards, and horticulture fruit from trees, bushes, and fruit herbs; and aquatic biomass comes from hydrophytic plants or hydrophytes (Tumuluru, 2020). The primary marine biomass sources are algae, water hyacinth, lakes, and seaweed (Tumuluru, 2020).

Woody and harvested biomasses have variable physical properties and chemical composition, limiting biomass use for large-scale applications. The physical properties, such as bulk density, irregular size, and irregular shape, create transportation, storage, handling, and flow challenges. In addition, harvested biomass has higher moisture content and a lower calorific value, reducing conversion efficiency. The conversion process requires more volume for the lower bulk density and energy content. Moisture in the biomass is another major limitation as it creates preprocessing challenges. During grinding, the higher moisture content in the biomass increases the energy consumption significantly, producing high variability in the particle size distribution. In addition, many of the grinding studies conducted by various researchers have indicated that the biomass's fibrous nature and moisture content increases the grinding energy. The major limitations of the biomass in terms of its physical, chemical, and rheological properties create challenges in handling, transportation, feeding, and conversion, which limit the biorefineries from operating at their designed capacities.

6.2. Torrefaction

Torrefaction is a thermal treatment of biomass in the temperature range of about 200–300°C in an inert environment (Tumuluru *et al.*, 2021, Tumuluru, 2015). During torrefaction, biomass loses mass due to the loss of moisture and low energy content volatiles. This causes increased energy content of the final product (Tumuluru, 2015). The final energy content depends on the severity of the thermal treatment. Figure 6.1 indicates the impact of the torrefaction temperature regime on the various biomass components, such as cellulose, hemicellulose, and lignin, in terms of their chemical composition and color change. Figure 6.1 indicates that the biomass undergoes devolatilization and carbonization reactions which result in a change of biomass color to dark brown (Tumuluru *et al.*, 2011).

Biomass torrefaction increase the porosity and decrease the bulk density (Tumuluru *et al.*, 2011). According to Chem *et al.* (2015), the final density depends on the initial density of the non-torrefied biomass. In addition, a high amount of fine particle dust is formed due to torrefaction, which results in handling issues. During torrefaction, bulk density decreases as energy density increases. Oliveira-Rodrigues and Rousset (2009) conducted the torrefaction tests on *Eucalyptus grandis* and indicated a significant loss in bulk

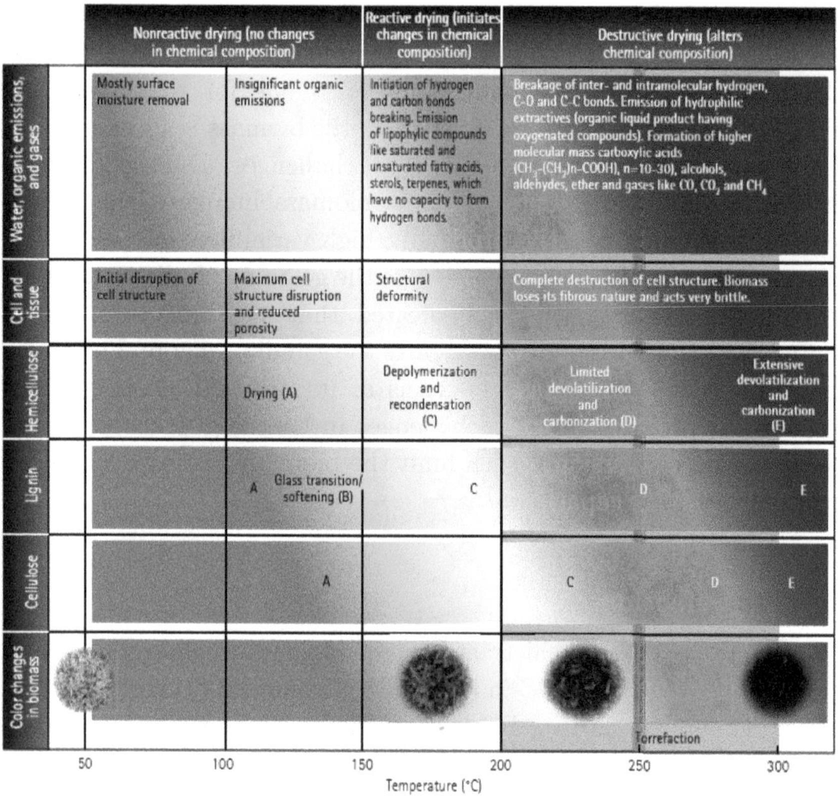

Figure 6.1. Biomass changes during torrefaction at different temperature regimes (Tumuluru *et al.*, 2011, 2021).

density when torrefied at 280°C for 30 min. This study showed that increasing the torrefaction temperature caused more mass loss, decreasing *Eucalyptus grandis* bulk density.

The other advantage of torrefaction is the increase in energy density. Many researchers have observed the increase in energy content of the biomass when torrefied at different temperatures and residence times (Bergman *et al.*, 2005; Ribeiro *et al.*, 2018; Tumuluru, 2015, 2016; Tumuluru *et al.*, 2010, 2011, 2012a, 2012b). Tumuluru (2016) observed that the calorific value of pine woodchips increased from 19 MJ/kg to 22.3 MJ/kg and in Douglas fir from 19.45 MJ/kg to 23.5 MJ/kg at 280°C. The application of temperature causes cellulose crystallization but creates cracks in its structure. The woody biomass has a decreased plastic and viscoelastic behavior at this stage. In addition, the high-temperature treatment causes modification to cell walls. The removal of the mass and active groups eases access into cell walls, decreasing or even deactivating the capability of the reverse osmose flow process into cells. This modification removes the swelling property of the wood structure.

Studies by Bergman and Kiel (2005) on the grinding of torrefied biomass indicated that power consumption reduces significantly after torrefaction (about 70–90% reduction), and throughput increases by about 7.5–15%. Similarly, Phanphanich and Mani (2011) observed similar trends where the grinding energy decreased from 237.7 to 37.6 kWh/t when torrefied at 300°C for 30 min; the dehydration reactions during torrefaction cause the shrinkage. Among the hemicellulose, the xylans degrade first at lower temperatures.

The torrefaction of woody, herbaceous, and other waste biomass changes proximate and ultimate composition and energy content (Tumuluru *et al.*, 2011). Studies by Phusunti *et al.* (2018) on microalgae indicated that mass yield and changes in the microalgae's properties are more influenced by torrefaction temperature than residence time. The calorific value increased to about 19.48 MJ/kg at a lower torrefaction temperature of 200°C for 30 min. In addition, the various components in the microalgae, such as protein, fat, and carbohydrates, have different degradation behaviors. Torrefaction studies on lodgepole pine at 270°C and 120 min indicated that volatile content decreased from 80% to about 45%, and the ash content increased from 0.77% to about 1.91% (Tumuluru, 2016). The other chemical components, such as hydrogen, oxygen, and sulfur content, decreased to 3%, 28.24%, and 0.01%, respectively. The higher heating value (HHV) increased to 23.67 MJ/kg at 270°C and 120 min by increasing the carbon content. The elemental ratios of hydrogen to carbon and oxygen to carbon (H/C and O/C) were about 0.56 and 0.47. Bajcar *et al.* (2018) studied the relationship between torrefaction parameters and physicochemical properties of torrefied products from plant-based biomass at torrefaction temperature of 200–300°C and 60 min residence time. The torrefaction process increased the calorific value of the biomass by more than 20%. This study also showed increased total carbon and ash content, whereas the hydrogen, oxygen, and moisture content decreased. Ramos-Carmona *et al.* (2017) studies on torrefaction of patula pine indicated that torrefied pine has higher chemical energy due to reducing O/C and H/C ratios and heating value. Compared to raw biomass, the torrefied material at 200–240°C did not significantly change physical properties and chemical composition, whereas, at 300°C, both the chemical composition and thermal behavior changed significantly. An analysis of torrefied material using Py-GC indicated that hemicellulose, cellulose, and lignin suffer a progressive thermal degradation with increased torrefaction temperature. A study was conducted by Zheng *et al.* (2017) on the effect of torrefaction temperature on the structural properties and pyrolysis behavior of biomass at different torrefaction temperatures of 210–300°C and a residence time of 30 min. The results indicated that higher torrefaction temperature increased the ash content and fixed carbon content and reduced the volatiles content. The oxygen content decreased from 41.77% to 32.62%, and the highest calorific value of 24.34 MJ/kg was achieved. This study indicated that torrefaction destroyed the cellulose and hemicellulose, which reduced the OH and C-O content and increased the C-C content. Zheng *et al.* (2015) studied the impact of torrefaction on the chemical structure and catalytic fast pyrolysis behavior of hemicellulose, lignin, and cellulose. These authors indicated a torrefaction temperature of 210–300°C and about 20–60 min of residence time brought structural changes in the hemicellulose, cellulose, and lignin. This study showed that the thermal stability of cellulose was the highest, followed by lignin and hemicellulose. Research by Wang *et al.* (2017) on the woody biomass properties indicated that the hemicellulose contents of the torrefied stem wood and stump decreased with an increase in torrefaction temperature and residence time. At 300°C, only a trace amount of hemicellulose was left. In contrast, the cellulose content in the bark decreased at a torrefaction temperature of 275°C.

Tumuluru *et al.* (2021) recently discussed the factors that impact the torrefied material properties and various reactors commonly used for the torrefaction process. These authors have developed a thermodynamic model to design a torrefaction reactor. Torrefied biomass has similar combustion characteristics; thus, biomass can be easily co-fired with coal in existing power generation reactors. Torrefied biomass becomes a good feedstock for gasification and pyrolysis because its moisture and light or low energy content volatiles have been removed. The impact of several combinations of coal and torrefied material has been tested on reactor performance and feeders. Co-firing will allow coal-fired utilities to meet 2020 standards without spending huge capital:

- Replacing coal with torrefied biomass will reduce plant carbon output by up to 2.4 tons per ton of torrefied mass, earning about $72 in carbon credits (Sheikh *et al.*, 2013).
- Torrefied wood can be stored and handled the same way as coal.
- Due to its hydrophobicity, it can be stored outside like coal.
- The torrefied biomass generates lower NO_x and SO_x than coal via combustion; hence it will decrease pollution and its associated costs.
- Due to the similarity of torrefied material with coal, no or little modification is required for existing utilities to feed the torrefied biomass (biocoal).

6.3. Steam Explosion

In 1926, Mason designed the steam treatment process for biomass (Mason, 1926), where wood chips were initially steam-treated in a Masonite gun (Brownell *et al.*, 1986). Since then, several authors have reported on systems that apply saturated steam in their pretreatment processes (Bach *et al.*, 2013; Chen *et al.*, 2012; Yan *et al.*, 2009, 2010). In general, biomass is treated at 180–240°C under 10–40 bar for 5–10 min. After treatment, the material decompresses explosively, resulting in the biomass fibers rupturing (Tang *et al.*, 2018). The severity of treatment depends on the temperature and residence time of the material in the reactor. The treatment effect can vary from small cracks in the wood structure to total defibrillation of the wood fibers (Tanahashi, 1990). It might also be due to the release of acetic acid during the treatment process, where the produced acid partially works as a catalyst and hydrolysis cell wall components (Glasser and Wright, 1998).

6.4. Chemical and Physical Changes of the Biomass

As previously stated, the wood fiber structure is cellular and composed of three major components: lignin, cellulose, and hemicellulose. Layers of cellulose fibers are embedded in a matrix of hemicellulose and lignin. Steam explosion helps release hemicellulose from the cell wall composite and makes the fibers and components accessible for chemical and physical separation and degradation. Hemicellulose is responsible for viscoelastic properties in wood. Its degradation makes the wood more brittle and rigid (Fengel and Wegener, 1989). During steam treatment, the released materials carry a big part of existing hydroxyl groups in the wood structure. Removing OH groups (e.g., dehydration reactions) results in a more hydrophobic surface. Brittleness (e.g., better grindability) and higher moisture resistance are important properties. Steam pretreatment leads to improved mechanical strength, hydrophobicity, and energy density of the wood pellets (Lam *et al.*, 2011; Reza *et al.*, 2012; Shaw *et al.*, 2009). The previous studies also showed that the moisture content of the produced solid increased up to two times after steam pretreatment (Lam *et al.*, 2011; Reza *et al.*, 2012).

During the last decade, much research was done to investigate steam pretreatment's effect on woody and agricultural biomass densification. Steam explosion treatment has been suggested as a pretreatment process for the wood pellet industry to produce more durable pellets (Obernberger and Thek, 2010). During the steam explosion, the lignin component that plays a role as an adhesive in holding the wood fibers together gets soft and breaks downs into parts with a smaller molecular weight. The softened lignin molecules get distributed on the surface of the fibers. Although not 100% documented, lignin can replay its role as a binder and re-bond the fiber pieces together through densification. As the concentration of lignin on the surface of the fibers increases, the chance exists for chemical binding. The fibers generated via the steam explosion process are more brittle. Once it gets compressed, there is less opportunity for particle bridging. As a result, the porosity of the produced pellets will be low compared to the untreated pellets. In addition, the steam-treated materials have a harsher surface than the untreated materials, which helps the particles interlock. The qualities that the biomass particles gain via steam treatment are the primary reasons that the pellets produced from steam-treated material have a higher density and durability.

The conventional explosion method is defined as a sudden drop in pressure in the reactor once the material discharge valve is opened and steam is released with the material. However, one problem associated with this method is that the released steam associated with the treated materials condenses on the materials. As a result, the discharged material has a much higher moisture content of 50–80%, depending on the bed temperature and steam saturation. Considering several experimental and process simulations (Shahrukh *et al.*, 2015), drying the wet exploded material consumes a huge amount of energy, suggesting whether the process is economical. Due to limitations in applying different apparatuses and methods for steam treatment of material, alternative methods exist to avoid the collection of discharged material in wet conditions. The steam explosion can be replaced with steam refining. The treated biomass is passed through a refiner in this method instead of discharging in an explosion. This method is suitable for continued steam treatment and is also available commercially.

6.5. Torrefaction Impact on Binding Characteristics of Biomass

Biomass loses mass while torrefiying. The material shrinks slightly, causing a minor decrease in volume. As a result, the torrefied materials have an even lower bulk density than untreated biomass (Tumuluru *et al.*, 2011). The handling and transporting of such low-density materials are of concern. The torrefied materials are densified by compressing the ground bulk to pellets or briquettes. Several studies have been completed on developing methods and conditions for the densification of torrefied materials (Tapasyi *et al.*, 2012).

However, the commercial technology of densifying torrefied materials has not been developed yet, especially on an industrial scale. Several issues still need to be addressed.

The torrefaction treatment reduces the binding characteristics of biomass, including its cohesiveness. Torrefied wood, with less cohesive characteristics, showed poor compressibility compared to untreated wood. Loss of cohesiveness resulted in a higher energy requirement for compression. Torrefaction increases the lignin content percentage in the biomass. The bonding was expected to be strongly related to the higher lignin percentage in torrefied products. The excess content of lignin partially helps, but due to structural changes of lignin during torrefaction, the modified lignin does not have the binding characteristics of untreated lignin.

Torrefied wood is not an ideal feedstock for densification in compaction characteristics. Increasing pelletizing temperature, moisture, and foreign binders may improve the pelletizing performance and make densification feasible. As the thermal treatment increases the hydrophobicity property of the treated material, the conditioning of torrefied material becomes complicated. The equilibrium moisture content of torrefied woody biomass is in the range of 3–2% (w.b.). The torrefied material will not be conditioned homogeneously at a high moisture content except by applying a binder.

Due to the thermal conductivity property of woody materials, particle size (distribution) of the input raw material is critical for all reactor technologies. It strongly influences the product quality and degree of torrefaction. The particle size distribution must be in a small range to have homogeneous torrefaction. Fire and explosion are risks when handling very small particles like dust. Torrefied material, especially in powder form, is highly reactive. Therefore, there is a high risk of explosion during milling, handling, overloading, and pelletizing. Most torrefaction pellet producers use additives (e.g., lignin, glycerin, and moisture) to combat this problem.

6.6. Densification of Torrefied Biomass

Ghiasi (2019) discussed two pelletization scenarios for the torrefied materials (Fig. 6.2). The pelletization process involves torrefaction

Figure 6.2. Two process configurations. Pathway I consists of torrefying wood chips and then turning the wood chips into pellets. In Pathway II, commercial pellets are heat-treated to make pellets. In this research, a steam pretreatment of wood chips before pelletization (grinding) is added to Pathway II to increase the density and durability of torrefied pellets.

before pelletization (i.e., raw material is dried, torrefied, ground, and pelletized) (Ghiasi, 2019). This pathway can be considered a conventional process scheme as most of today's torrefaction plants are using this pathway to produce torrefied pellets. High die temperature and the use of binders are often necessary to effectively pelletize torrefied biomass in the conventional process scheme. The alternative pathway to produce torrefied pellets is pelletizing first and further torrefaction of the pelletized biomass. The wood chips are dried, ground, and pelletized in this pathway. These regular pellets are torrefied in the final step to make treated pellets. This author conducted pelleting of torrefied biomass using both methods and compared the quality in terms of physical properties and chemical composition. Table 6.1 gives the major advantages and disadvantages of two pelletization scenarios for the torrefied material.

According to Tumuluru *et al.* (2011), the variability in feedstock quality due to differences in raw materials, tree species, climatic and seasonal variations, storage conditions, and time significantly influences pellet quality. However, torrefying the biomass after pelletization produces a uniform feedstock with consistent quality.

Table 6.1. Pros and cons of two pathways to produce torrefied pellets.

Process	Pros	Cons
Torrefy — Pelletize Biomass is ground, dried, torrefied, and then pelletized.	Using binders, it may be possible to produce durable pellets with a bulk density equivalent to regular pellets.	Deeply torrefied biomass tends not to bind; extreme die temperatures or binders may be needed. Carbonized biomasses wear the die.
Pelletize — Torrefy Ground biomass is pelletized in a similar way to the existing commercial processes. Regular pellets are then torrefied.	Pelleting white wood is well known. As a feedstock to torrefier, pellets have low moisture with known properties. The torrefier can be controlled more easily than torrefying the wood chips.	Depending upon the severity of torrefaction, pellets lose a degree of density and durability. Torrefied pellets may require gentler handling than regular pellets.

In their review on torrefaction, these authors concluded that compared to raw biomass pelletization, the energy consumption could be increased by about 1.3–2 when the torrefied material is densified at a temperature of 180–225°C in the absence of any binder.

Table 6.2 shows the physical properties of untreated Douglas fir pellets and torrefied Douglas fir pellets produced through both pathways. The equilibrium moisture content at ambient temperature for untreated pellets is noticeable, for the equilibrium moisture content is 8%. It was expected that torrefaction increases hydrophobicity and, as a result, reduces the equilibrium moisture content and the final moisture content to <3% (w.b.). This improvement resulted in a new pathway. As it has been reported, the severity of treatment causes the treated materials to hold less water. The addition of a binder depends on its chemical structure but usually increases the equilibrium moisture content of the torrefied pellets. Table 6.2 shows that the torrefied pellet with a binder carries 8% moisture content. Several types of research show that increasing moisture content increases the possibility of self-healing and off-gassing (Guo, 2013).

Table 6.2. The physical property of torrefied pellets.

Sample	Density (single pellet) (kg/m^3)	Moisture content (%)	Durability	
			Dural	Tumbler
Raw material pelletized	1158.5	8	82	95.2
Raw material, pelletized and subsequently torrefied	1031.2	3	84	98.2
Raw material, torrefied and subsequently pelletized	1207.3	8	85	98.2

Untreated laboratory-made pellets had an average unit density of 1160 kg/m^3, comparable with the commercial white pellet density. As mentioned, untreated wood chips and pellets lose the same percentage of their solid dry mass during torrefaction. Although it does lose mass in the case of the pellets, the pellet does not shrink considerably in relation to the mass loss. As a result, the density of the pellets decreases via the torrefaction process. Table 6.2 shows that the torrefied pellet density has decreased to 1031 kg/m^3. This happens for wood chips during torrefaction as well. In the conventional method, the torrefied wood chips grind and densify. As a result, the torrefied material gains its density through densification. In fact, the nature of the binder being used affects the torrefied pellet density. The density of the torrefied pellets produced through the conventional method is typically about 1207 kg/m^3.

The durability of the produced pellets is presented in Table 6.2. These laboratory-made untreated pellets have low durability considering the European durability standard. An interesting note is that pellet durability increased via torrefaction. Untreated pellets have 95.2% durability but improve to 98.2% by thermal treatment. This means that some chemical bonding and crystallization happen due to high-temperature treatment. The torrefied ground material hardly gets bonded to each other and requires high pressure and temperature or the addition of a binder. The results also indicated that the torrefied pellets produced via conventional methods with the addition of a binder have 98.6% durability. It is good to note that these pellets were produced in the laboratory using a laboratory-scale pelletizer. The use of a commercial pelletizer will result in much more durable pellets.

Untreated white pellets are produced commercially. The pelletization industry for these pellets has matured greatly over the last 20 years. These pellets are made in several million tons throughout the US and Canada. Considering the new torrefaction pathway, it looks more practical to torrefy the produced white pellet by adding a torrefaction unit to existing pellet production plants. The positive of using this method is that white pellets are readily available as raw materials extruded from a pelletizer at about 85–100°C and that hot steam is exhausting from the dryer at about 240–280°C. This steam can be used as a heat source and inert gas, preventing oxygen access to the torrefaction reactor and avoiding oxidation reactions.

The torrefaction of commercial white pellets has been investigated at the University of British Columbia's Biomass and Bioenergy Group on a laboratory scale since 2012. In these studies, commercial white pellets produced in British Columbia were made from about 90–92% mountain pine and 8–10% Douglas fir as raw materials and torrefied at different temperatures ranging from 200–300°C. The torrefaction proceeded under two other methods: an inert chamber and one carrying gas, steam, or N_2.

Bed temperature has a considerable effect on the severity of thermal treatment. All types of woody feedstock have three main components that build up more than 95% of a biomasses structure. These three components, hemicellulose, lignin, and cellulose, behave differently through torrefaction. Cellulose is the last component that starts thermal degradation, while hemicellulose starts degradation at a lower temperature (Stelte, 2012). The experimental results and data from the literature review show that all three components resist thermal degradation under 200–220°C. Under 200°C, the materials lose their moisture and some extractives. Although pure hemicellulose is expected to degrade at 190–200°C, degradation of the other components starts at temperatures higher than 220°C (Bergman, 2005). This happens when materials are treated under atmospheric pressure and in a dry chamber. As the structure of the cell walls is a composite of the three components, it stays firm as long as they have not started decomposition. When the material gets treated under different conditions, the decomposition behavior changes. For example, when the material is exposed to hot steam at 4–5 bar pressure, the hemicellulose degrades at 170–180°C (Pirraglia *et al.*, 2013). In atmospheric pressure, woody materials lose their moisture and light volatile organic materials (VOCs) under the thermal treatment at temperatures below 200°C. Although this treatment removes all bonded and free water from the material in a certain residence time, the treatment condition is not severe enough to damage the main component structures of the woody materials.

In most cases, the cell walls remain firm and have the same behavior in the absorption and desorption of water. Removing water helps the materials lose their elasticity and become more rigid. This improves the grindability of the woody biomass, especially with the compaction force used for size-reduction (Shang *et al.*, 2012). The woody materials have less conductivity, and heat transfer via conduction has a low rate. This requires a longer residence time when exposed to a certain temperature. In addition, particle size has a major effect on the residence time of the materials in the reactor. Another parameter that can play a major role is the conductivity of the hot gas chamber.

The following study investigates the effect of temperature, chamber gas composition, and material residence time on the severity of woody biomass thermal treatment. First, the woody biomass (or commercial white pellet) is thermally treated in several different temperature ranges. Although the untreated material quality has a baseline for comparison, the quality of the pellets treated at 200°C was considered as a second baseline. In this condition, the pellets are fully dried and free from any moisture content on their durability, grindability, and density. Next, batches of these same pellets are thermally treated in different temperature ranges from 200–300°C in two different inert chambers.

6.7. Physical Properties of Torrefied and Densified Biomass

6.7.1. *Mass Loss and Calorific Value*

Figure 6.3 shows the calorific value of the torrefied wood pellets in nitrogen (N_2) and steam environments. The torrefaction of pellets at 200°C removes moisture and light VOCs. A jump of calorific value from 19 to 20 MJ/kg was observed. When the torrefaction temperature is increased to 300°C, the calorific value increases to more than 26 MJ/kg. Steam was more effective than the N_2 gas in increasing the calorific value of the treated pellets at 260°C, 280°C, and 300°C. This agrees with the mass loss and pellet density ratio discussed in the following paragraphs.

Figure 6.4 provides a sample plot of mass vs. time during torrefaction at 280°C. The mass of the biomass reduced about 30% from its initial 600 g. At about 200°C, the moisture content and some light VOCs were lost. As the chamber temperature gradually

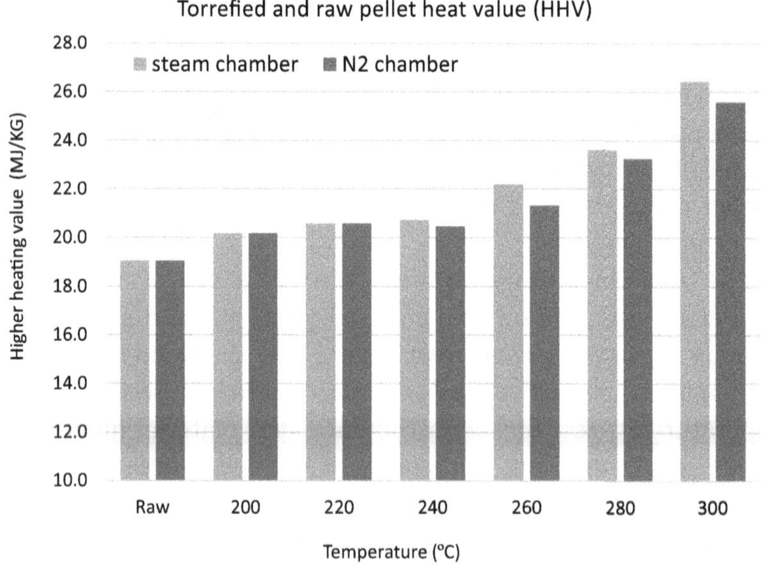

Figure 6.3. Calorific value changes by torrefying a commercial pellet in different temperature ranges and at two-reactor chamber inert gas. All pellets were torrefied at the set temperature for a residence time of 30 min.

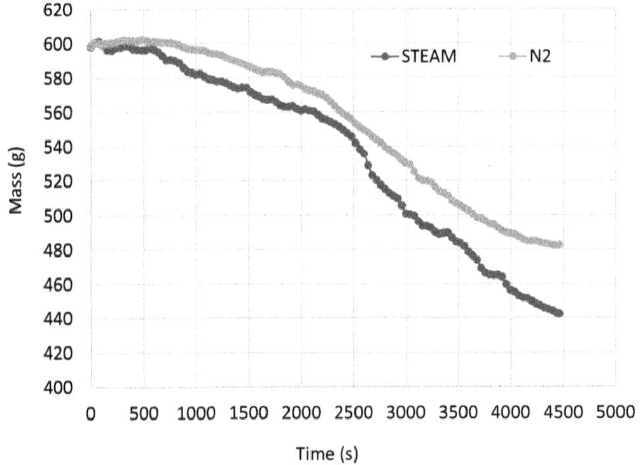

Figure 6.4. Pellet mass loss during torrefaction at 280°C treated in the chamber with N_2 or steam as the carrier gas. The decrease in mass was gradual up to around 2400 s (40 min). Treatment for a longer time accelerated the mass loss.

increased, the decrease in mass increased as well. Treatment at a longer time at 280°C causes an increasing mass loss rate, which was observed to be larger when using steam as an inert gas than for N_2 treatment conditions. It is plausible that the convective heat exchange between the pellets and their immediate environment was affected by the inert gas, steam, or N_2.

Figure 6.5 shows the percentage of mass loss in torrefied wood pellets treated at 200–300°C in steam or an N_2 chamber. Mass loss ranges from 5% to 38% for steam and 5% to 34% for an N_2 chamber, depending on the treatment temperature. By increasing treatment temperature, mass loss increased as well. In torrefaction at 240°C, pellets in a steam chamber lost more mass as compared to torrefaction in an N_2 chamber at the same temperature (e.g., 11.94% vs. 8.39%), while their calorific value remained almost unchanged (e.g., 20.73 MJ/kg vs. 20.50 MJ/kg). The final higher calorific value of pellets treated in the steam chamber was 18.26 MJ. The total heat content of a torrefied pellet in an N_2 chamber at 240°C was 18.78 MJ. This shows that even the high calorific value of pellets treated in a steam chamber is higher than that of torrefied pellets in an N_2 chamber despite losing more mass.

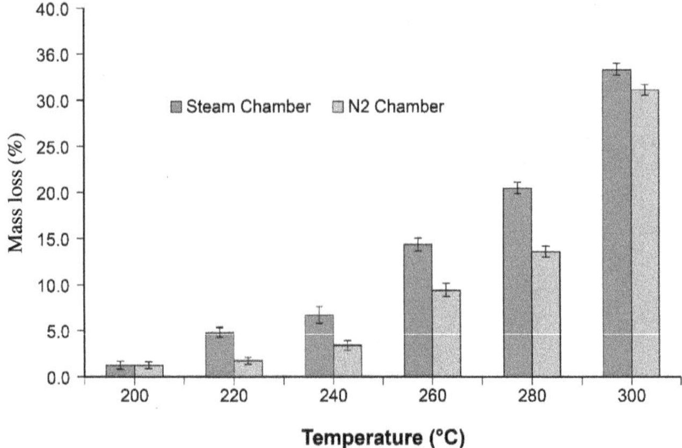

Figure 6.5. Percent mass loss changes via torrefying a commercial pellet in several different temperature ranges and at two-reactor chamber inert gas. All pellets were torrefied at the set temperature for a residence time of 30 min.

6.7.2. *Single Pellet Density*

During torrefaction, pellets lose their mass. By increasing the treatment temperature, the mass loss percentage increases considerably. Although the pellets shrink slightly during torrefaction, the pellet volume does not decrease relatively. As a result, single pellet density decreases during torrefaction. Figure 6.6 shows the individual density of the torrefied pellets. The treated pellet's density (single pellet) decreased by almost 13.8% when these pellets were torrefied at 280°C for 30 min in an N_2 chamber. The decrease in pellet density was smaller for pellets treated at 260°C (6.35%) and less at lower temperatures. The drop in pellet density was larger for pellets treated in a steam chamber. For example, the density of torrefied pellets in a steam chamber at 280°C decreased by 17.3%. The corresponding decrease in the N_2 chamber was 13.8%. The pellet's length did not reduce considerably. The average pellet diameter before and after torrefaction showed that the diameter of the pellets decreased slightly when torrefied.

6.7.3. *Pellet Durability*

Figure 6.7 shows the drop in pellet durability after torrefaction. The average durability of untreated pellets was more than 99.7%. The

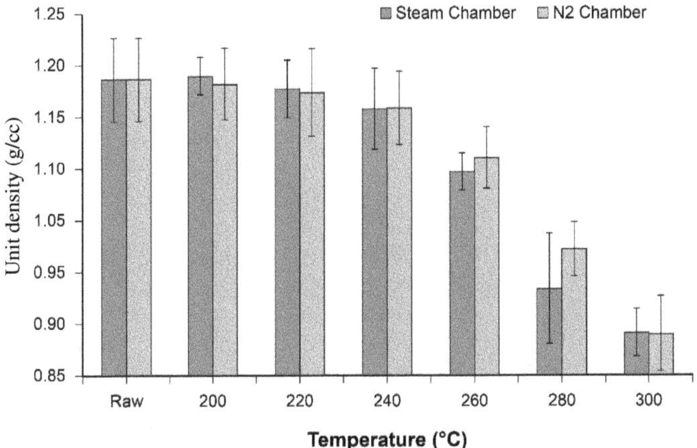

Figure 6.6. Single pellet density changes via a torrefying commercial pellet in different temperature ranges and at two-reactor chamber inert gas. All pellets were torrefied at the set temperature for a residence time of 30 min.

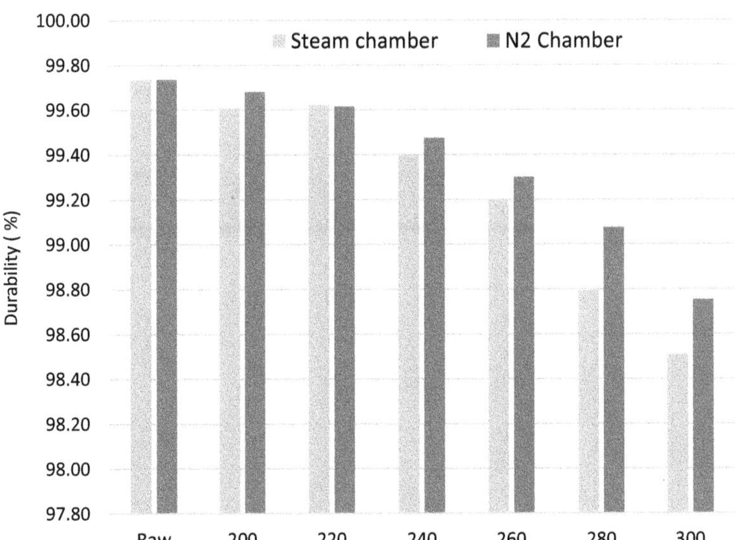

Figure 6.7. The durability of untreated and torrefied commercial pellets was measured using a tumbler. These commercial pellets were torrefied both in steam and an N_2 chamber. In addition, the effect of chamber gas composition has been presented.

average durability of torrefied pellets at 280°C was 98.7%, which is in the range of durable pellets according to ISO standard 17226. Pellet durability did not decrease substantially considering the pellet's mass and density loss at 280°C. After treatment, the high durability indicates that new chemical bonds were shaped among particles, preventing the pellets from disintegrating despite losing some mass. Pellets treated in a steam chamber lost more durability than pellets treated in an N_2 chamber due to higher mass loss.

6.7.4. *Ash Content*

Torrefaction does not affect the mass value of the ash. However, as the pellets lose a portion of their organic compound, the ratio of minerals to organic compounds increases slightly. Figure 6.8 shows the ash content of untreated pellets and the pellets torrefied at different temperature ranges in steam or an N_2 chamber. The ash content of pellets increased due to torrefaction. The proportion of ash content increased with an increased torrefaction temperature. Nevertheless,

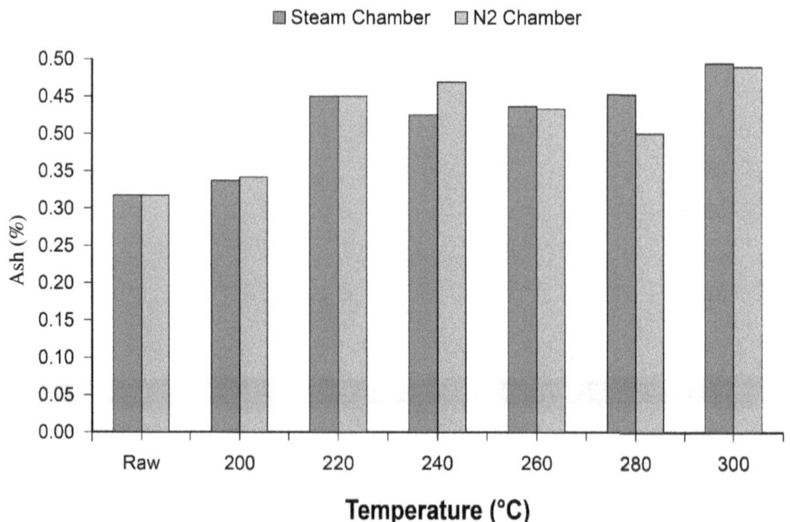

Figure 6.8. Ash content of untreated and torrefied pellets. The ash content percentage of the pellets increases with a corresponding increase in the severity of the thermal treatment.

torrefaction of untreated pellets at 300°C did not cause the pellets to have an ash content of over 0.6%.

6.8. Discussion

Steam has a higher thermal conductivity than nitrogen and increases torrefaction severity. Another reason for applying steam as heat compared to inert gas is that a huge volume of hot steam is available in an industrial pellet mill exhausted from this dryer and the torrefaction unit. Maintaining the steam at a high temperature will make the torrefaction process safe and more auto-thermal. The advantage of using hot steam over other alternatives has been presented in detail by showing comparative results from using both a steam chamber and a nitrogen chamber. In this study, the feasibility and effect of different treatments have been investigated. Increasing the temperature increases the rate of the reaction, but the nature of the reactants also changes in different temperature ranges. At different temperatures, the weight loss analysis shows that material loses a certain amount of mass at each temperature setting and then reaches a steady state.

Increasing the process's residence time will not considerably affect the mass loss ratio. This means that the concentration of the reactants that could react and transfer to a gas phase was reduced significantly. The reaction is in the termination step or the equilibrium stage. When exposed to the higher temperature, not only does the slope of the weight loss graph of the same materials changes, presenting an increase in reaction rate, but the mass loss ratio of the pellets increases as well. This means that some new components that react at a higher temperature were not reactive at lower temperatures. This is not a regular reaction rate increase with a temperature increase through the Arrhenius equation. Several models have been introduced to estimate the material's mass loss and reaction rate at different temperatures (Huang *et al.*, 2012; Lee *et al.*, 2012; Li *et al.*, 2012). The problem with the woody and agricultural biomass feedstocks is that these materials have multi-component structures and compositions that differ from specie to specie. Even in the same species, the plant's age and the location of the sample in the plant impact the torrefaction process.

Most of the reactions, including evaporation of free and bonding water, emission of light VOCs, and some breakdown of the components, are endothermic reactions. The heat value of the material increases through the removal of these masses. Most of these reactants have non-carbon groups or have low carbon content. The chamber temperature provides enough activation energy for components undergoing exothermic reactions, like oxidation, but the limitation of access to free oxygen prevents such reactions. One major reason for torrefying woody and agricultural biomass is to increase the energy density per unit mass of biomass. As stated previously, removing moisture content, light VOCs, and oxygen-containing groups help increase the carbon to oxygen ratio in a treated material structure. As a result, it increases the heat value of the biomass. For example, if we have 600 g of white pellets with a 19.04 MJ/kg heat value after torrefaction at 260°C for 30 min, the pellets lose 114/g of their initial mass. The remaining torrefied pellets can generate 10.87 MJ of heat, which, compared to 600 g white pellets, will generate 11.42 MJ of lost heat content via the torrefaction process. This observation means that the materials lose 4.8% heat content while losing 19.5% initial mass. Although it is a small percentage of the energy content, it has shown that the lost mass has a heat value that can be recycled. According to the research by ECN Netherlands (Bergman *et al.*, 2010) and some other process simulation works, condensable and non-condensable emitted gases carry the released heat. Bergman *et al.* report that through the combustion of released gas (e.g., Tar, CO, CH_4, and H_2), it is possible to provide the required heat for the torrefaction process and heating side units. This is the amount of energy that is spent on upgrading biomass via the torrefaction process.

During torrefaction processes, materials lose their moisture content. The equilibrium moisture content decrease is in the range of 2–3%. The structural analysis of the torrefied materials shows that the percentage of the insoluble components in acid increases due to an increase in the crystalline structure of biomass. This structural change causes the material to be more rigid and firm. This is an important improvement in material grindability since rigid materials become less elastic and brittle. Thus, grinding will be easier by applying compact or shear force.

6.8.1. *Pellet Quality*

Besides all the torrefaction's advantages after pelletization, the major concern is pellet quality. The regular untreated commercial pellets lose their mass via the torrefaction process. Although pellets slightly shrink through thermal treatment, pellet volume does not change considerably and in the same ratio as the mass changes. As a result, the torrefied pellets have lower bulk and single pellet density. This change does not affect the pellet's bed voidage but increases the torrefied pellet's porosity. To overcome this decrease and produce high-quality pellets, it will be more efficient to go through the production of very dense raw pellets before torrefaction by applying different methods and equipment even though one or two steps. The two-step process includes a steam explosion of raw materials before densification. During the steam explosion process, the particles lose their strength, and the fibers get separated via the displacement of lignin and the partial removal of hemicellulose (Stelte, 2012). As a result, fibers become flexible and more compressible. This pretreatment technique has been studied in detail on the laboratory scale as a part of high-quality torrefied pellet production research (Bergman, 2005).

Densification of agri-biomass is very complicated. The agri-biomass does not pelletize in the condition that woody biomass easily densifies. Usually, due to the low bulk density and elasticity of the agri-biomass, it needs to be compressed more than the woody biomass. As a result, the material needs to reside in the pelletization die more than the woody materials do. Several studies have shown that as the material requires a higher residence time in the die and higher compression, that material will need to have a higher initial moisture content before pelletization. Some researchers have called the densification method of agri-biomass "wet densification or high moisture pelleting" (Tumuluru, 2014).

Steam explosion of agricultural and woody biomasses eases their densification, resulting in durable pellets with a higher density and a slightly higher heat value. The steam explosion process is required for biofuel production through bio-fermentation. Several studies have shown that the steam-exploded pellet keeps the same quality as the undensified steam-exploded materials. In addition, due to the high density of the steam-exploded pellets, the storage and transportation of the steam-exploded pellet are more economical than the loss of the steam-exploded materials. Lam (2011) and Tooyserkani (2013) have studied the steam explosion of woody biomass and the densification of steam-exploded materials. They have worked at the laboratory scale producing single pellets and studying the single pellet properties. Following their works, Ghiasi (2019) produced steam-exploded pellets on a semi-laboratory scale resulting in several kilos of woody and agricultural steam-exploded pellets. The pellets improved quality, durability, density, heat value, and hydrophobicity compared to those produced from untreated pellets of the same materials. The steam explosion effect on biomass structure proves that it is suitable for ethanol.

6.8.2. *Single Pellet Density of Different Thermally Treated Wood and Agricultural Biomass*

Figure 6.9 shows the untreated pellet density change after different treatment methods. The graph shows that the untreated pellet loses much of its density through torrefaction. The average density of untreated Douglas fir pellets changes from 1.17 g/cm^3 to 1.07 g/cm^3 via torrefaction. This change happens severely for agricultural biomass (switchgrass) from 1.09 g/cm^3 to 0.92 g/cm^3 after torrefaction at 280°C for 15 min. Steam treatment before densification increases the average pellet density to 1.31 g/cm^3 for woody biomass (e.g., Douglass fir) and 1.34 g/cm^3 for agricultural samples (e.g., switchgrass). Even after torrefaction, the steam-treated pellets have a higher density than the torrefied untreated

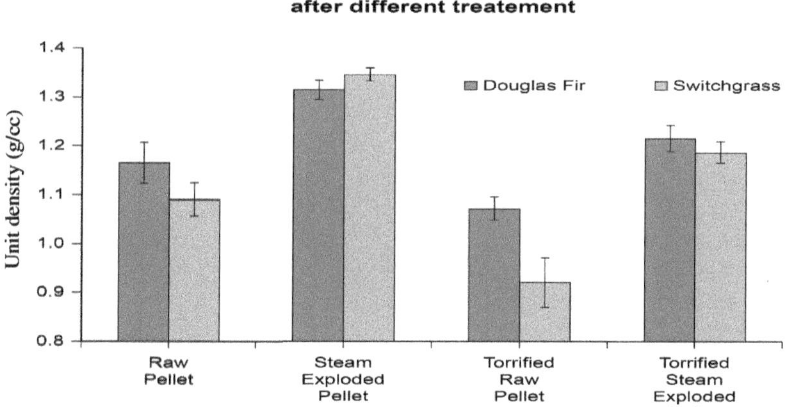

Figure 6.9. The effect of several different thermal treatments on pellet density. The steam-treated pellets have the highest density. Even after torrefaction, these pellets have an even higher density than the untreated pellets.

pellets. Torrefied steam-treated Douglas fir pellets have an average density of 1.21 g/cm^3, while for switchgrass, it is 1.18 g/cm^3. The calculation shows an 11.5% increase in density for a torrefied woody pellet and 22.5% for agricultural torrefied pellets. While steam treatment tests were run at 210°C for five minutes, the pellets were further torrefied at 280°C for 15 min.

6.8.3. *Pellet Durability*

Interestingly, torrefaction after pelletization improves the durability of the pellets depending on their initial durability, as shown in Fig. 6.10. Even though torrefaction helps ease the grinding process, the durability results indicate that the fine particles are more tightly bound to each other during the torrefaction process. These bonds are strong enough to resist the compact force via the tumbling of the pellets. The bonds give more rigidity to the pellets. In this case, the elasticity of woody particles decreases. This improves the grindability of the material as well. In the case of Douglas fir pellets, the untreated pellets have an average of 95% durability, while after torrefaction, they reach an average of 98.5%. Steam-treated material bonds to

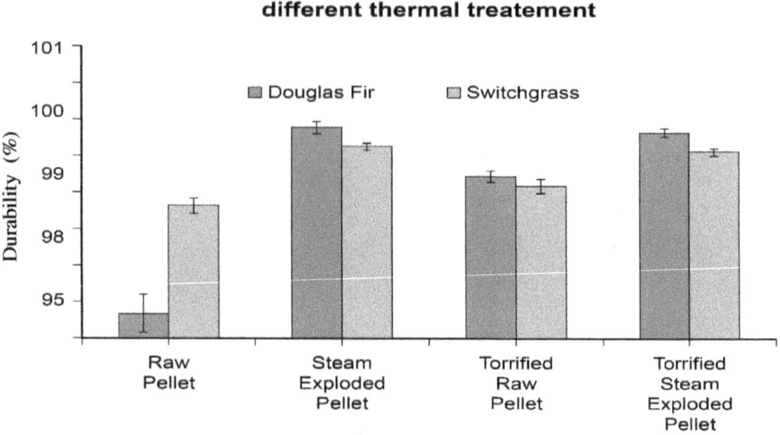

Figure 6.10. The untreated and thermal treated woody and agricultural biomass pellet durability. The torrefaction of the pellets binds the particles to each other and reduces the percentage of fines generation during durability tests using a tumbler.

itself efficiently via the densification process. As a result, they have high density and durability. Because the particle compresses and interlocks to each other, the steam-treated pellet's thermal treatment results in a higher dense and durable torrefied pellet. As Figs. 6.9 and 6.10 show, these qualities improve both the woody and agricultural pellets.

6.9. Application of Torrefied and Densified Biomass

6.9.1. *Gasification*

Sarkar *et al.* (2013) studied the impact of torrefaction and densification on gasification. Devolatilization kinetics was determined at three heating rates (e.g., 10°C, 30°C, and 50°C for 1 min) in an inert (nitrogen) and oxidizing (air) atmospheres using a thermogravimetric analyzer. Gasification performance was evaluated at three temperatures (e.g., 700°C, 800°C, and 900°C) using an externally heated fixed-bed reactor with air at an equivalence ratio (ER) of 0.3. A devolatilization study showed that switchgrass torrefied at

270°C had the highest carbon (C) and the lowest hydrogen (H) and oxygen (O) contents of 59.16%, 4.67%, and 34.53% (d.b.), respectively. This resulted in the lowest atomic O/C (0.44) and H/C (0.95) ratios and the highest heating value of 27.11 MJ/kg (d.b.). Combined torrefaction and densification of switchgrass resulted in the least volatile and highest ash and fixed carbon contents of 62.63, 5.91, and 31.45% (d.b.), respectively. Combined torrefied and densified switchgrass had the highest rate of devolatilization in both atmospheres. The gasification study showed that the bulk density of combined torrefied and densified switchgrass was the highest (598.17 kg/m^3, d.b.), requiring less space to store and transport. Pretreatments of switchgrass and gasification temperatures had significant effects on gasification performance. The results indicated that combined torrefaction and densification resulted in the highest yields of H_2 (0.03 kg/kg biomass) and CO (0.72 kg/kg biomass), highest syngas LHV (5.08 MJ/Nm^3), CCE (92.53%), and CGE (68.40%) at the gasification temperature of 900°C. This study has also indicated that the rate of devolatilization follows the following order: combined torrefied and densified switchgrass > densified switchgrass > torrefied at 230°C > switchgrass torrefied at 270°C > raw switchgrass (Table 6.3).

Table 6.3. Effects of pretreatments on average H_2, CO, CH_4 yields, CCE and CGE (Sarkar *et al.*, 2013).

Pretreatment	H_2 yield (kg/kg biomass)	CO yield (kg/kg biomass)	CH_4 yield (kg/kg biomass)	LHV (MJ/Nm^3)	CCE (%)	CGE (%)
No pretreatment	0.013	0.41	0.10	4.46	87.61	58.28
Torrefaction at 230°C	0.01	0.23	0.11	4.06	66.86	42.65
Torrefaction at 270°C	0.007	0.16	0.12	3.89	58.12	34.12
Densification	0.02	0.49	0.08	4.48	89.58	63.98
Torrefaction and densification	0.03	0.62	0.07	4.98	90.69	64.76

6.9.2. *Pyrolysis*

Yang *et al.* (2014) studied the impact of torrefaction and densification on switchgrass pyrolysis products. The pyrolysis of pretreated switchgrass (torrefied at 230°C and 270°C, densification, and torrefaction at 270°C followed by densification) was studied at three temperatures (e.g., 500°C, 600°C, and 700°C) using gas chromatogram mass spectroscopy (Py-GC/MS). The torrefaction of switchgrass improved the energy content and oxygen to carbon (O/C) ratio. The anhydrous sugars and phenols in pyrolysis products of torrefied switchgrass were higher than the pyrolysis products of raw switchgrass. Increasing the torrefaction temperature from 230°C to 270°C, the anhydrous sugars and phenols contents in pyrolysis products increased, whereas the content of guaiacols decreased. High pyrolysis temperatures (600°C and 700°C as compared to 500°C) increased lignin and anhydrous sugar decomposition, increasing phenols, aromatics, and furans. Densification improved the depolymerization of cellulose and hemicellulose during pyrolysis. The effects of torrefaction and densification on pyrolysis products of switchgrass indicated that anhydrous sugars and phenols obtained were higher than those obtained from raw switchgrass. The contents of anhydrous sugars and phenols also increased as torrefaction temperature increased. This study indicated that densification enhanced the depolymerization of cellulose and hemicellulose, resulting in production of small molecules, such as furans, ketones, and acids.

6.10. Conclusions

Thermal pretreatment of biomass impacts the physical properties, chemical composition, and energy content. Thermal pretreatment results in mass loss which lowers the bulk density. Densifying the thermally pretreated biomass increases the bulk density by 5–6 times and also increases energy density. Improving the density also helps improve transportation logistics. During torrefaction processes, materials lose their moisture content. The equilibrium moisture content decrease is in the range of 2–3%. Also, densifying the torrefied biomass increases the calorific value of the biomass. Steam explosion of agricultural and woody biomass eases their densification and results in durable pellets with high density and slightly high heat value. In the case of torrefaction, the densification is a challenge as the glass transition temperature of the lignin is changed. Gasification of the torrefied and densified switchgrass resulted in the highest yields of H2 (0.03 kg/kg biomass) and CO (0.72 kg/kg biomass). The syngas also had the highest LHV (5.08 MJ/Nm^3), CCE (92.53%), and CGE (68.40%) values at the gasification temperature of 900°C. Torrefaction and densification on pyrolysis products of switchgrass indicated that anhydrous sugars and phenols obtained from torrefied switchgrass were higher than those obtained from raw switchgrass. The densification of the torrefied switchgrass enhanced the depolymerization of cellulose and hemicellulose, promoting the production of small molecules, such as furans, ketones, and acids. Higher pyrolysis temperature favored the decomposition of lignin and anhydrous sugars, resulting in increased yields of phenols, aromatics, and furans.

References

Bach, Q. V., Tran, K. Q., Khalil, R. A., Skreiberg, Ø., & Seisenbaeva, G. (2013). Comparative assessment of wet torrefaction. *Energ. Fuels*, *27*, 6743–6753.

Baicar, M., Zagula, G., Saletnik, B., Tarapatskyy, M., & Puchalski, C. (2018). Relationship between torrefaction parameters and physicochemical properties of torrefied products obtained from selected plant biomass. *Energies*, *11*, 2919.

Bergman, P. C. A. & Kiel, J. H. A. (2005). Torrefaction for biomass upgrading. Proceedings of the 14th European Biomass Conference & Exhibition, Paris, France. pp. 17–21. Retrieved from https://citeseerx.ist.psu.edu/document?repid=rep1&type=pdf&doi=666ee7ddfcbf1bf59a60d4f4b98ac9ee05217efe.

Bergman, P. C. A., Boersma, A. R., Zwart, R. W. H., & Kiel, J. H. A. (2010). BIOCOAL. ECN Report #ENC-C-05-013, Torrefaction for Biomass Co-firing in Existing Coal-Fired Power Stations.

Bergman, P. C. A. (2005). Combined torrefaction and pelletization: The TOP process. Energy Research Centre of the Netherlands (ECN), ECN Biomass, July 2005.

Brownell, H. H., Yu, E., & Saddler, J. (1986). *Steam-Explosion Pretreatment of Wood: Effect of Chip Size, Acid, Moisture Content and Pressure Drop*. John Wiley & Sons Inc. doi:10.1002/bit.260280604.

Chen, W.-H., Peng, J., & Bi, X. T. (2015). A state-of-the-art review of biomass torrefaction, densification and applications. *Renew. Sust. Energ. Rev.*, *44*, 847–866.

Chen, W. H., Ye, S. C., & Sheen, H. K. (2012). Hydrothermal carbonization of sugarcane bagasse via wet torrefaction in association with microwave heating. *Bioresour Technol.*, *118*, 195–203.

Fengel, D. & Wegener, G. (1989). *Wood: Chemistry, Ultrastructure, Reactions.* Berlin: Walter de Gruyter.

Ghiasi, B. (2019). Steam assisted pelletization and torrefaction of lignocellulosic biomass. PhD, Thesis submitted to the University of British Columbia, Vancouver, British Columbia, Canada.

Glasser, W. G. & Wright, R. S. (1998). Steam-assisted biomass fractionation. II. fractionation behavior of various biomass resources. *Biomass Bioenergy*, *14*(3), 219–235.

Guo, W. (2013). Self-heating and spontaneous combustion of wood pellets during storage. PhD, Thesis, Chemical and Biological Engineering Department, The University of British Columbia, Vancouver, Canada.

Huang, Y. F. Chen, W. R. Chiueh, P. T. Kuan, W. H., & Lo, S. L. (2012). Microwave torrefaction of rice straw and pennisetum. *Bioresour. Technol.*, *123*, 1–7.

Lam, P. S., Sokhansanj, S., Bi, X. Lim, C. J., & Melin, S. (2011). Energy input and quality of pellets made from steamtreated Douglas fir (Pseudotsuga menziesii). *Energy Fuels*, *25*(4), 1521–1528.

Lee, J., Kim, Y., Lee, S., & Lee, H. (2012). Optimizing the torrefaction of mixed softwood by response surface methodology for biomass upgrading to high energy density. *Bioresour. Technol.*, *116*, 471–476.

Li, J., Brzdekiewicz, A., Yang, W., Blasiak, W. (2012). Co-firing based on biomass torrefaction in a pulverized coal boiler with aim of 100% fuel switching. *Appl. Energy*, *99*, 344–354.

Mason, W. H. (1926). Low-temperature explosion process of disintegrating wood and the like. United States Patent, 1, 586, 159.

Meng, J., Park, J., Tilotta, D., & Park, S. (2012). The effect of torrefaction on the chemistry of fast pyrolysis bio-oil. *Bioresour. Technol.*, *111*, 439–446.

Obernberger, I. & Thek, G. (2010). *The Pellet Handbook, the Production and Thermal Utilization of Biomass Pellets.* Washington, DC: IEA Bioenergy, Earthscan LLC.

Oliveira-Rodrigues, T. & Rousset, P. L. A. (2009). Effects of torrefaction on energy properties of eucalyptus grandis wood. *Cerne 15*(4), 446–452.

Phanphanich, M. & Mani, S. (2011). Impact of torrefaction on the grindability and fuel characteristics of forest biomass. *Bioresour. Tech.*, *102*(2), 1246–1253. doi:10.1016/j.biortech.2010.08.028.

Phusunti, N., Phetwarotai, W., & Tekasakul, S. (2018). Effects of torrefaction on physical properties, chemical composition and reactivity of microalgae. *Korean J. Chem. Eng.*, *35*(2), 503–510. doi:10.1007/s11814-017-0297-5.

Pirraglia, A., Gonzalez, R., Denig, J., *et al.* (2013). Technical and economic modeling for the production of torrefied lignocellulosic biomass for the US densified fuel industry. *Bioenerg. Res.*, *6*, 263–275. https://doi.org/10.1007/s12155-012-9255-6.

Ramos-Carmona, S., Pérez, J. F., Pelaez-Samaniego, M. R., Barrera, R., & Garcia-Perez, M. (2017). Effect of torrefaction temperature on properties of *Patula pine. Maderas, Cienc. Tecnol., 19*(1), 39–50. doi:10.4067/S0718-221X2017005000004.

Reza, M. T., Lynam, J. G., Vasquez, V. R., & Coronella, C. J. (2012). Pelletization of biochar from hydrothermally carbonized wood. *Environ. Prog. Sust. Energy, 31*(2), 225–234.

Ribeiro, J., Godina, R., Matias, J., & Nunes, L. (2018). Future perspectives of biomass torrefaction: review of the current state-of-the-art and research development. *Sustainability, 10*(7), 2323. doi:10.3390/su10072323.

Rijal, B., Igathinathane, C., Karki, B., Yu, M., & Pryor, S. W. (2012). Combined effect of pelleting and pretreatment on enzymatic hydrolysis of switchgrass. *Bioresour. Technol., 116*, 36–41.

Sarkar, M. (2013). Effect of torrefaction and densification on devolatilization kinetics and gasification performance of switchgrass. Master of Science Thesis, Submitted to the Faculty of the Graduate College of the Oklahoma State University.

Sarkar, M., Kumar, A., Tumuluru, J. S., Patil, K. N., & Bellmer, D. D. (2014). Gasification performance of switchgrass pretreated with torrefaction and densification. *Appl. Energy, 127*, 194–201.

Shahrukh, H., Oyedun, A. O., & Kumar, A. Ghiasi, B. Kumar, L. Sokhansanj, S. (2015). Net energy ratio for the production of steam pretreated biomass-based pellets. *Biomass Bioenergy, 80*, 286–297.

Shang, L., Niels, P., Nielsen, K., Dahl, J., Stelte, W., Ahrenfeldt, J., Holm, J. K., Thomsen, T., Henriksen, U. B. (2012). Quality effects caused by torrefaction of pellets made from Scots pine. *Fuel Process. Technol., 101*, 23–28.

Shaw, M. D., Karunakaran, C., & Tabil, L. G. (2009). Physicochemical characteristics of densified untreated and steam treated poplar wood and wheat straw grinds. *Biosyst. Eng., 103*(2), 198–207.

Sheikh, M. M. I., Kim, C., Park, H., Kim, S., Kim, G., Lee, J., Kim, J. W. (2013). Influence of torrefaction pretreatment for ethanol fermentation from waste money bills. *Biotechnol. Appl. Biochem., 60*(2), 203–209. doi:10.1002/bab.1070.

Stelte, W. (2012 December). Torrefaction of unutilized biomass resources and characterization of torrefaction gasses. In *Energy & Climate Centre for Renewable Energy and Transport Section for Biomass.* Danish Technological Institute, Gregersensvej 2C, DK-2630 Taastrup, Denmark.

Svanberg, M., Olofsson, I., Floden, J., & Nordin, A. (2013). Analysing biomass torrefaction supply chain costs. *Bioresour. Technol., 142*, 287–296.

Tanahashi M. (1990). Characterization and Degradation Mechanisms of Wood Components by Steam Explosion and Utilization of Exploded Wood. Wood research: bulletin of the Wood Research Institute Kyoto University, 77, 49–117. https://repository.kulib.kyoto-u.ac.jp/dspace/handle/2433/53271.

Tang, Y., Chandra, R. P., Sokhansanj, S., & Saddler, J. N. (2018). Influence of steam explosion processes on the durability and enzymatic digestibility of wood pellets. *Fuel, 211*, 87–94.

Tapasvi, D., Khalil, R., Skreiberg, O., Khanh-Quang, T., & Gronli, M. (2012). Torrefaction of Norwegian birch and spruce: An experimental study using macro-TGA. *Energy Fuels*, *26*(8), 5232–5240.

Thangalazhy-Gopakumar, S., Adhikari, S., & Gupta, R. B. (2012). Catalytic pyrolysis of biomass over H^+ZSM-5 under hydrogen pressure. *Energy Fuels*, *26*(8), 5300–5306.

Tooyserkani, Z. (2013). Hydrothermal pretreatment of softwood biomass and bark for pelletization. PhD, Thesis, Chemical and Biological Engineering Department, The University of British Columbia, Vancouver, Canada.

Tumuluru J. S. (2014). Effect of process variables on the density and durability of the pellets made from high moisture corn stover. *Biosyst. Eng.*, *119*, 44–57.

Tumuluru, J. S. (2015). Comparison of chemical composition and energy property of Torrefied Switchgrass and Corn Stover. *Front. Energ. Res.*, *3*, 46. doi:10.3389/fenrg.2015.00046.

Tumuluru, J. S. (2016). Effect of deep drying and torrefaction temperature on proximate, ultimate composition, and heating value of 2-mm Lodgepole Pine (*Pinus contorta*) grind. *Bioengineering*, *3*(2), 16. doi:10.3390/bioengineering3020016.

Tumuluru, J. S. (2020). Bioenergy feedstock types and properties. In J. S. Tumuluru (Ed.), *Biomass Densification* (pp. 1–21). New York, NY, USA: Springer. doi:10.1007/978-3-030-62888-8_1.

Tumuluru, J. S., Sokhansanj, S., Hess, J. R., Wright, C. T., & Boardman, R. D. (2010). *Biomass Torrefaction Process Review and Moving Bed Torrefaction System Model Development*, INL/EXT-10-19569. Idaho Falls, ID, USA: Idaho National Laboratory.

Tumuluru, J. S., Sokhansanj, S., Hess, J. R., Wright, C. T., & Boardman, R. D. (2011). Review: A review on biomass torrefaction process and product properties for energy applications. *Indust. Biotech.*, *7*(5), 384–401. doi:10.1089/ind.2011.7.384.

Tumuluru, J., Boardman, R., Wright, C., & Hess, J. (2012a). Some chemical compositional changes in miscanthus and white oak sawdust samples during torrefaction. *Energies*, *5*, 3928–3947. doi:10.3390/en5103928.

Tumuluru, J. S., Kremer, T., Wright, C. T., & Boardman, R. D. (2012b). Proximate and ultimate compositional changes in corn stover during torrefaction using thermogravimetric analyzer and microwaves. ASABE Annual International Meeting. 29 July–1 August 2012, Dallas, TX, USA, Paper Number 121337398.

Tumuluru, J. S., Ghiasi, B., Soelberg, N. R., & Sokhansanj, S. (2021). Biomass torrefaction process, product properties, reactor types, and moving bed reactor design concepts. *Front. Energy Res.*, *9*, 728140. doi: 10.3389/fenrg.2021.728140.

U.S. Department of Energy (DOE) (2016). Billion-ton report: Advancing domestic resources for a thriving bioeconomy — Volume 1: Economic availability of feedstocks. ORNL/TM-2016/160, Oak Ridge National Laboratory, Oak Ridge, TN, USA.

U.S. Energy Information Administration (EIA) (2018). Biomass explained. EIA, Washington, DC, USA. Retrieved from https://www.eia.gov/energy explained/biomass/.

Wang, L., Barta-Rajnai, E., Skreiberg, Ø., Khalil, R., Czégény, Z., Jakab, E., *et al.* (2017). Impact of torrefaction on woody biomass properties. *Energ. Proced.*, *105*, 1149–1154. doi:10.1016/j.egypro.2017.03.486.

Yan, W., Acharjee, T. C., Coronella, C. J., & Vásquez, V. R. (2009). Thermal pretreatment of lignocellulosic biomass. *Environ. Prog. Sustain Energy*, *28*, 435–440.

Yan, W., Hastings, J. T., Acharjee, T. C., Coronella, C. J., & Vásquez, V. R. (2010). Mass and energy balances of wet torrefaction of lignocellulosic biomass. *Energy Fuels*, *24*, 4738–4742.

Yang, Z., Sarkar, M., Kumar, A., Tumuluru, J. S., & Huhnke, R. L. (2014). Effects of torrefaction and densification on switchgrass pyrolysis products. *Bioresour. Technol.*, *174*, 266–273.

Zheng, A., Jiang, L., Zhao, Z., Huang, Z., Zhao, K., Wei, G., *et al.* (2015). Impact of torrefaction on the chemical structure and catalytic fast pyrolysis behavior of hemicellulose, lignin, and cellulose. *Energy Fuels*, *29*(12), 8027–8034. doi:10.1021/acs.energyfuels.5b01765.

Zheng, Y., Tao, L., Yang, X., Huang, Y., Liu, C., Gu, J., *et al.* (2017). Effect of the torrefaction temperature on the structural properties and pyrolysis behavior of biomass. *BioResources*, *12*(2), 3425–3447. doi:10.15376/biores.12.2.3425-3447.

https://doi.org/10.1142/9781800613799_0007

Chapter 7

Impact of Densified Biomass Feedstocks on Biofuel Production from Biomass Gasification

Joseph D. Smith
Missouri University of Science and Technology
Rolla, Missouri, USA
smithjose@mst.edu

Abstract

Economic production of transportation fuels from biomass sources is an ongoing challenge and opportunity. Hybrid energy systems combine various forms of energy into resilient supplies. This chapter describes biofuel production using biomass gasification. A key step in optimizing the biomass gasification step is densification of various forms of biomass including lignocellulosic biomass. An emphasis of this chapter is providing examples of the impact of feedstock preparation on biomass gasification. These examples show the beneficial impact that mechanical, chemical, and thermal preprocessing technologies can have on feedstock quality which also impacts the quality and quantity of biofuel production. This chapter also includes a discussion of the economics associated with biofuels production from multiple biomass types. Finally, this chapter discusses the key analytical methods used to quantify biofuel quality which also helps demonstrate the feasibility of substituting biofuels generated by biomass gasification for current fossil fuels. This chapter demonstrates that the technology to sustainably generate biofuels currently exists and that the key issue is whether biofuels can be produced in a cost-competitive process as a drop-in compatible fuel using the existing refining and distribution infrastructure currently used by petroleum-based refineries. It's also important to note that the current petroleum industry evolved over the past one hundred years and has seen several generations of technology evolution to get to where it is today. Biomass densification is an example of how the biofuels industry is evolving today into a cost-competitive environmentally friendly substitute for current fossil fuels.

Keywords: Biomass gasification, hybrid energy, feedstock densification, feedstock logistics, biofuel generation, climate change mitigation, green energy

7.1. Introduction

Economic production of transportation fuels from biomass sources (i.e., biofuels) is an ongoing challenge and opportunity. For developing countries, coal, natural gas, and crude oil (fossil fuels) represent the cheapest way to bring their citizens out of energy poverty.[1] At the same time, fossil fuels are a significant source of greenhouse gas (GHG) emissions which impact climate change. Biofuels represent a way to reduce climate impact of fossil fuels while reducing energy poverty in developing countries. Also, as shown in Fig. 7.1, biofuels have been shown to reduce combustion particulate matter by 10%, CO emissions by 11%, and unburned hydrocarbon emissions by 21%

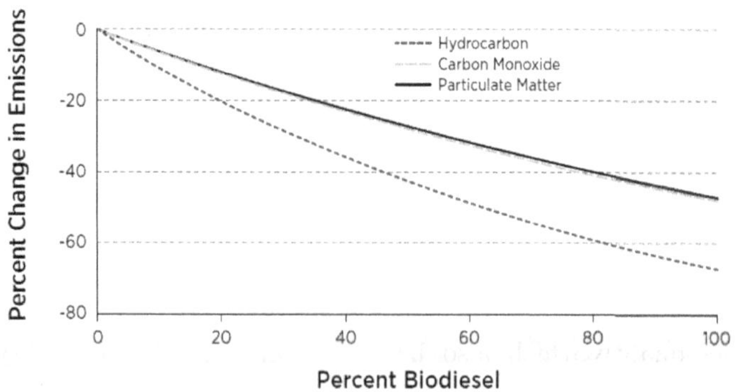

Figure 7.1. Comparison of combustion emissions from biodiesel compared to fossil diesel (Vehicle Technologies Program, 2011).

[1]Bill Gates' discussion on energy poverty provides strong motivation for developing biofuels: see https://www.gatesnotes.com/energy.

when compared to traditional diesel (Vehicle Technologies Program, 2011). However, the logistical challenges associated with transporting biomass from where it is grown and harvested to a bio-refinery are significant. For example, the USA produces 1556 million gallons of biodiesel in 2016 and 75% of the biodiesel production came from soybean. In addition, designing a bio-refinery that can efficiently convert a heterogeneous biomass feedstock into a useable drop-in compatible liquid biofuel and producing and selling biofuels in a market dominated by low-cost fossil fuels are key challenges to establishing a vigorous biofuels industry. Although many books and technical papers have been written on biofuels, the following chapter discusses work related to biodiesel production from vegetable oils including Waste Cooking Oil (WCO), biomass gasification to produce biocrude and syngas for conversion to biofuels, and bio-digestion of wastewater to produce biofuels.

Thermal-chemical conversion of biomass into producer gas or syngas is accomplished using less oxygen than required for complete combustion of the biomass feed (biomass gasification). Biomass gasification occurs between 1112°F and 2732°F and produces a low- to medium-energy gas, depending upon the gasifier design and operating conditions. Biomass gasification is currently used to produce electric power via gas-fired turbines. Significant work has also been done to develop economical and sustainable gasification technology to generate "green"-hydrogen (as opposed to hydrogen produced by steam-methane reforming) to power fuel cell systems, to produce plastics, fertilizers, and drugs and to produce alcohols for transportation fuels.

Syngas produced by biomass gasification is a mixture of gases, mainly composed of carbon monoxide and hydrogen. Recovering energy from waste materials generated in sawmills, paper mills, and landfills via biomass gasification is a cost-effective and reliable process that provides both electric power and clean transportation fuels. Currently, biomass covers approximately 10% of the global energy supply (Schill, 2009). Among renewable energy resources, biomass and renewable waste account for just under two-thirds (64.2%) (Renewable Energy Statistics, 2015). Two common biomass gasifiers are generally based on updraft and downdraft designs.

7.2. Biomass Gasification

One of the oldest and simplest biomass gasifier designs feeds biomass through the top of the reactor in a countercurrent to the air fed through the bottom of the reactor. This design is referred to as an updraft gasifier. In this design, biomass entering the gasifier is subjected to the highest gas temperature created in the combustion zone. As the dried biomass moves further down the reactor, it undergoes gasification with volatile gases evolved from the biomass burned below the gasification zone in the combustion zone. The solid biochar is removed from the bottom of the gasifier while the syngas generated in the gasification zone exits near the top of the gasifier (see Fig. 7.2). This type of gasifier is very simple and low cost, a key disadvantage to the countercurrent design is the high amount of tar produced in the process (Deshpande *et al.*, 2013).

Another design, referred to as a downdraft gasifier, feeds air and biomass through the top of the reactor. This design is the most commonly used style of gasifier for commercial operation. In the

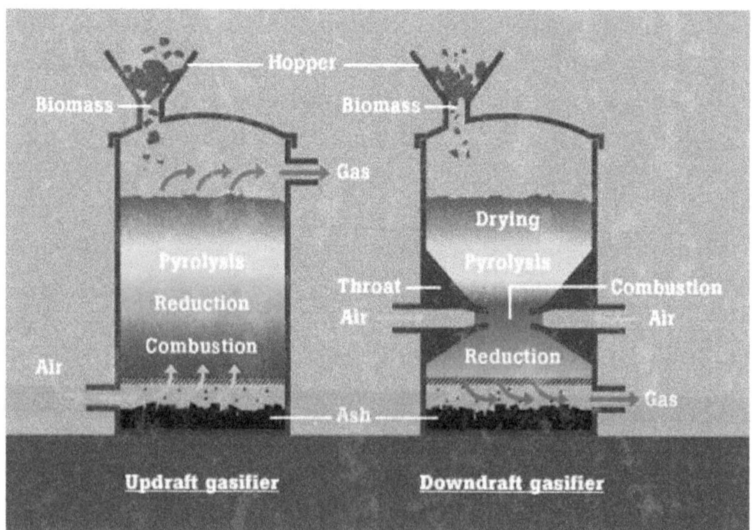

Figure 7.2. Updraft versus downdraft fixed-bed gasifiers (https://www.researchgate.net/figure/Schematic-of-updraft-and-downdraft-fixed-bed-gasifiers-5_fig1_323927618).

downdraft gasifier, air fed near the reactor top moves downward with the biomass feedstock also fed through the top of the gasifier. In this design, a gasification zone is established below the combustion zone where heat is generated to drive the drying process above the combustion zone and the gasification zone below the combustion zone (see Fig. 7.2). This simple gasifier design is also low cost and easy to operate but generates much less tar compared to updraft gasifiers (Patra and Sheth, 2015; Sastry, 2011).

7.2.1. *Densification of Gasifier Feed Materials*

The feed properties and the experimental procedure used to run a gasifier are critically important to what is produced in the gasifier and the safe operation of the gasifier. Therefore, the densification process is discussed next including how various types of feedstocks impact gasifier performance.

7.2.2. *Biomass Feedstock Characteristics*

In the previous work (Smith *et al.*, 2019; Golpour *et al.*, 2017; Yu and Smith, 2018), gasification experiments were carried out with three different types of feed stocks including processed hardwood pellets as well as waste wood materials from a paper mill. The waste wood feed materials included pine flakes and chips with varying moisture content, heating value, and bulk density. As shown in Fig. 7.3, the wood chips were raw unprocessed high moisture material obtained from a Canadian paper mill, and flakes and pellets were low moisture-processed material. Flakes were processed wood without bark and cut into small uniform sizes while pellets were made from compacted oak sawdust. Five samples of each type shown in Fig. 7.3 were collected randomly and Proximate and Ultimate analyses for each were measured using a thermogravimetric analyzer and a CHN elemental analyzer, respectively. These analyses were carried out on dry basis by placing samples in a vacuum oven for eight hours at a temperature of 300°F. The average Proximate Analysis, Ultimate Analysis, and Heating value of these materials are shown in Tables 7.1 and 7.2, respectively.

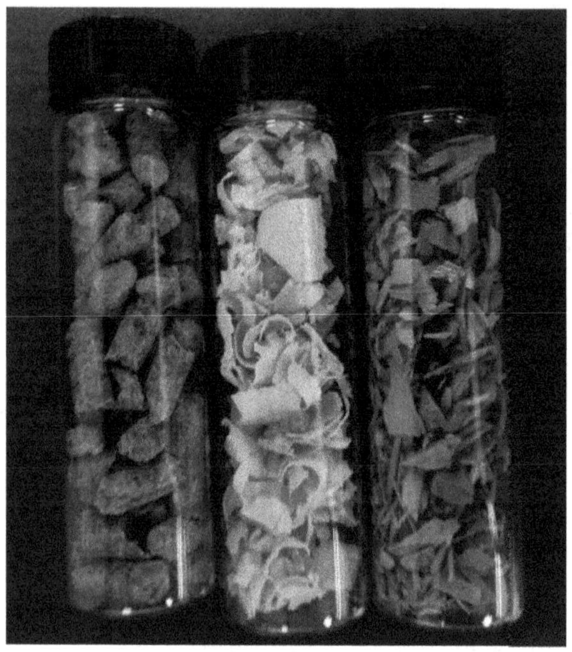

Figure 7.3. Left to right: pellets, flakes, and chips.

Table 7.1. Proximate and ultimate analysis of biomass feedstocks.

Proximate analysis	Chips	Flakes	Pellets	Ultimate analysis (wt %)	Chips	Flakes	Pellets
Moisture (%)	35.19	11.01	7.56	Carbon	48.81	48.24	49.03
Volatile dry (%)	82.28	86.15	87.23	Hydrogen	5.96	6.15	5.58
Fixed carbon dry (%)	17.26	13.32	12.39	Oxygen	44.98	45.55	45.33
Ash dry (%)	0.46	0.53	0.38	Nitrogen	0.26	0.06	0.06

Table 7.2. Heating value of biomass feedstocks.

Heating value	Chips	Flakes	Pellets
Cal/g	4510	4562	4622
Btu/lb	8118	8212	8319

7.3. Gasification Testing with Different Feedstocks

To illustrate the impact that feedstock densification has on biofuel production, several gasification experiments were performed using pre-processed pellets and raw wood flakes and chips. These tests were conducted in a downdraft biomass gasifier similar to that shown in Fig. 7.2. The following section presents the results of these tests. The optimum operating condition for each feedstock was controlled by feedstock composition and transportability within the reactor.

7.3.1. *Pellets*

Fuel pellets made from compacted sawdust with a binder had the lowest moisture content and highest heating value as recorded in Tables 7.1 and 7.2. The pellets were approximately shaped like a cylinder with an $L/D = 4$ (1/4″ long by 1/16″ diameter). The pellets were found to flow freely in the gasifier during each test with no bridging observed in the bed. The pellets also ignited most easily and produced the highest quality syngas compared to wood flakes and chips.

In addition, air flow into the bed above the combustion zone in the gasifier appeared to be impacted by the feedstock. For each test, inlet feedstock flow rate was directly related to the optimal amount of air flow into the reactor required to establish a stable gasification process in the reactor. Too much air increased the combustion zone temperature which reduced the amount of syngas produced in the gasification zone. On the other hand, too little air reduced the combustion zone temperature which also reduced syngas production in the gasification zone. Temperature profiles inside the reactor are shown in Fig. 7.4.

As shown in Fig. 7.4, the experiment began by igniting the feedstock. The temperatures for the three distinct zones in the gasifier were monitored to establish when the gasifier had reached quasi-steady operation (approximately 30 min after startup). Efficient and safe gasifier operation is possible by tracking temperatures in the combustion and gasification zones along with oxygen concentration in the syngas leaving the gasifier. Normal gasifier operating conditions

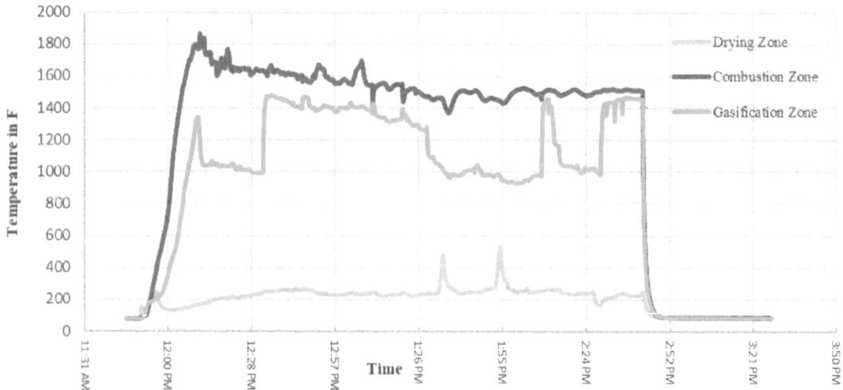

Figure 7.4. Gasifier temperature profile when pellets used.

for the combustion and gasification zones are 1500°F and 1200°F, respectively. Once the gasifier reaches steady operation, additional feedstock is added to the top of the reactor. For this test, the gasifier was operated for approximately 2 h with shutdown commencing at 2:43 PM. The shutdown process involves isolating the combustion/gasification zones inside the gasifier and adding pure nitrogen to the top of the bed which flows through the bed to quench the combustion process which shuts down the gasification process. The primary Induced-Draft (ID) fan remains on during shutdown to sweep nitrogen through the reactor and purge the syngas from the system. Bed temperatures are monitored during shutdown to confirm bed temperature is below reaction temperature which completes the reactor shutdown process.

During operation, syngas leaving the gasifier is cooled to below 200°F before it flows through the primary ID fan. In addition, syngas leaving the reactor passes through a tar separation and collection section to clean the syngas before it enters the ID fan. Bio-oil, also referred to as tar (see Fig. 7.5), is collected and recycled to the gasifier during operation to minimize waste production during the gasification process. The effluent syngas temperature along with ID fan inlet and outlet temperatures are monitored during gasifier operation (see Fig. 7.6).

Figure 7.5. Bio-oil (tar) produced during biomass gasification.

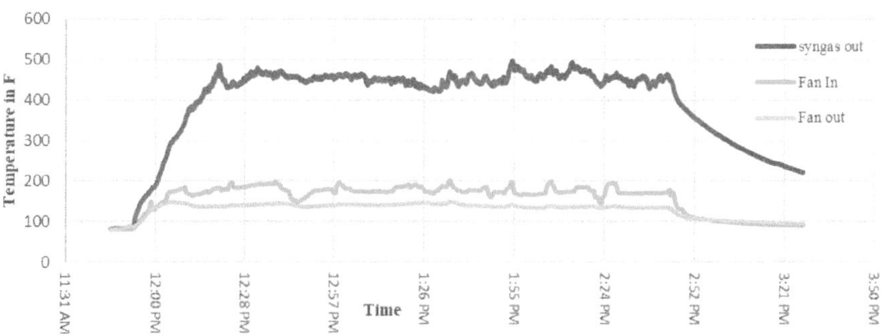

Figure 7.6. Exit syngas temperature when feeding pellets to the gasifier.

7.3.2. *Flakes*

Like wood pellets, flakes of wood or wood shavings were also tested in the gasifier. This feedstock had a slightly higher moisture content compared to the pellets. Plus, wood flakes were also found to have more difficulty flowing through the reactor during the gasification process. To assist the flow of wood flakes, a large vibrator was attached to the inner core of the reactor where the gasification process took place. Although the volumetric flow of flakes to the gasifier was approximately the same as pellets, given the lower

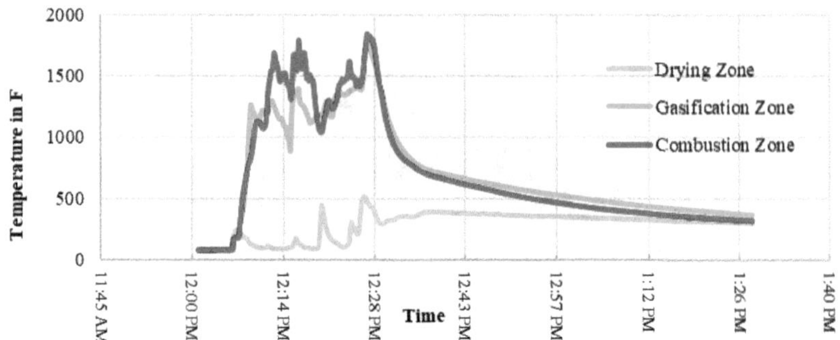

Figure 7.7. Gasifier temperature profile when wood flakes are fed to the gasifier.

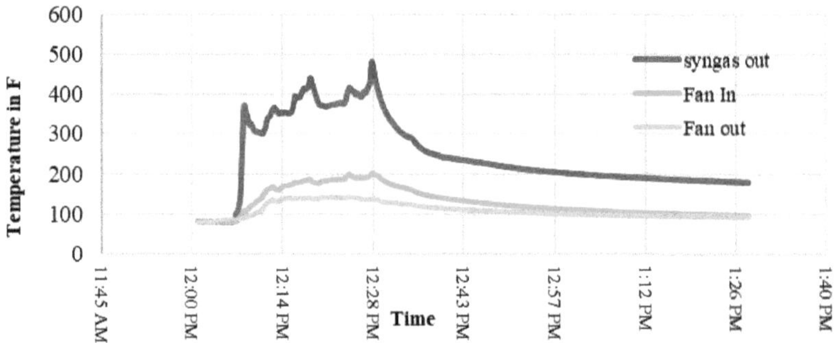

Figure 7.8. Exit syngas temperature when feeding wood flakes to the gasifier.

density of wood flakes, the overall mass flow to the gasifier was much lower than with pellets. Therefore, the overall production of syngas and biochar was less with flakes. The relative temperature profiles in the reactor, at the syngas outlet and through the ID fan, are shown in Figs. 7.7 and 7.8.

7.3.3. *Chips*

The gasifier was also operated when feeding chips of wood to it. These chips were cylindrical in shape approximately 1–2″ (2.5–5 cm) long and approximately 1/16″ in diameter (0.16 cm), much longer and thinner than pellets. The wood chips also had a much higher moisture content than the pellets (35% vs. 7.5%). Due to these two main differences, gasifier startup and operation were much more difficult. The high moisture content required more energy input into the gasifier to start it up and establish a continuous combustion zone which drives the gasification zone. The long thin shape of the wood chips led to excessive bridging inside the gasifier during normal operation which required the operator to frequently stir the bed inside the reactor to keep it flowing.

In this test, pellets were initially used to start up the reactor and once a stable combustion zone was achieved, wood chips were added to the gasifier. Due to the bridging tendency of the wood chips, the vibrator was operated but visual observation of the combustion zone from above the reactor was also used to monitor bridging inside the reactor. When bridging occurred, the bed would form void spaces and the oxygen content of the effluent gas would start to increase. When this was observed, the operator would insert a long metal rod into the bed and stir it to break the bridge and create a more homogenous bed confirmed by a more uniform temperature profile inside the reactor. The temperature profiles inside the bed and in the reactor effluent are shown in Figs. 7.9 and 7.10.

To establish continuous gasifier operation, a much lower mass flow of wood chips was used during this test. The impact of bridging

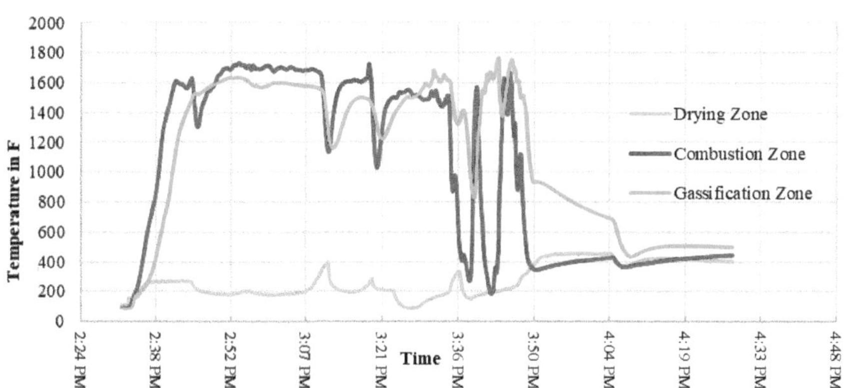

Figure 7.9. Gasifier temperature profile when wood chips were fed to the gasifier.

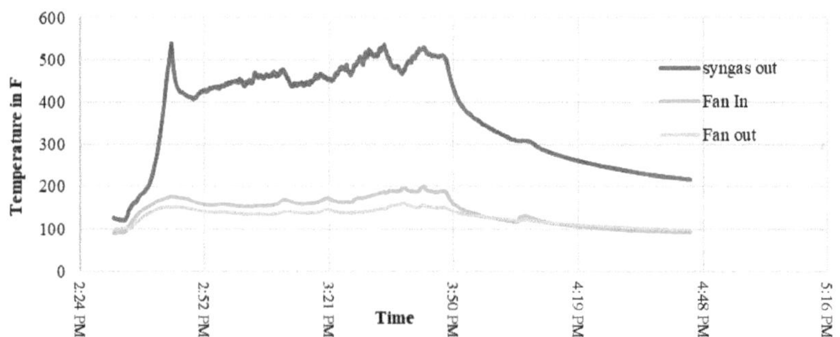

Figure 7.10. Exit syngas temperature when feeding wood chips to the gasifier.

on gasifier operation is shown at times 3:34 PM and 3:40 PM in Fig. 7.9. This unsteady bed temperature was the result of a void forming in the bed due to bridging with the operator stirring the bed to break the bridge and the bed re-establishing continuous flow. This phenomenon was observed using a camera mounted above the reactor with a view of the bed during gasifier operation. When bridging occurred, a glowing zone was observed. Also, the syngas flame in the gas flare used to burn the effluent syngas was much less intense and less stable compared to pellets or wood flakes. This also indicated a lower heating value of the produced syngas when using wood chips due to the higher moisture content.

7.4. Syngas Composition and Bio-Oil (Tar) Production

To quantify how feedstock quality impacted syngas production, the composition of syngas produced in each of these steps was measured. Gas samples of syngas were collected in sample bags during each gasifier test and the composition of each sample was determined using an FTIR instrument. Results of the syngas composition collected during the pellet gasification test are shown in Table 7.3. The high concentration of nitrogen in the syngas was due to using air as the oxidizer during the gasification tests. Syngas quality (i.e., heating value) would be much higher if pure oxygen was used instead of air in the gasification process. However, as expected based on previous work, the main components of the produced

Table 7.3. Syngas composition for pellet gasification.

Component	Volume (%)
Hydrogen	18
Carbon monoxide	21
Carbon dioxide	16
Methane	2
C2 + hydrocarbons	2
Nitrogen	41

syngas for pellets, wood flakes, and wood chips were H2 and CO which accounted for approximately 40% by volume of the produced syngas.

Besides syngas, liquid bio-oil (tar) was also produced during each gasification test. This liquid was a dark color mostly during the beginning of each gasifier test. During the pyrolysis reaction, carbonaceous biomass feedstock underwent thermal degradation to form volatiles, char, and ash. During low-temperature operation (i.e., during the first 10 min of each test when the gasifier was heating up), the low gas temperature condensed the gaseous volatiles into this brownish-black thick liquid (see Fig. 7.5) often referred to as wood vinegar or bio-oil. Liquid composition was approximately 70% carbon, 7–9% hydrogen, and 20% oxygen but also contained up to as much as 20–30% water for wood chips with their higher moisture content. The quality and composition of this bio-oil were directly related to the feedstock quality. This bio-oil (wood vinegar) is considered environmentally friendly given the lower CO_2 emissions compared to fossil fuels. The bio-oil has an approximate heating value of 25–30 MJ/kg (25–35% lower compared to crude oil) (Chhiti and Kemiha, 2013; Mohan *et al.*, 2006). More discussion of this bio-oil is given in the following.

As shown, feedstock quality greatly affects syngas and bio-oil (tar) production which thus affects biofuels production. Thus, biofuels production is directly tied to feedstock densification. The following discussion on biofuels production via biomass gasification is intended to illustrate this point.

7.5. Biofuels from Gasification

The gasification process generates syngas plus bio-oil, bio-char, and ash (Rauch, 2006). Typical syngas composition is mainly made up of carbon monoxide and hydrogen with smaller quantities of carbon dioxide, methane, and water. If the gasification process generates syngas with significant amounts of non-combustibles such as carbon dioxide and nitrogen from the air, the effluent gas is alternatively referred to as producer gas (http://www.enggcyclopedia.com/2012/01/syngas-producer-gas/). As a renewable energy resource, syngas has many advantages as it is ecologically benign in nature and can be stored, transported, or converted to other higher-value chemicals and liquid fuels. Syngas forms the basis for producing many products including chemicals, plastics, fertilizers, and electric power (see Fig. 7.11).

Syngas can also be used to produce naphtha, waxes, and lubes via the Fischer–Tropsch process. It also represents a sustainable route to liquid biofuels which becomes more critical as cheap oil becomes more expensive due to increasing demand, implementation of climate change policies (i.e., carbon tax), and reduction of easily obtainable

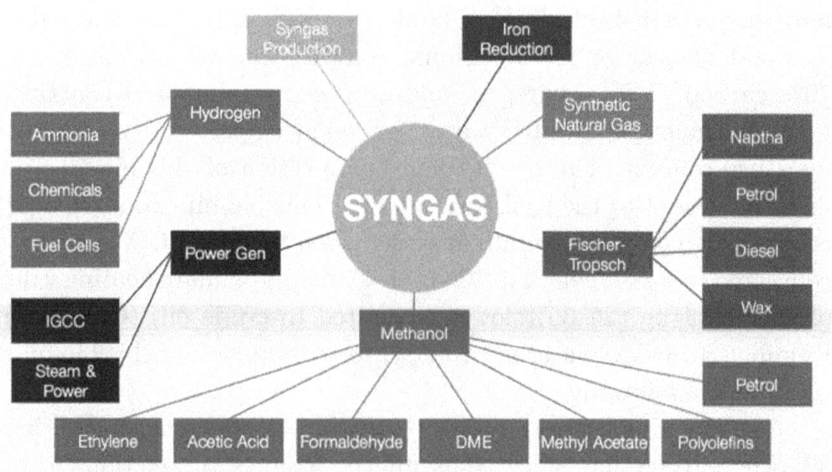

Figure 7.11. Applications of syngas (http://wastetoenergysystems.com/syn-gas-a-versatile-and-renewable-fuel/).

crude oil supplies. Syngas also represents the best feedstock for methanol production that itself is a biofuel and can be blended with standard diesel to reduce NO_x and soot (PM) emissions (Yusaf *et al.*, 2013). Other uses include carbon-neutral electric power generation via combined cycle gas-turbine systems and hydrogen-powered fuel cells, ammonia/urea fertilizer production, methanol and associated chemicals production, and hydrogen for upgrading heavy petrochemicals in existing refineries. Taken together, electricity power generation and Fischer–Tropsch liquid fuels represent approximately 50% of the total Syngas use (Rauch, 2006). The heat produced when syngas is burned can also be used in desalination plants. According to the US Department of Energy, in 2009, the potential of biomass usage for electricity generation was projected as 22 GW by 2022 (Bain, 2012).

Besides using the syngas as a biofuel, the liquid product can be efficiently utilized using conversion technologies to produce mixed alcohols that could be blended with petroleum gasoil for co-cracking to produce bio gasoline, diesel, and fuel oil. Previous work demonstrated bio-oils (tar) produced in biomass gasification could be processed via conventional petrochemical refinery operations (Agblevor *et al.*, 2010; Agblevor and Mante, 2010). A simulated high-temperature distillation curve for the bio-oil predicted 100% distillation with no residues (Fig. 7.12). The work showed that blends of 15 wt% bio-oil and 85 wt% gas-oil could produce distillate fractions without loss of conversion (Table 7.4). This work also showed that cracked products from bio-oil/gas-oil blends contained no oxygenated compounds and that they contained high amounts of aromatic compounds characteristic of high-octane fuel (see Fig. 7.13). This suggests that the bio-oil/gas-oil blends are best classified as "synthetic" sweet crude which would bring a premium price per barrel.

Certainly, biofuel costs depend on many factors including supply and preparation of biomass feedstock to the refinery. Work by Hess *et al.* showed for a biorefinery producing 4000 barrels per standard day of liquid biofuels from wood, switchgrass, or corn stover, between three and six 4000 ft^3 hopper cars per hour would be required.

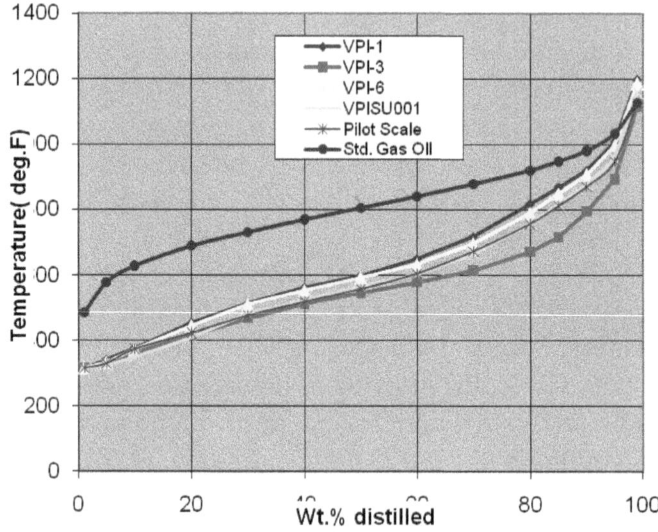

Figure 7.12. High-temperature distillation curves of SGO and BPO.

Table 7.4. Distillate fractions of standard gas-oil/stable bio-oil blends produced from different catalysts compared to standard gas oil (VPI-4, VPI-4ST, and VPISU001 are proprietary catalysts).

Fraction	Standard gas oil	VPI-4	VPI-4ST	VPISU001
H2 (%)	0.61	0.53	0.44	0.56
Total C2	2.98	2.99	2.92	2.94
LPG	16.00	16.19	16.00	15.95
Gasoline	43.97	44.01	44.44	44.35
LCO	17.06	16.93	17.23	17.23
HCO	12.94	13.07	12.77	12.77
Coke	7.06	6.81	6.64	6.76
Conversion (%)	70.00	70.00	70.00	70.00
Cat/oil	6.00	6.08	5.96	5.81

Current railroad infrastructure could not support and would require a bio-refinery to develop its own marshaling yards and secure its own hopper cars and switching trains. Supply via river barges would require a bio-refinery to provide a dock with unloading/loading equipment. The capital investments associated with either railroad

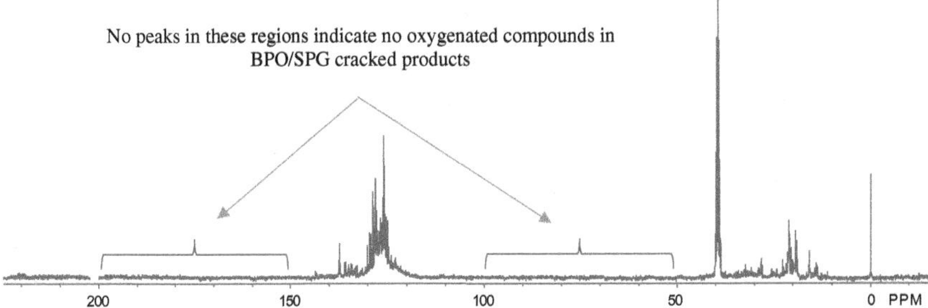

Figure 7.13. ^{13}C-NMR spectrum of blends of stabilized bio-oil/gas-oil distillates (note absence of oxygenated peaks between 50–100 ppm and 150–200 ppm).

or river badge shipping would make the feedstock logistics supply costs prohibitively expensive for biofuels production. The best way to reduce/minimize feedstock costs would be to pre-process the biomass feedstock before shipping. The standard "Advanced Uniform Format Feedstock Supply Concept" shown in Fig. 7.14 involves collecting (harvesting) biomass from the farm and accumulating it in a central "Shipping Terminal". The proposed approach would be to densify the biomass at this step before sending it to a bio-refinery next to or as part of a large standard petro refinery. This would reduce shipping costs and facilitate processing in a bio-refinery. Hess *et al.* (2009) have shown that densified feedstock pellets support highly efficient biomass gasification and subsequent syngas production which could be used to produce bio-oil for upgrading and blending with gas-oil in a large regional petro-refinery (see Fig. 7.15).

The business model proposed by Hess assumed the produced bio-oil would be sold to the refiner at a discounted price based on conventional benchmark crude prices. He proposed co-locating a pre-processing facility near a secure biomass source which involved shipping biomass from the shipping terminal to the pre-processing bio-refinery. Collecting and storing biomass at a centralized depot is an essential step in Hess' bio-fuel production model. Adding the densification step at the depot would further reduce the shipping and processing cost so the only step required at the bio-refinery would be related to producing bio-oil ready for blending to generate a superior

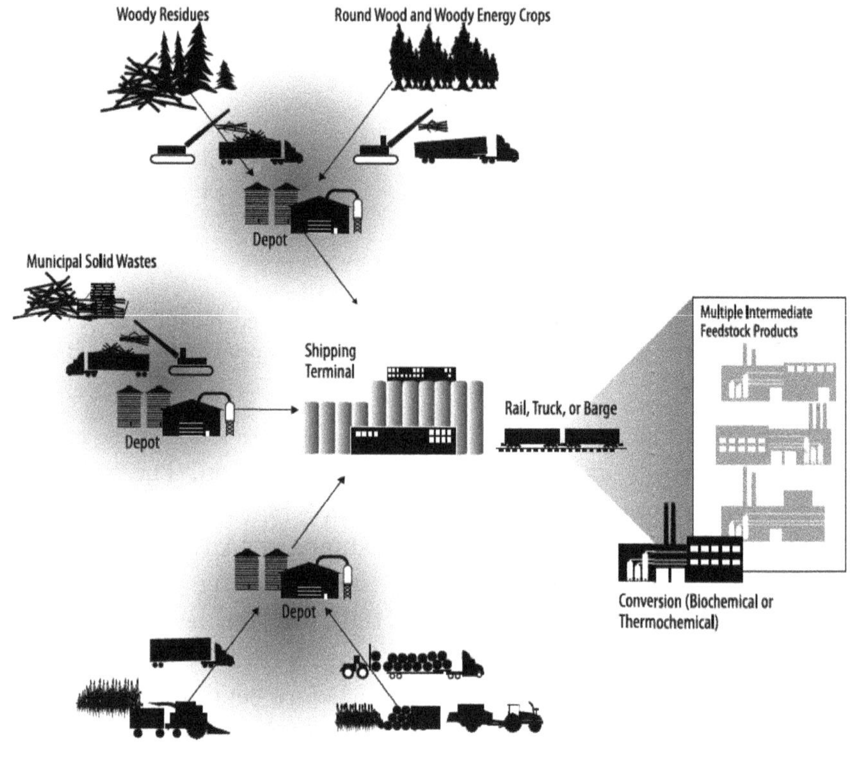

Figure 7.14. Advanced uninform format feedstock supply concept.

biofuel which has very low sulfur content and a high-octane number. In addition, bio-oil is generally hydrogen deficient due to its high aromaticity, so they also require hydrogen upgrading. Blending with gas-oil reduces the bio-oil viscosity from between 20 and 8.6 API (SG $\sim 0.93 - 1.01$) to between 30 and 40 API (SP $\sim 0.876 - 0.825$).[2] To minimize the impact on a standard petroleum refinery, a bio-refinery "across the fence" would blend and hydrotreat the Bio-oil before it is cracked and blended with gas-oil in the standard petro-refinery.

To illustrate how feedstock quality affects the biorefinery design and operation, three feedstocks are considered including Virginia

[2]API is defined as API = 141.5/SG − 131.5.

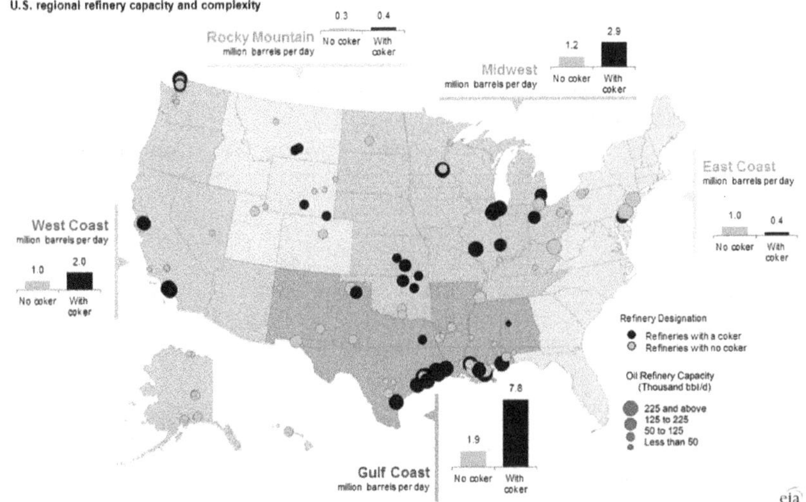

Figure 7.15. Relative location of existing US regional petroleum refineries for potential collocation of future bio-refineries.

Note: According to EIA, as of January 1, 2014, there were 133 operating refineries with atmospheric crude oil distillation units (ACDU) totaling capacity of 18.9 million barrels per stream day. Heavy capacity denotes refineries with coking capacity; light capacity denotes refineries without coking capacity.

Source: U.S. Energy Information Administration (2015).

poplar, switchgrass, and corn stover. The representative elemental analysis of these feedstocks is given in Table 7.5 with bio-oil yields given in Table 7.6.

These yields are based on catalytic pyrolysis of the feedstocks in a fluidized bed. The resulting bio-oil remains liquid on heating which was then processed in a lab-scale cracking reactor with success with gas-oil blends up to 25%. Hess used these yields to develop his economics of bio-fuel production using a catalytic pyrolysis process.

Pricing for raw materials and various products is provided in Table 7.7. A material and energy balance was performed around the catalytic pyrolysis unit considering the expected product streams. This analysis considered standard unit operations employed in a general petro-refinery. The unit operations and associated equipment

Table 7.5. Biomass feedstock analysis (moisture and ash free basis).

Biomass feedstock (wt%)	Virginia poplar	Switchgrass	Corn stover
Carbon	49.9	50.2	50.3
Hydrogen	6.1	5.6	5.4
Oxygen	43.7	43.6	43.2
Nitrogen	0.2	0.4	0.9
Sulfur	0.1	0.1	0.2

Table 7.6. Product yields for three biomass feedstocks.

Biomass feedstock (wt%)	Virginia poplar	Switchgrass	Corn stover
Light gases	21.5	12.7	11.6
Organic liquids	34.6	31.4	35.8
Char	23.7	39.0	33.5
Water	20.2	20.9	19.0

Table 7.7. Pricing and product disposition.

Product or raw material	Pricing	Disposition
Biocrude	\$65/barrel	To adjoining refinery
Gases	No pricing (\$4/MMBTU can be inferred) from steam pricing	To process and gas turbine use
Biochar	\$60/t as soil amendment	Farming
Water	Process water (\$0.1/1000 gallons)	Cooling water, or water made is treated and sent to river
Boiler feed water	\$1.50/1000 gallons	Producing net steam for export (150 psig superheated)
Catalyst	\$3000/t	—
Chemical and misc.	At 0.15% of investment	Water treatment, caustic, ammonia, etc.
Power	10 cents per KW	Same price consuming or generation of excess

Table 7.8. Required processing units and estimated costs for each operation (Vehicle Technologies Program, 2011).

Raw material	Virginia poplar ($)	Switchgrass ($)	Corn stover ($)
Catalytic pyrolysis unit	50.00	50.00	50.00
Gas recovery	16.00	11.00	10.00
Amine treating	12.00	8.00	7.00
Wastewater treatment	10.00	10.00	10.00
Cooling tower	1.90	1.30	1.30
Grinding mill	6.00	6.00	6.00
Screens for char	0.60	0.60	0.60
Rotary dryers	6.90	6.90	6.90
Gas turbine	11.40	9.60	5.60
Tankage	15.60	15.60	15.60
Subtotal	130.00	119.00	113.00
Offsite (50%)	65.00	60.00	57.00
Engineering (30%)	59.00	54.00	51.00
Total cost ($MM)	254.00	232.00	220.00

in the standard refinery helped determine the required capital investment (see Table 7.8) required plus a 25% contingency based on 2010 dollars.

An assumed discount of 10 cents per gallon of bio-oil provided to the petro-refinery is based on a biocrude with 20% oxygen and lab-scale blending results shown in Table 7.9. The lab-scale blending yields were assumed representative for what would be achieved in a full-scale refinery. Based on this analysis, Biofuels derived from biomass gasification are a viable alternative to standard liquid petroleum-based fuels. The biggest challenge in providing the biofuel based on blended bio-oil and gas-oil remains cost associated with shipping logistics of the biomass feedstock. Our recommended modification to the Advanced Uniform Format concept is the add the densification step in the collection depot to reduce shipping and handling costs. Integrating the bio-refinery with a standard petro-refinery further reduces the production costs. Also, solutions to the technical challenges of producing bio-oil/gas-oil blended biofuels makes biomass gasification a viable option for large-scale sustainable

Table 7.9. Yields for biocrude.

Product	Yield (wt%) SGO	Yield (wt%) SGO + 25% BPO	Derived yields on BPO
H2 thru C2	5.5	4.4	1.2
C3's	9.6	8.9	6.8
Butenes	4.3	4.5	5.0
Isobutane	8.0	8.1	8.6
N-butane	2.8	2.2	0.6
Gasoline	45.7	46.0	47.1
Heating oil	8.6	10.0	14.4
Residual fuel	3.0	3.2	3.8
Coke	12.6	12.6	12.6
Gasoline octane	98.8	99.3	100.6

production of liquid transportation fuels using the existing refining infrastructure.

7.6. Biofuel Quality Characterization

Fuel quality must also be considered when developing a biodiesel process. Biofuel quality can be quantified using various methods including Thermogravimetric Analysis (TGA) and Gas Chromatography/Mass Spectrometry (GC/MS). Standard tests developed by the American Society for Testing and Materials (ASTM) are normally used to quantify biofuel quality. Two tests are normally used including the Cetane number (CN) test and the Pour Point test. These methods and tests will briefly be reviewed but more details are available elsewhere if required.

7.7. Thermogravimetric Analysis

Thermogravimetric analyzers (TGAs) measure changes in the physical and chemical properties of a material as a function of increasing temperature (with constant heating rate) or as a function of time (with constant temperature and/or constant mass loss). For biofuel production, the blending process can best be quantified by comparing the boiling point curves for pure components. A TGA instrument can be used to easily and efficiently monitor bio-fuel blend quality.

In general, the TGA has been used to characterize material performance in terms of mass loss due to decomposition, oxidation, or loss of volatiles and/or moisture. Common applications of TGA include the following:

- materials characterization through analysis of characteristic decomposition patterns,
- studies of degradation mechanisms and reaction kinetics,
- determination of organic content in a sample, and
- determination of inorganic (e.g., ash) content in a sample, which may be useful for corroborating predicted material structures or simply used for chemical analysis.

To determine bio-fuel yield, the blending process can be followed using a TGA as illustrated in the sample TGA curve shown in Fig. 7.16. Here, curves for the wt% of pure bio-fuel and pure gas-oil are shown with a 50/50 blend of both as a function of temperature.

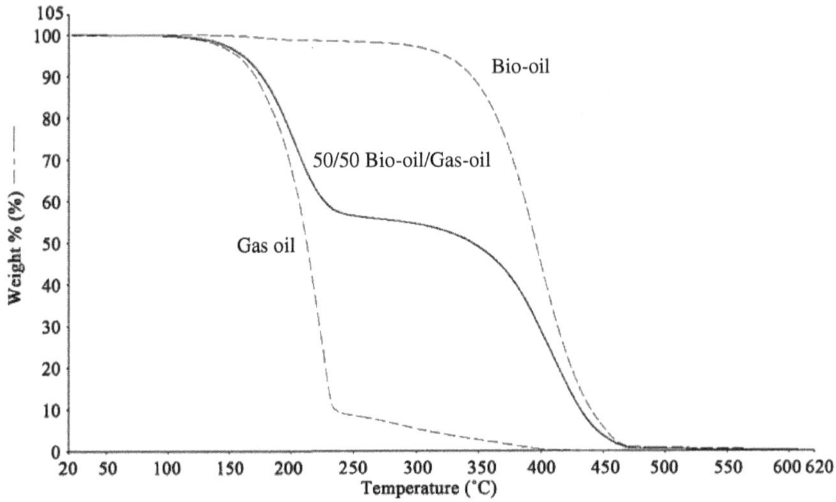

Figure 7.16. Representative TGA curve (20°C/min heating rate).

In addition, the derivative of the TGA trace (referred to as DTGA) can be used to determine weight loss temperature more closely.

7.7.1. *TGA Analysis*

As illustrated in Fig. 7.16, the TGA analysis covered a temperature range between 20°C and 550°C and shows a decrease in bio-fuel mass beginning between 125°C and 130°C. This decrease is attributed to the light fraction of bio-fuel vaporizing first. The second step shown in the TGA curve starts around 232°C which is likely due to heavier components in the bio-fuel. No further changes are observed in the bio-fuel TGA curve which suggests the bio-fuel composition is uniform beyond the initial light fraction. By comparison, the TGA for gas-oil indicates evaporation begins at approximately 325°C while the mixture of bio-fuel/gas-oil shows the combined effect with the mixture with a 0.9% loss of sample at 125°C, 41.54% loss at 232°C, and 98% loss at 460°C. The bio-oil mass loss at 125°C correlates to the wt% bio-oil in the sample while the mass loss associated with gas-oil at 325°C correlates to the wt% gas-oil in the sample. Since these temperatures vary by approximately 200°C, this method is quite effective in distinguishing bio-oil and gas-oil.

7.8. Gas Chromatography Mass Spectrometry

Gas chromatography–mass spectrometry (GC/MS) combines the features of gas chromatography (GC) and mass spectrometry (MS) to identify specific substances in a sample. This instrument identifies molecules based on characteristic fragmentation patterns at specific retention times in a specific packed column. The difference in chemical properties between different molecules in a mixture and their relative affinity for the stationary phase (packing) in the column promotes separation as the sample moves through the column driven by capillary forces. Molecules are retained in the column and separate throughout the column at different times (retention times) which allows the MS to capture, ionize, accelerate, deflect, and detect ionized molecules individually. The MS accomplishes this by cracking each molecule into ionized fragments and detecting these fragments using their mass-to-charge ratio. Used together, these instruments provide a much more accurate identification of particular molecules not possible in the individual devices. The MS relies on a very pure sample to differentiate between multiple molecules that may take the same amount of time to travel through the GC column. Sometimes, two molecules might have similar fragment ionization patterns in the MS, so GC is used to differentiate. Combining both processes reduces analysis error but this complex dual system is cost prohibitive for normal applications in practical petro-chemical refining.

7.9. Standard Biofuel Quality Tests

Biofuel product quality can be quantified using six standard ASTM definitions including the following:

1. Standard Specification — defines requirements to be satisfied by subject,
2. Standard Test Method — defines the way a test is performed and the precision of the result, test result may be used to assess compliance with Standard Specification,
3. Standard Practice — defines a sequence of operations that, unlike a Standard Test Method, does not produce a result,
4. Standard Guide — provides an organized collection of information or series of options that does not recommend a specific course of action,
5. Standard Classification — provides an arrangement or division of materials, products, systems, or services into groups based on similar characteristics such as origin, composition, properties, or use, and
6. Terminology Standard — provides agreed definitions of terms used in other standards.

Bio-fuel/gas-oil blend quality can be quantified using standard ASTM tests including the Cetane Number and Pour Point as described in the following.

7.9.1. *Cetane Number*

Cetane number (CN) of a fuel measures the combustion quality of biodiesel during compression ignition (CI). This number is critical for CI-based engines including high compression diesel and gasoline engines. The CN measures a fuel's ignition delay (time period between the start of injection and the first identifiable pressure increase during fuel combustion). For a particular CI engine, higher CN fuels have shorter ignition delays than lower CN fuels. In short, the higher the CN, the more easily the biodiesel will combust under compression (such as in a diesel engine).

7.9.2. *Pour Point*

The pour point is an indication of the lowest temperature at which a fuel is capable of flowing under very low forces. The pour point is prescribed in accordance with the conditions of storage and use. Higher pour point fuels are permissible where heated storage and adequate piping facilities are provided. An increase in pour point can occur when residual fuel oils are subjected to cyclic temperature variations that can occur in the course of storage or when the fuel is preheated and returned to storage tanks.

Table 7.10 compares properties of regular diesel and biodiesel.

Table 7.10. Properties of fossil-based diesel compared to biodiesel.

Properties	Biodiesel standard ASTM D6751-09	Regular diesel	Testing method used for regular diesel
Specific gravity	0.86–0.9	0.85	ASTM D4052
Viscosity (mm^2/s) at 400°C	1.9–6	2.6	ASTM D445
Calorific value (MJ/kg)	Report	42	ASTM D240
Cetane number	47	46	ASTM D613
Pour point (0°C)	Report	−20	ASTM D97

7.10. Conclusions and Recommendations

A 1 ton per day downdraft biomass gasifier was designed, fabricated, and operated to produce syngas and bio-oil from various feedstock including pellets, wood flakes, and high moisture wood chips. Experiments were recorded with temperature profiles through the gasifier and at the exit and through the ID fan. Several tests were conducted to evaluate the effect of temperature, air flow rate, and feed moisture content on syngas and bio-oil (tar) production.

Biomass feed rate and air flow rate into the reactor were found to be critical to maintaining steady gasifier operation. An optimum biomass feed rate together with the best air flow rate required to retain the steady operation was found for each feedstock. Feedstock quality was shown to have a dramatic impact in terms of combustion intensity and gasification rate. Higher density, higher heating value, and lower moisture cylindrically shaped pellets were easier to use in the gasifier and resulted in a more stable combustion and gasification zone plus they produced higher quality syngas compared to wood chips and flakes. The effect of reactor size syngas production is expected not to nonlinear which was also shown through a scaling study considering a 4″ gasifier, an 8″ gasifier, and a 12″ gasifier. Each reactor size showed much the same behavior in terms of syngas and bio-oil production, but the larger gasifier appeared to be more steady and easier to operate. Plus, heat loss through the reactor walls was much less which also contributed to a more steady higher quality syngas and bio-oil production.

In summary, energy is key to economic prosperity, and biofuels are an important component of resilient sustainable energy. In this chapter, biomass gasification of various feedstocks was used to produce syngas and bio-oil has been discussed. Although several gasification technologies have been developed around the world in various research laboratories, the "downdraft" gasifier design appears to be the most widely used design. As discussed, many high-value products can be produced from syngas and bio-oil due to the ultralow sulfur content of most biomass. This makes biomass gasification a key technology in the biofuels industry. The main challenge in growing the biofuels industry has been feedstock logistics needed to supply a bio refinery with ample biomass to sustain the production of a cost-competitive bio-fuel. Previous economic analysis by Hess identified the breakeven point of $65/bbl for petroleum crude and $90/bbl for bio-oil from which the biofuel industry based on biomass gasification would be feasible.

It is clear from this discussion that the technology to sustainably generate biofuels currently exists. It is also clear that the key issue is whether the biofuels can be produced in a cost-competitive process to supply a drop-in compatible fuel using the extensive existing refining and distribution infrastructure currently used by petroleum-based refineries. It's important to note that the petroleum industry was not developed in a few short years or even decades but took several generations to get to where it is today. The same can be said for the biofuels industry — developing and implementing this industry will take time and money before it can truly become a significant energy resource.

References

Agblevor, F. A. & Mante, O. (2010). VTIP #10-080. Multimedia fluidized bed pyrolysis production of biogasoil. Patent application # 61/305,188.

Agblevor, F. A., Mante, O. Abdoulmoumine, N., & McClung, R. (2010). Production of stable biomass pyrolysis oils using fractional catalytic pyrolysis. *Energy Fuels*. doi:10.1021/ef1004144.

Bain, R. L. (2012). USA biomass gasification status. National Renewable Energy Laboratory, April 18. Retrieved from http://www.ieatask33.org/app/webroot/files/file/2012/USA.pdf.

Bhavanam, A. & Sastry, R. C. (2011 December). Biomass gasification processes in downdraft fixed bed reactors: A review. *Int. J. Chem. Eng. Appl.*, *2*(6), 425–433.

Chhiti, Y. & Kemiha, M. (2013). Thermal conversion of biomass, pyrolysis and gasification: A review. *Int. J. Eng. Sci.*, *2*(3), 75–85.

Deshpande, A. D. P., Shailesh, L. P., Ghadge, A. G., & Raibhole, V. N. (2013). Testing and parametric analysis of an updraft biomass gasifier. *Int. J. Chem. Tech Res.*, *5*(2), 753–760.

Encyclopedia, Syngas/Producer gas. Retrieved from http://www.enggcyclopedia.com/2012/01/syngas-producer-gas/. http://wastetoenergysystems.com/syn-gas-a-versatile-and-renewable-fuel/.

Golpour, H., Boravelli, T., Smith, J. D., & Safarpour, H. R. (2017 June). Production of Syngas from biomass using a downdraft Gasifier. *Int. J. Eng. Res. Appl.*, *7*(6, (Part-2)), 61–71. https://www.ijera.com/papers/Vol7_issue6/Part-2/J0706026171.pdf.

Hess, J. R., Kenney, K. L., Ovard, L. P., Searcy, E. M., & Wright, C. T. (2009 April). Commodity-scale production of an infrastructure-compatible bulk solid from Herbaceous Lignocellulosic biomass. INL/EXT-09-14752, Uniform-format bioenergy feedstock supply system design report series.

Mohan, D., Pittman, C. U., & Steele, P. H. (2006). Pyrolysis of wood/biomass for bio-oil: A critical review. *Energy Fuels*, *20*(3), 848–889. https://doi.org/10.1021/ef0502397.

Patra, T. K., & Sheth, P. N. (2015). Biomass gasification models for downdraft gasifier: A state-of-the-art review. *Renewable Sust. Energy Rev.*, *50*, 583–593.

Rauch, H. B. R. (2006). Review applications of gases from biomass gasification: Syngas production and utilization. In *Handbook Biomass Gasification.* Biomass Technology Group.

Renewable Energy Statistics (2015 May). EuroStat statistics explained. Retrieved from http://ec.europa.eu/eurostat/statistics-explained/index.php/Renewable_energy_statistics.

Schill, S. R. (2009 September 19). IEA task40: Biomass provides 10 percent of global energy use. *Biomass Magazine.* Retrieved from http://biomassmagazine.com/authors/view/Sue_Retka%20Schill.

Smith, J. D., Alembath, A., Al-Rubaye, H., Yu, J., Gao, X., & Golpour, H. (2019). Validation and application of a kinetic model for downdraft biomass gasification simulation. *Chem. Eng. Technol.*, *42*(12), 2505–2519. https://doi.org/10.1002/ceat.201900304.

U. S. Energy Information Administration (EIA). (2015). Regional refinery trend evolve to accommodate increased domestic crude oil production. Retrieved from https://www.eia.gov/todayinenergy/detail.php?id=19591 (Accessed February 2, 2023).

Vehicle Technologies Program, U.S. Department of Energy. (2011). *Energy Efficiency and Renewable Energy*, DOE/GO-102011-3001.

Yu, J. & Smith, J. D. (2018). Validation and application of a kinetic model for biomass gasification simulation and optimization in updraft gasifiers. *Chem. Eng. Process. Process Intens.* https://doi.org/10.1016/j.cep.2018.02.003.

Yusaf, T., Hamawand, I., Baker, P., & Najafi, G. (2013). The effect of methanol-diesel blended ratio on CI engine performance. *Int. J. Automot. Mech. Eng.*, *8*, 1385–1395.

https://doi.org/10.1142/9781800613799_0008

Chapter 8

Densification of Agricultural Residue for Hydrothermal Liquefaction

Bharathkiran Maddipudi*, Anuradha R. Shende*,‡, Vinod Amar*, Khang Huynh*, Jaya Shankar Tumuluru†, and Rajesh V. Shende*,§

*Karen M. Swindler Department of Chemical and Biological Engineering, South Dakota Mines Rapid City, South Dakota, USA

†Southwestern Cotton Ginning Laboratory United States Department of Agriculture Agriculture Research Service Las Cruces, New Mexico, USA

‡anuradha.shende@sdsmt.edu

§rajesh.shende@sdsmt.edu

Abstract

Crop harvest leaves behind agricultural waste with a significant amount of moisture content. Transporting high moisture agriculture waste biomass to a biorefinery is highly uneconomical. Achieving dry biomass (<10%) for subsequent thermochemical processing such as pyrolysis or gasification is cost-intensive. Hydrothermal liquefaction (HTL) technology is more suited for agricultural waste conversion because it can accept high moisture feedstock to produce biofuels and bioproducts. HTL technology has demonstrated many more advantages over other thermochemical technologies at the laboratory scale. Yet, there is no commercial HTL plant to process agricultural residue or lignocellulosic biomass. Transporting wet biomass and making it HTL-ready are the two significant obstacles to commercializing the HTL technology. Biomass densification may reduce the transportation cost and allow processability at higher biomass loadings. Although densification via pelletization has been reported in the literature, currently, there is no understanding of the influence of densified biomass on HTL product distribution or yield. This chapter covers essential information on HTL technology, reported yields of bioproducts (bio-oil and hydrochar), and the parameters to consider in configuring an HTL-ready feedstock with emphasis on corn stover feedstock densification.

Keywords: Hydrothermal liquefaction, agricultural residue, corn stover, lignocellulosic biomass, densification, bio-oil, biochar

8.1. Background

The International Energy Outlook Narrative published in October 2021 projects that global renewable energy consumption will increase from 15% to 27% by 2050. Figure 8.1(a) shows that renewable energy consumption will be almost equal to petroleum by 2050. In 2020, biomass generated 39% of energy from all renewable feedstock (Fig. 8.1(b)).

The US Department of Energy (DOE) aims to replace 30% of current US petroleum consumption with bioenergy. To achieve this goal, it is estimated that ~1 billion tons of biomass must be available annually from agricultural and non-agricultural sources. A joint US Department of Agriculture (USDA)-DOE report suggests biomass residues from forestland, agricultural land, and perennial bioenergy crops cultivated on idle pasture will have to be included in the feedstock supply chain (Perlack *et al.*, 2005). Therefore, there is an

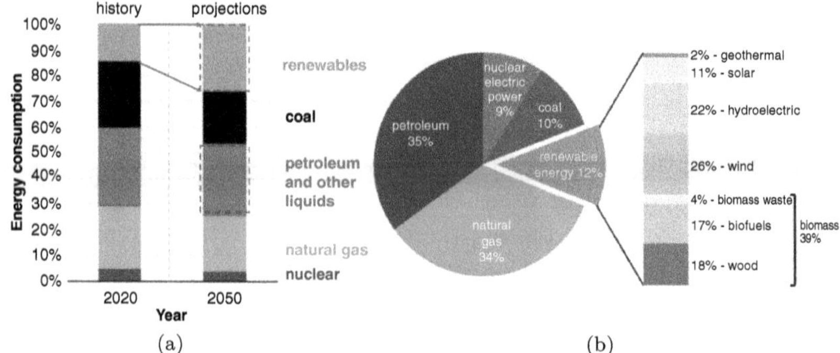

Figure 8.1. (a) Primary world energy consumption by source and (b) biomass utilization among renewables in 2020 (EIA, 2021).

urgent need to modify feedstock transportation, configure conversion-ready feedstock, and improve conversion technologies to meet the target bioenergy production.

Currently, thermochemical and biochemical are the two actively investigated conversion routes to valorize lignocellulosic biomass (LCB), including agricultural residue to bioenergy and bioproducts. Thermochemical processing accepts a broad range of feedstock, reforms almost all feedstock carbon, and is more tolerant to metal, plastic, or toxic contaminants than the biochemical routes. The four thermochemical technologies for biomass conversion are combustion, pyrolysis, gasification, hydrothermal liquefaction (HTL) and hydrothermal carbonization (HTC). Presently, commercial plants employing combustion, pyrolysis, and gasification technologies successfully convert LCB to heat, electricity, transport fuel, and value-added products as independent entities or integrated biorefineries. However, these conversion technology platforms require pre-dried biomass, a cost-intensive pretreatment. The HTL technology eliminates the pre-drying process. However, HTL has yet to overcome several technical challenges to establish itself as an economically viable technology. Irrespective of conversion technologies, high moisture agricultural residue transportation from farm to storage facilities and generating conversion-ready feedstock are logistic and costly challenges associated with all biomass valorization technologies. This chapter focuses on the HTL conversion of agricultural residue, which is available in bales (e.g., corn stover) and pulverized powder (Fig. 8.2). The agricultural residue can be preprocessed and pelletized into a densified form for transport and storage. The challenges in generating HTL-ready agricultural residue for optimal bio-oil and char production with references to feedstock composition, particle size reduction, and densification for scaled-up HTL processing are addressed in the following.

8.2. Hydrothermal Liquefaction

Since the first report in 1960 by Appell *et al.* (1971) on liquefaction of cellulosic and waste biomass, laboratory-scale HTL processing

Figure 8.2. (a) Agricultural waste, (b) corn stover bales, (c) pulverized corn stover, and (d) corn stover pellets.

of feedstock such as wastepaper, pinewood, municipal solid waste, and corn stover has been widely investigated for biofuel and char products (Shende *et al.*, 2017; Nan *et al.*, 2016; Tungal and Shende, 2014, 2013). The hydrothermal process employs a temperature range of 200–370°C and pressure between 40 and 200 bar to convert biomass to liquid fuel (Fig. 8.3). Water behaves like a supercritical fluid after reaching the so-called "critical point" at 374°C and 220 bar. Hydrothermal processing carried out between 180°C and 250°C is called hydrothermal carbonization (HTC), while hydrothermal liquefaction (HTL) typically process operates in the higher temperature range of 200–290°C (Lachos-Perez *et al.*, 2022; de Caprariis *et al.*, 2017)). Products from HTL are bio-oil, aqueous co-product, solid char, and a mixture of gases. In the HTC process, most carbon from the biomass feedstock forms the solid product called hydrochar. For simplicity, we have used the term HTL for all hydrothermal processes referred to in this article.

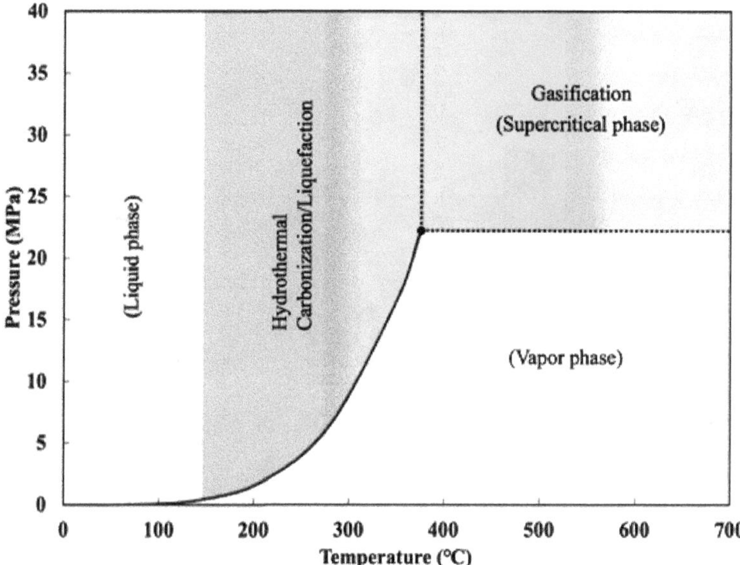

Figure 8.3. Typical T and P ranges are for biomass carbonization/liquefaction (Peterson *et al.*, 2008).

Under subcritical conditions, the dielectric constant of water is significantly lower, and its ionization constant is approximately three orders of magnitude higher than under ambient conditions. These altered properties of water create an acidic medium supporting hydrolysis of biomass components. Subcritical water serves as a reaction medium and acts as a reactant leading to liquefaction of biomass. The liquefaction proceeds through chemical modifications involving depolymerization, solvolysis, and the chemical and thermal decomposition of monomers into smaller molecules. Several competing reaction pathways ultimately liquefy biomass to biocrude and generate solid biochar and gaseous hydrocarbons. A schematic of a typical laboratory HTL reactor setup is shown in Fig. 8.4. The main equipment is the high-temperature and high-pressure reactor (e.g., PARR) fitted with a stirrer and thermocouple. Pulverized biomass (15–20 wt% in water) is loaded with or without a catalyst in the reactor. The reactor is heated to a set point temperature. The reaction is performed for residence times ranging between 15 min

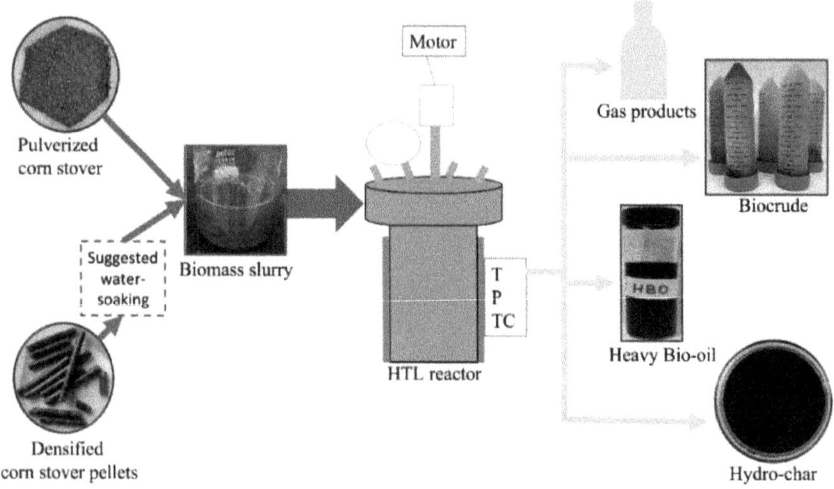

Figure 8.4. A schematic of a bench-scale HTL batch reactor setup with the recovery of gaseous products, oil, aqueous biocrude, and hydrochar.

and 4 h. Gaseous and liquid samples can be collected while HTL is in progress and analyzed by GC/GCMS/HPLC techniques. At the end of the reaction, the reactor contents are cooled and filtered to recover the hydrochar. Bio-oil is extracted from hydrochar with acetone. The bio-oil and char products are washed with acetone and dried (Tungal and Shende, 2014).

The HTL process has some outstanding advantages, such as (i) it uses water (a green solvent) which serves as a low-cost reaction medium and reactant, (ii) unlike biochemical processing, HTL does not require elaborate feedstock pre-treatment, (iii) it eliminates the feedstock drying step which is critical for combustion, gasification, and pyrolysis, (iv) it is amenable to mixed feedstock streams, (v) closed and sealed HTL reactor eliminates secondary environment pollution during the conversion process, (vi) aqueous waste co-product can be recycled for water conservation, and (vii) waste stream such as biosolids are sterilized at high temperature and pressure reaction conditions.

Table 8.1 summarizes HTL reaction conditions and yields of bio-oil and hydrochar from various forest and agricultural residues.

Table 8.1. Bio-oil and biochar yields from HTL processing of representative agricultural and forest residue.

Lignocellulosic feedstock	Reaction conditions	Bio-oil yield (wt%) and HHV (MJ/kg)	Biochar yield (wt%) and HHV (MJ/kg)	References
Soybean straw	*T*: 320°C, *t*: 60 min B:S — 3:8 Catalyst: None	Yield = 15.8 HHV = 32.98	NA	Tian *et al.* (2020)
Pine wood	*T*: 200–275°C, *t*: 30–120 min B:S — 1:10 Catalyst: $Ni(NO_3)_2$	Yield = HHV = 27.20	Yield = NA HHV = 31.83	Tungal and Shende (2014)
Corn stover	*T*: 250–375°C, *t*: 0–60 min B:S — 1:6 Catalyst: None	Yield = 29.25 HHV = 35.13	Yield = 30.21 HHV = 24.7	Mathanker *et al.* (2020)
Poplar	*T*: 220–280°C, *t*: 30 min B:S — 1:10 Catalyst: None	Yield=19.88 HHV = 27.97	Yield = 18.56 HHV = 25.07	Wu *et al.* (2018)
Oakwood	*T*: 260–320°C, *t*: 15 min B:S — 1:5 Catalyst: Fe	Yield = 37 HHV = 32.37	Yield = 21 HHV = NA	De Caprariis *et al.* (2019)
Pinewood sawdust	*T*: 300°C, *t*: 30 min B:S — 1:10 Catalyst: Na_2CO_3	Yield = 31 HHV = 27.95	Yield = 44 HHV = 27.76	Hu *et al.* (2020)
Barley straw	*T*: 280–400°C, *t*: 15 min B:S — 3:20 Catalyst: K_2CO_3	Yield = 34.85 HHV = 27.29	Yield = 8.10 HHV = 17.19	Zhu *et al.* (2015)
Rice husk	*T*: 240–360°C, *t*: 20 min B:S — 3:20 Catalyst: NaOH	Yield = 28 HHV = 24.81	Yield = 32 HHV = 20.15	Huang *et al.* (2017)
Pine sawdust	*T*: 300°C, *t*: 60 min B:S — 1:8 Catalyst: Zn/HZSM-5	Yield = 59.09 HHV = 33	Yield = 13 HHV = NA	Cheng *et al.* (2017b)
Pine sawdust (PSD)	*T*: 300°C, *t*: 60 min B:S — 2:11 Catalyst: Ni/HZSM-5	Yield = 63 HHV = 28.38	Yield = 22 HHV = NA	Cheng *et al.* (2017a)

Note: *T*: temperature, *t*: time, B:S: biomass-to-solvent ratio, and NA: not available.

Researchers are attempting to improve product and co-product qualities to target high-end applications for increased revenue from the HTL process. Shende *et al.* have reported co-product H2 (143 MJ/kg) with the concomitant bio-oil product (Tungal and Shende, 2014). Researchers are pursuing valorization of char to high surface area porous carbon for electrical charge/energy storage, biocompatible slow-release soil amendment, or as adsorbents for nutrients and pollutants (Houck *et al.*, 2020; Amar *et al.*, 2021; Luutu, 2022; Wu *et al.*, 2021; Yu *et al.*, 2021).

Most data presented in the literature are for batch HTL processing of LCB. Semi-continuous or continuous HTL processing is mostly reported only for algal feedstock. Figure 8.5 shows a block flow diagram for a continuous HTL process. Either pulverized or densified biomass can be made into a slurry and pumped into a digester tank. The preheated slurry can be continuously fed into a tubular reactor at the desired set temperature and pressure. Adjusting the mass flow rate can provide essential dwell time to complete the reaction. HTL processed slurry can be separated through a filter to recover the hydrochar and aqueous biocrude. Char extraction with acetone yields heavy bio-oil (HBO).

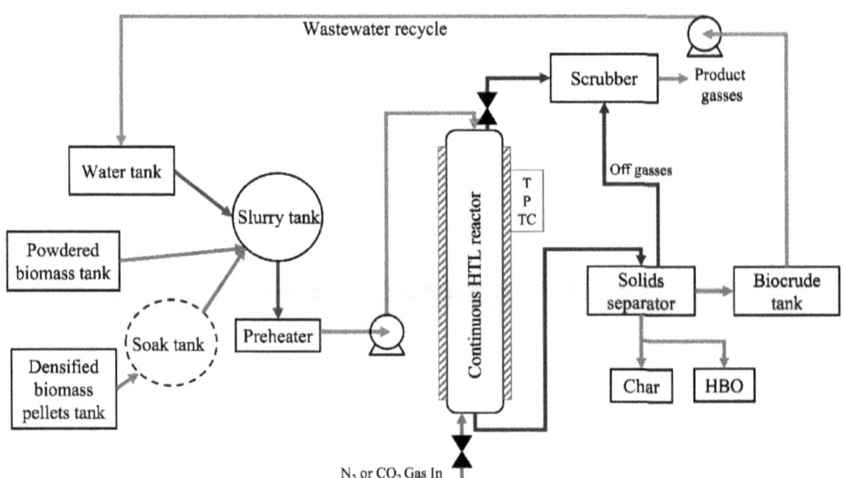

Figure 8.5. A block flow diagram for a continuous HTL process.

Aqueous biocrude can be extracted with dichloromethane to achieve recovery of light bio-oil (LBO). After the recovery of LBO, the aqueous biocrude can be recycled into a water tank and reused for HTL processing. Based on T-P swing operations and slurry characteristics (chemical and physical), process intensification efforts will still be required to achieve the commercial-scale energy-efficient HTL biomass processing system. HTL technology is most suited to generate bioenergy from agricultural residue. And yet, there is not a single HTL plant available for commercial processing of lignocellulosic biomass. Some of the challenges to commercializing this technology lie in generating HTL-ready feedstock for optimal bio-oil and char products production. In the lab-scale HTL process, the influences of feedstock particle size, biochemical properties, and density on HTL product distribution and yield are well documented. Very little is known of the influence of these feedstock attributes from the viewpoint of scaling up the HTL process. Feedstock characteristics projected to play a significant role in designing a continuous or semi-continuous HTL reactor are its composition, particle size, density, and pumpability/flowability.

8.3. Feedstock Composition

A typical HTL plant may receive a range of LCB, such as switchgrass, corn stover, miscanthus, or lodgepole pine. These feedstocks vary widely in sizes, bulk densities, biochemical compositions, and hydrothermal recalcitrance. The feedstock's innate morphological and compositional diversity would require re-optimizing (and re-engineering) HTL reactor design and process for each different type of biomass, thus increasing processing costs. Complicating this further is that the same feedstock varies markedly from region to region (Shi *et al.*, 2013), and each feedstock within a region varies from year to year based on weather conditions, handling, storage, and crop variety (Lubowski *et al.*, 2005; Schmer *et al.*, 2010). Some of the feedstock biochemical characteristics (e.g., lignin and ash contents) significantly affect bio-oil and char production. An HTL plant may receive mixed feedstock or consider mixing feedstock to improve pumpability or product yield. Prior knowledge about synergistic and antagonistic interactions in mixed feedstock and the effect on product yield and quality is necessary to prepare the feedstock. This is an area of active research where researchers have used prediction models and actual feedstock or model compounds to evaluate the product yield from co-liquefaction of mixed feedstock (Madesen *et al.*, 2019; Teri *et al.*, 2014). A consensus of a decrease in bio-oil yields following the order of lipid > protein > carbohydrate > lignin is evident from the literature. Researchers have combined high dry matter and fibrous feedstock with oleaginous biomass to increase product yields and feedstock pumpability. Yang *et al.* (2018) reported a synergistic effect on bio-oil yield from cellulose–lipid and hemicellulose–lipid interactions, while cellulose–lignin had an antagonistic effect on char quantities. Quantitative and qualitative analyses of bio-oil and biochar products from co-liquefaction experiments have improved our understanding of synergistic and antagonistic HTL feedstock combinations (Brilman *et al.*, 2017; Sharma *et al.*, 2021; Pederson, 2016; Sintamarean, 2017). For instance, Leng *et al.* (2018) reported that co-liquefaction of wood sawdust, rice straw, and sewage sludge increased the bio-oil yield from 22% to 32%. Liu *et al.* (2021) performed fast HTL of corn stover and cow manure to produce bio-oil and biochar with HHV of 34 MJ/Kg and 27.3 MJ/kg, respectively. An additional benefit of improved feedstock pumpability was reported from a synergistic combination of swine manure and sewage sludge by Shah *et al.* (2021). More feedstock combinations and co-liquefaction studies are required to construct a database of synergistic combinations. The database will aid in developing software to prepare synergistic mixed HTL-ready feedstock on-site.

8.4. Particle Size

The lignin-carbohydrate super polymeric combination does not pose a barrier to HTL conversion. Therefore, energy and cost-consuming LCB pre-treatments generic to biochemical processing are not required to prepare HTL-ready feedstock. However, feedstock particle size is a crucial factor to consider when generating HTL-ready feedstock. Particle size reduction facilitates continuous feeding and higher bio-oil and hydrochar yields. Physical methods such as mechanical extrusion, milling, and chemical processes have been explored to prepare HTL-ready feedstock with specific dimensions. However, a complete life cycle and techno-economic evaluation of these methods are missing. Chipping, milling, and grinding are the most used mechanical methods to reduce particle size. Chipping reduces particle size to 10–30 mm. Milling and grinding reduce the particle size to 0.2 mm. Compared with other particle size reduction methods, milling is energy and time efficient. Feedstock initial size, moisture content, HTL reactor design, and feeding rate dictate the selection of the milling method. Hammer milling, two-roll milling, vibratory milling, and colloid milling are commonly used milling methods. It has been proposed that a 4–10 mm particle size favors optimal HTL product distribution (Xue *et al.*, 2016). Compared with cellulose, lignin hydrolysis is much more affected by particle size (Madhawa *et al.*, 2020).

8.5. Densification

Low density limits most LCB, such as corn stover, for long-distance transportation. Therefore, following particle size reduction, densification is required to compress loose feedstock into a pellet, briquette, cube, or another high-density form. Densification also helps reduce the moisture content of the biomass, making it more suitable for long-term storage (Tumuluru *et al.*, 2011). The advantages of densification include the following: (i) improved handling and conveyance efficiencies throughout the supply system and biorefinery infeed, (ii) controlled particle size distribution for improved feedstock uniformity and density, (iii) fractionated structural components for improved compositional quality, and (iv) conformance to predetermined conversion technology and supply system specifications. Standard biomass densification systems have been adapted from other highly efficient processing industries, such as feed, food, and

Table 8.2. Hydrothermal processing of biomass pellets.

Feedstock	Reaction conditions	Biochar yield (wt%) and HHV (MJ/kg)	References
Bagasse from plants *Grindelia squarrosa*	*T*: 200°C, *t*: 5 min B:S — 1:5 Catalyst: None	Yield = 59 HHV = 19.9	Reza *et al.* (2016)
Bagasse from plants *Ericameria nauseosa*	*T*: 200°C, *t*: 5 min B:S — 1:5 Catalyst: None	Yield = 79 HHV = 23.2	Reza *et al.* (2016)
Alnus incana (Gray alder)	*T*: 220°C, *t*: 90–240 min B:S — 1:4 Catalyst: None	NA	Raghu *et al.* (2017)
Softwood	*T*: 200–225°C, *t*: 60 min B:S — 1:4 Catalyst: None	Yield = 63.3–66.62 HHV =	Stirling *et al.* (2018)
Corn stover	*T*: 210°C, *t*: 60 min B:S — 1:15 Surfactant: CS/SA/SP	Yield = 54.5 HHV = 27.98	Tu *et al.* (2019)
Microalgae-fungal *Penicillium* sp. and *Chlorella* sp.	*T*: 220°C, *t*: 240 min B:S — 1:10 Catalyst: None	Yield = 28.03 HHV = 27.70	Chen *et al.* (2020)

Note: *T*: temperature, *t*: time, B:S: biomass-to-solvent ratio, and NA: not available.

pharmaceutical manufacturing processes, including (i) pellet mills, (ii) cubers, (iii) briquette presses, (iv) screw extruders, (v) tabletizers, and (vi) agglomerators (Tumuluru *et al.*, 2011). Among these, pellet mills produce cylindrical pellets with typical length and diameter dimensions (i.e., 13–19 mm and 6.3–6.4 mm and bulk densities of 600–700 kg/m^3). Pellet quality plays a significant role in handling, storage, and transportation properties. Densification is considered critical to producing a feedstock material suitable as a commodity product. Low-temperature (200–260°C) hydrothermal processing, referred to as hydrothermal carbonization (HTC) of pelletized biomass, has been reported in the literature. The results are summarized in Table 8.2, which shows microalgae yielding the lowest biochar compared to other biomass substrates.

8.6. HTL Technology-Ready Corn Stover

One of the most interesting agricultural residue feedstock for HTL conversion is corn stover (Zea mays). Cornstalk, cob, leaves, sheath, husks, and silk after corn harvesting constitute corn stover. It is inexpensive, fairly uniform, and abundantly available. Corn stover consists of 32.4–35.5% cellulose, 18.5–21.8% hemicellulose, and 11.2–14.9% lignin (Weiss *et al.*, 2010). With more than 90 million acres of land dedicated to corn plantations, the US is the largest producer of corn globally. However, only about 5% of corn stover finds application while the remainder is burned or left on the fields (Alavijeh and Karimi, 2019). The high volume, low bulk density, and moisture content of corn stover pose a logistical challenge to the corn stover supply chain. Harvested corn stover is bailed into mostly round bales, although square bales can be more efficiently packaged and transported.

Corn stover bailing significantly influences storage, transportation, and process economics. After the bales are transported to a preprocessing facility, they can be made into a densified form. Figure 8.6 indicates the flow diagram for the high moisture pelleting process tested for the corn stover bales. Table 8.3 provides the energy consumption for various unit operations based on configuration-1 (stage-2 grinder fitted with 1/4 inch screen) for corn stover bales at 25% (w.b.) moisture content. We believe particle size reduction and densification will offer economical transport and storage of corn stover. Additionally, HTL processing of densified pellets will result in higher product yields. More studies and techno-economic analyses are required to confirm the energy and costs associated with the size reduction and pelletization processes. Densification of corn stover with two or more synergistic feedstocks requires process development and optimization.

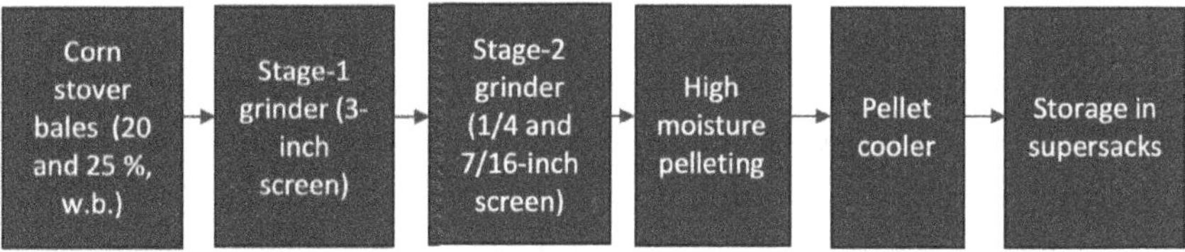

Figure 8.6. A schematic representation to prepare conversion-ready densified corn stover feedstock.

Table 8.3. Energy consumption and dry throughput for high moisture pelleting process.

Unit operations	Specifications	Energy consumption (KWh/dry ton)	Throughput (dry ton/h)
Bale moisture (%, w.b.)	25	—	—
Stage-1 screen size (inches)	3	12.864	3.34
Stage-2 screen size (inches)	1/4	73.11	1.02
Pellet mill (die and L/D ratio)	6 mm die and L/D 8	36.0	3.3
Pellet cooler	NA	3.02	5

Note: NA: not applicable.

8.7. Summary

Significant quantities of agricultural residue are available globally, which can be utilized for thermochemical processing in an integrated biorefinery to manufacture biofuels and bioproducts. Contrasting to these technologies, in HTL technology, water acts as a reactant and reaction medium contributing to depolymerization, solvolysis, and the chemical and thermal decomposition of monomers into smaller molecules. Although HTL has many advantages, the technology has not been commercialized. Variations in feedstock particle sizes, bulk densities, and biochemical compositions (ash content) are some of the challenges to achieving conversion-ready feedstock for HTL. Additional challenges also exist as studies related to the effects of particle size and densification of agricultural waste feedstock for pilot scale HTL conversion are missing in the literature. There appears to be a trade-off between the cost associated with the initial densification step and HTL-derived product yields and quality from the densified pellets compared to pulverized feedstock, which needs to be thoroughly analyzed for technology commercialization.

References

Alavijeh, M. K. & Karimi, K. (2019). Biobutanol production from corn stover in the US. *Ind. Crops Prod.*, *129*, 641–653.

Amar, V., Houck, J., Maddipudi, B., Penrod, T., Shell, K., Thakkar, A., Shende, A., Hernandez, S., Kumar, S., & Gupta, R. (2021). Hydrothermal liquefaction (HTL) processing of unhydrolyzed solids (UHS) for hydrochar and its use for asymmetric supercapacitors with mixed (Mn, Ti)-Perovskite oxides. *Renewable Energy*, *173*, 329–341.

Appell, H. R. (1971). *Converting Organic Wastes to Oil: A Replenishable Energy Source*. US Department of Interior, Bureau of Mines.

Brilman, D., Drabik, N., & Wądrzyk, M. (2017). Hydrothermal co-liquefaction of microalgae, wood, and sugar beet pulp. *Biomass Convers. Biorefinery*, *7*(4), 445–454.

Chen, J., Ding, L., Liu, R., Xu, S., Li, L., Gao, L., Wei, L., Leng, S., Li, J., Li, J., Leng, L., & Zhou, W. (2020). Hydrothermal carbonization of microalgae-fungal pellets: removal of nutrients from the aqueous phase fungi and microalgae cultivation, *ACS Sustainable Chem. Eng.*, *8*(45), 16823–16832.

Cheng, S., Wei, L., Alsowij, M., Corbin, F., Boakye, E., Gu, Z., & Raynie, D. (2017a). Catalytic hydrothermal liquefaction (HTL) of biomass for bio-crude production using Ni/HZSM-5 catalysts. *AIMS Environ. Sci.*, *4*(3), 417–430.

Cheng, S., Wei, L., Julson, J., Kharel, P. R., Cao, Y., & Gu, Z. (2017b). Catalytic liquefaction of pine sawdust for biofuel development on bifunctional Zn/HZSM-5 catalyst in supercritical ethanol. *J. Anal. Appl. Pyrolysis*, *126*, 257–266.

de Caprariis, B., Bavasso, I., Bracciale, M. P., Damizia, M., De Filippis, P., & Scarsella, M. (2019). Enhanced bio-crude yield and quality by reductive hydrothermal liquefaction of oak wood biomass: Effect of iron addition. *J. Anal. Appl. Pyrolysis*, *139*, 123–130.

de Caprariis, B., De Filippis, P., Petrullo, A., & Scarsella, M. (2017). Hydrothermal liquefaction of biomass: Influence of temperature and biomass composition on bio-oil production. *Fuel*, *208*, 618–625.

Houck, J. D., Amar, V. S., & Shende, R. V. (2020). Sol-gel derived mixed phase (Mn, Ti)-oxides/graphene nanoplatelets for hybrid supercapacitors. *Int. J. Energy Res.*, *44*(15), 12474–12484.

Hu, Y., Gu, Z., Li, W., & Xu, C. C. (2020). Alkali-catalyzed liquefaction of pinewood sawdust in ethanol/water co-solvents. *Biomass Bioenergy*, *134*, 105485.

Lachos-Perez, D., César Torres-Mayanga, P., Abaide, E. R., Zabot, G. L., & De Castilhos, F. (2022). Hydrothermal carbonization, and Liquefaction: Differences, progress, challenges, and opportunities. *Bioresour. Technol.*, *343*, 126084, ISSN 0960-8524. https://doi.org/10.1016/j.biortech.2021.126084.

Leng, L., Li, J., Yuan, X., Li, J., Han, P., Hong, Y., Wei, F., & Zhou, W. (2018). Beneficial synergistic effect on bio-oil production from co-liquefaction of sewage sludge and lignocellulosic biomass. *Bioresour. Technol.*, *251*, 49–56.

Liu, Q., Xu, R., Yan, C., Han, L., Lei, H., Ruan, R., & Zhang, X. (2021). Fast hydrothermal co-liquefaction of corn stover and cow manure for biocrude and hydrochar production. *Bioresour. Technol.*, *340*, 125630.

Luutu, H., Rose, M. T., McIntosh, S. *et al.* (2022). Plant growth responses to soil-applied hydrothermally-carbonised waste amendments: A meta-analysis. *Plant Soil*, *472*, 1–15. https://doi.org/10.1007/s11104-021-05185-4.

Madsen, R. B. & Glasius, M. (2019). How do hydrothermal liquefaction conditions and feedstock type influence product distribution and elemental composition? *Ind. Eng. Chem. Res.*, *58*(37), 17583–17600.

Mathanker, A., Pudasainee, D., Kumar, A., & Gupta, R. (2020). Hydrothermal liquefaction of lignocellulosic biomass feedstock to produce biofuels: Parametric study and product characterization. *Fuel*, *271*, 117534.

Nan, W., Shende, A. R., Shannon, J., & Shende, R. V. (2016). Insight into catalytic hydrothermal liquefaction of cardboard for biofuels production. *Energy Fuels*, *30*(6), 4933–4944.

Patel, M., Oyedun, A. O., Kumar, A., & Gupta, R. (2019). What is the production cost of renewable diesel from woody biomass and agricultural residue based on experimentation? A comparative assessment. *Fuel Process. Technol.*, *191*, 79–92.

Pedersen, T. H., Grigoras, I., Hoffmann, J., Toor, S. S., Daraban, I. M., Jensen, C. U., Iversen, S., Madsen, R., Glasius, M., & Arturi, K. R. (2016). Continuous hydrothermal co-liquefaction of aspen wood and glycerol with water phase recirculation. *Appl. Energy*, *162*, 1034–1041.

Perlack, R. D., Wright, L. L., Turhollow, A. F., Graham, R. L., Stokes, B. J., & Erbach, D. C. (2005). Biomass as feedstock for a bioenergy and bioproducts industry: The technical feasibility of a billion-ton annual supply. US Department of Energy. doi:10.2172/885984.

Peterson, A. A., Vogel, F., Lachance, R. P., Fröling, M., Antal Jr, M. J., & Tester, J. W. (2008). Thermochemical biofuel production in hydrothermal media: A review of sub-and supercritical water technologies. *Energy Environ. Sci.*, *1*(1), 32–65.

Raghu, K. C., Babu, I., Alatalo, S., Föhr, J., Ranta, T., & Tiihonen, I. (2017). Hydrothermal carbonization of deciduous biomass (*Alnus incana*) and pelletization prospects, *J. Sustainable Bioenergy Syst.*, *7*(3), 138–148.

Reza, M. T., Yang, X., Coronella, C. J., Lin, H., Hathwaik, U., Shintani, D. K., Neupane, B. P., & Miller, G. C. (2016). Hydrothermal carbonization (HTC) and pelletization of two arid land plants bagasse for energy densification, *ACS Sustainable Chem. Eng.*, *4*(3), 1106–1114.

Sharma, K., Shah, A. A., Toor, S. S., Seehar, T. H., Pedersen, T. H., & Rosendahl, L. A. (2021). Co-hydrothermal liquefaction of lignocellulosic biomass in supercritical water. *Energies*, *14*(6), 1708.

Shell, K. M., Rodene, D. D., Amar, V., Thakkar, A., Maddipudi, B., Kumar, S., Shende, R., & Gupta, R. B. (2021). Supercapacitor performance of corn stover-derived biocarbon produced from the solid co-products of a hydrothermal liquefaction process. *Bioresour. Technol. Rep.*, *13*, 100625.

Shende, A., Nan, W., Kodzomoyo, E., Shannon, J., Nicpon, J., & Shende, R. (2017). Evaluation of aqueous product from hydrothermal liquefaction of cardboard as bacterial growth medium: Co-liquefaction of cardboard and bacteria for higher bio-oil production. *J. Sustainable Bioenergy Syst.*, *7*(2), 51–64.

Shende, A., Tungal, R., Jaswal, R., & Shende, R. (2015). A novel integrated hydrothermal liquefaction and solar catalytic reforming method for enhanced hydrogen generation from biomass. *Am. J. Energy Res.*, *3*(1), 1–7.

Sintamarean, I. M., Pedersen, T. H., Zhao, X., Kruse, A., & Rosendahl, L. A. (2017). Application of algae as cosubstrate to enhance the processability of willow wood for continuous hydrothermal liquefaction. *Ind. Eng. Chem. Res.*, *56*(15), 4562–4571.

Stirling R. J. Snape, C. E., & Meredith, W. (2018). The impact of hydrothermal carbonization on the char reactivity of biomass. *Fuel Process. Technol.*, *177*, 152–158.

Teri, G., Luo, L., & Savage, P. E. (2014). Hydrothermal treatment of protein, polysaccharide, and lipids alone and in mixtures. *Energy Fuels*, *28*(12), 7501–7509.

The EIA (2021). Energy information administration. International energy statistics: Petroleum-International Energy Administration (IEA). Annual Reporting on Renewables.

Tian, Y., Wang, F., Djandja, J. O., Zhang, S.-L., Xu, Y.-P., & Duan, P.-G. (2020). Hydrothermal liquefaction of crop straws: Effect of feedstock composition. *Fuel*, *265*, 116946.

Toor, S. S., Rosendahl, L., Nielsen, M. P., Glasius, M., Rudolf, A., & Iversen, S. B. (2012). Continuous production of bio-oil by catalytic liquefaction from wet distiller's grain with solubles (WDGS) from bio-ethanol production. *Biomass Bioenergy*, *36*, 327–332.

Tu, R., Sun, Y., Wu, Y., Fan, X., Wang, J., Cheng, S., Jia, Z., Jiang, E., & Xu, S. (2019). Improvement of corn stover fuel properties via hydrothermal carbonization combined with surfactant, *Biotechnol. Biofuels*, *12*, 249.

Tumuluru, J. S., Wright, C. T., Hess, J. R., & Kenney, K. L. (2011). A review of biomass densification systems to develop uniform feedstock commodities for bioenergy application. *Biofuels, Bioprod. Biorefin.*, *5*(6), 683–707.

Tungal, R. & Shende, R. V. (2013). Subcritical aqueous phase reforming of wastepaper for biocrude and H_2 generation. *Energy Fuels*, *27*(6), 3194–3203.

Tungal, R. & Shende, R. V. (2014). Hydrothermal liquefaction of pinewood (Pinus ponderosa) for H2, biocrude and bio-oil generation. *Appl. Energy*, *134*, 401–412.

Weiss, N. D., Farmer, J. D., & Schell, D. J. (2010). Impact of corn stover composition on hemicellulose conversion during dilute acid pretreatment and enzymatic cellulose digestibility of the pretreated solids. *Bioresour. Technol.*, *101*(2), 674–678.

Wu, L. Wei, W., Wang, D., & Ni, B.-J. (2021). Improving nutrients removal and energy recovery from wastes using hydrochar. *Sci. Total Environ.*, *783*, 146980. https://doi.org/10.1016/j.scitotenv.2021.146980

Wu, X.-F., Zhou, Q., Li, M.-F., Li, S.-X., Bian, J., & Peng, F. (2018). Conversion of poplar into bio-oil via subcritical hydrothermal liquefaction: Structure and antioxidant capacity. *Bioresour. Technol.*, *270*, 216–222.

Xue, Y., Chen, H., Zhao, W., Yang, C., Ma, P., & Han, S. (2016). A review on the operating conditions of producing bio-oil from hydrothermal liquefaction of biomass. *Int. J. Energy Res.*, *40*(7), 865–877.

Yang, J., Niu, H., Corscadden, K., & Astatkie, T. (2018). Hydrothermal liquefaction of biomass model components for product yield prediction and reaction pathways exploration. *Appl. Energy*, *228*, 1618–1628.

Yu, J., Tang, T., Cheng, F., Huang, D., Martin, J. L., Brewer, C. E., Grimm, R. L., Zhou, M., & Luo, H. (2021 February 7) Waste-to-wealth application of wastewater treatment algae-derived hydrochar for Pb(II) adsorption. *Methods*, *8*, 101263.

Zhang, B., von Keitz, M., & Valentas, K. (2008). Thermal effects on hydrothermal biomass liquefaction. In *Biotechnology for Fuels and Chemicals* (pp. 511–518). Springer.

Zhang, B., von Keitz, M., & Valentas, K. (2009). Thermochemical liquefaction of high-diversity grassland perennials. *J. Anal. Appl. Pyrolysis*, *84*(1), 18–24.

Zhu, Z., Toor, S. S., Rosendahl, L., Yu, D., & Chen, G. (2015). Influence of alkali catalyst on product yield and properties via hydrothermal liquefaction of barley straw. *Energy*, *80*, 284–292.

https://doi.org/10.1142/9781800613799_0009

Chapter 9

Densification of Microwave-Torrefied Oat Hull for Use As a Solid Fuel

Esteban Valdez[*,§], Lope G. Tabil[*,¶], Edmund Mupondwa[†,‖], Tim Dumonceaux[†,**], Duncan Cree[‡,††], and Hadi Moazed[*,‡‡]

[*]*Department of Chemical and Biological Engineering*
University of Saskatchewan
Saskatoon, Saskatchewan, Canada
[†]*Saskatoon Research and Development Centre*
Agriculture and Agri-Food Canada
Saskatoon, Saskatchewan, Canada
[‡]*Department of Mechanical Engineering*
University of Saskatchewan
Saskatoon, Saskatchewan, Canada
[§]*esteban.valdez@usask.ca*
[¶]*lope.tabil@usask.ca*
[‖]*edmund.mupondwa@agr.gc.ca*
[**]*tim.dumonceaux@agr.gc.ca*
[††]*duncan.cree@usask.ca*
[‡‡]*hmoazed955@yahoo.com*

Abstract

Quality properties of microwave-torrefied oat hull pellets were determined as a result of pelleting trials at different pellet die temperatures, sample particle sizes, and binder formulations. Microwave-torrefied ground biomass samples were compressed at two levels of pellet die temperatures (95°C and 145°C), two particle sizes (fine and coarse), and two levels of binders at different wt% (sodium lignosulfonate and sawdust). Quality of pellets was examined based on relaxed pellet density, pellet tensile strength, moisture absorption, moisture content, ash content, and higher heating value. Results showed that pellets made from fine particle size at high die temperature (145°C) have improved density and tensile strength. The best binder for density and tensile strength improvement was 10 wt% sodium lignosulfonate. However, since raw sawdust is abundantly available and cheaper than lignosulfonate, it is best recommended as a low-cost effective binder for densifying torrefied oat hull into pellets. Overall, all the densified pellets had a moisture content of 4–5% wet basis. Moisture absorption decreased when increasing torrefaction temperature, and pellets formulated with lignosulfonate presented higher moisture absorption and lower calorific value. Moreover, the addition of lignosulfonate significantly increased the pellet ash content, while sawdust significantly decreased the ash content of microwave torrefaction oat hull pellets.

Keywords: Microwave, torrefaction, pelleting, quality, binders

9.1. Introduction

It has been reported that torrefaction enhances biomass quality in means of developed hydrophobicity, lower moisture content, higher heating value, and better grindability (Arias *et al.*, 2008; Iroba *et al.*, 2017; Peng *et al.*, 2013a; Verhoeff *et al.*, 2011). However, the next challenge will be decreasing the cost of transportation through pelletization (Gilbert *et al.*, 2009) and reduction of pellet "fines" formation. Pelletization was described as the next challenge because torrefaction reduces the strength and density of pellets. Peng *et al.* (2013b) evaluated the pellet density of torrefied softwood residues and observed that torrefied biomass was more difficult to densify into strong pellets than control pellets under the same operating conditions. Iroba *et al.* (2017) described a decrease in tensile strength of severely torrefied pellets. Li *et al.* (2012) observed a lower density of torrefied sawdust pellets compared to untreated sawdust pellets due to the removal of most low melting or softening point components.

In order to improve pellet quality, resulting from high torrefaction treatment, binders can be added prior to densification. The fraction of added binders normally ranges from 0.5 to 20 wt%, and it is limited by the cost and emissions produced during the end use of pellets (Peng *et al.*, 2015). Soleimani *et al.* (2017) investigated the use of lignosulfonate as a binder for the pelletization of wheat straw at a die temperature of 90°C, 10% moisture content, and 4000 N compressive load concluding that it was one of the most effective binders, which increased pellet durability. Peng *et al.* (2015) also investigated the use of lignosulfonate and sawdust as binders to improve the pellet quality of torrefied pine wood chips and observed that both binders improved density and tensile strength. However, an increased moisture absorption content and decreased heating value were observed when compared to torrefied pellets with no binder.

Particle size is another important parameter in the pellet process, affecting contact between particles, compressibility, flow rate of the material, and friction in the die (Harun and Afzal, 2016). A study examining pelleting characteristics of canola and oat straw showed that a decrease in particle size using hammer mill screen size from 6.4 mm to 1.6 mm increased the particle density (Adapa *et al.*, 2010). Another study by Castellano *et al.* (2015) showed an increase in pellet mechanical durability from pellets produced at 2 mm screen size, compared to 4 mm screen size. Shaw *et al.* (2009) reported a higher mechanical durability of wheat straw and poplar wood pellets, when using a smaller particle size. Peng *et al.* (2013a) concluded that the hardness and density of torrefied pellets are dependent on torrefaction weight loss and compression die temperature, a higher die temperature resulted in a higher pellet quality. Peng *et al.* (2015) also reported the use of raw sawdust as a binder with die temperatures between 110°C and 170°C which increased the density and hardness of torrefied wood chip pellets.

In this work, the effect of microwave torrefaction severity, die temperature, particle size, and binder use on pellet quality was investigated. The main objective of the study was to identify optimal densification parameters for torrefied oat hulls based on the quality of processed pellets. The quality of pellets was determined in terms of final pellet density, tensile strength, moisture content, moisture absorption content, higher heating value, and ash content.

9.2. Materials and Methods

9.2.1. *Sample Preparation*

Microwave-torrefied oat hull produced at three reaction temperatures of 225°C, 255°C, and 285°C (3 min residence time) was used as feedstock in this work. Oat hull samples were torrefied via microwave torrefaction at 650 watts, and the residence time started once the biomass reached the reaction temperature. Oat hull samples were prepared by grinding in a knife mill (Model No. 3690604, Dietz-motoren GmbH & Co. KG, Dettingen, Germany) with a screen size of 3.2 mm and by a precision grinder (Model No. 117739, Huddinge, Sweden) with a screen size of 4.6 mm. Subsequently, the ground microwave torrefaction samples were conditioned to an average moisture content of 10% by spraying a calculated amount of water (based on mass balance) and stored at room temperature (25°C) for two weeks in hermetic bottles (Iroba *et al.*, 2017; Satpathy *et al.*, 2013). After the storage period, the binder weight percentage was added, and samples were densified. Sodium lignosulfonate was purchased from Lignotech USA Inc. (Belmont, WI), and sawdust forest product processing residue was acquired from Meadow Lake, SK. The reason for selecting sawdust and lignosulfonate as binders was because sawdust is a by-product of Saskatchewan's forest industry, and lignosulfonate has been considered to be an effective binder used in biomass pellets (Abedi and Dalai, 2017; Lu *et al.*, 2014; Soleimani *et al.*, 2017).

9.2.2. *Particle Size Analysis*

The particle size analysis of ground samples was determined in accordance with ANSI/ASAE standard S319.3 (2008), by placing a 100 g sample in a stack of sieves arranged from largest to smallest opening. For samples ground by knife mill and precision grinder, US sieve numbers 12, 16, 30, 40, 50, and 70 (sieve size: 1.68 mm, 1.19 mm, 0.595 mm, 0.420 mm, 0.297, and 0.210 mm, respectively) and 16, 30, 40, 50, 70, and 100 (1.19 mm, 0.595 mm, 0.420 mm, 0.297 mm, 0.210 mm and 0.149 mm, respectively) were used. A Ro-Tap sieve shaker (Tyler Industrial Products, Mentor, OH) held the set of sieves for 10 min, which was previously determined as optimal through trails. The geometric mean diameter (d_{gw}) and standard deviation of particle diameter (S_{gw}) were calculated based on ASABE standard S319 (2008) (Adapa *et al.*, 2011; Iroba *et al.*, 2017; Satpathy *et al.*, 2013).

9.2.3. *Pelleting Setup*

A single pellet unit mounted to an Instron tester (Model No. 3366, Instron Corp., Norwood, MA) produced the pellets. The Instron tester fitted with a 5000 N load cell compressed the sample with a steel rod into a heated die at a speed of 50 mm/min (Fig. 9.1). A heating cable surrounded the cylinder die to supply a specific die temperature. About 0.5 g of the biomass was loaded and compressed to a pre-set load of 4000 N for 60 s to prevent the spring back effect of the pellet (Mani *et al.*, 2006). The pellets were stored inside plastic bags at room temperature for further analysis. Twelve pellets were produced for each sample.

Independent variables for pellet formation were die temperature (95°C and 145°C), particle size (coarse and fine), and binder addition (sawdust and lignosulfonate). During the compression process, commercial and pilot plant pellet equipment heat the biomass by frictional forces, which caused the pellets to have an outlet

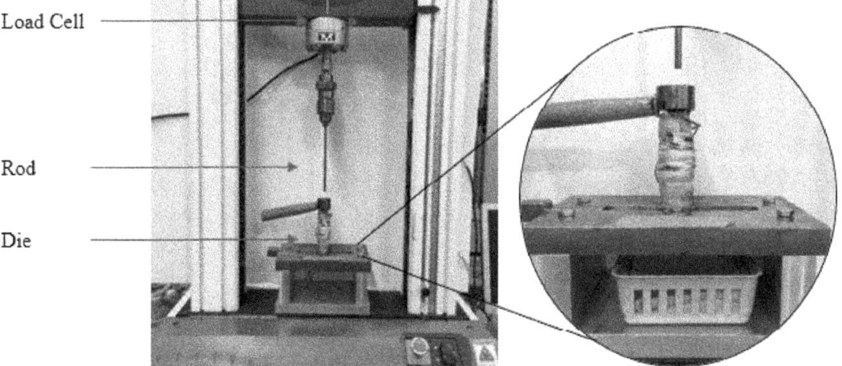

Figure 9.1. Single pelleting unit setup installed on Instron Model 3366 tester.

temperature of 95°C (Serrano *et al.*, 2011). Peng *et al.* (2015) recommended the use of binders and torrefied biomass with die temperatures between 110°C and 170°C to enhance pellet density and tensile strength. Therefore, in order to analyze the effect of high die temperature on pellet quality, a die temperature of 145°C was set.

Previous studies indicated that small biomass particle size led to an enhancement in pellet physical quality (Castellano *et al.*, 2015; Shaw *et al.*, 2009). Consequently, coarse and fine particle sizes, ground by a knife mill and precision grinder, were tested. Moreover, other studies stated that biomass torrefaction enhances physicochemical properties like grindability, hydrophobicity, and calorific value (Iroba *et al.*, 2017; Tumuluru *et al.*, 2011). However, it also decreases densification and strength of pellets (Iroba *et al.*, 2017; Peng *et al.*, 2013a). For these reasons, two binders (sawdust and lignosulfonate) were employed in this study.

9.2.4. *Unit Density of Pellet*

The mass, length, and diameter of pellets were measured after pellet formation to determine unit density. A digital caliper measured three different points of the pellet height and diameter to obtain an average volume. All processed pellets were stored in plastic bags at room temperature (25°C). After 14 days, the measurements were repeated to evaluate relaxed density (Iroba *et al.*, 2017; Shaw *et al.*, 2009).

9.2.5. *Pellet Tensile Strength*

Diametral compression test measured pellet tensile strength. Pellets were cut into 2.2 mm tablets using a diamond-cutting wheel attached to a Dremel rotary tool (Robert Bosch GmbH, Stuttgart, GER) and by scalpel. Each individual tablet, positioned on the lower metal plate, was compressed with a 5000 N load cell (Fig. 9.2) by the upper plunger at a speed of 1 mm/m until ideal tensile failure, considered as a crack dividing the tablet into two semi-circular parts (Fig. 9.3) (Emadi *et al.*, 2017; Tabil and Sokhansanj, 1996). Upon failure, fracture force was recorded. Twelve replicates were made for

Figure 9.2. Instron 3366 tester setup for pellet tensile strength tests.

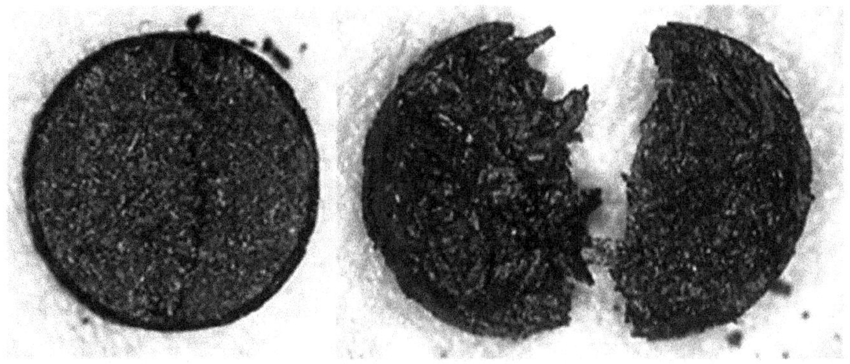

Figure 9.3. Ideal pellet tensile strength fracture after testing.

each sample, and the tensile strength of pellets was calculated as

$$\delta_x = 2F/\pi dl \tag{9.1}$$

where δ_x is the tensile strength (Pa), F is the fracture load (N), d is the tablet diameter (m), and l is the tablet thickness (mm).

9.2.6. *Moisture Absorption Test of Pellets*

The moisture absorption of non-torrefied and torrefied oat hull pellets was evaluated in a controlled environment chamber (Espec SH-641 benchtop chamber, ESPEC Corp., Osaka, Japan) by measuring the moisture uptake rate under 90% relative humidity and a temperature of 25°C for 72 h (Iroba *et al.*, 2017; Satpathy *et al.*, 2013). These conditions were tested to resemble the "worst case scenario" for storage conditions from rainy and humid weather in Canada. A THELCO laboratory PRECISION oven dried the pellets at 103°C for 24 h, before the moisture absorption test (Acharjee *et al.*, 2011; Emadi *et al.*, 2017; Peng *et al.*, 2013a; Satpathy *et al.*, 2013). Glass containers, stored inside the humidity chamber, held about 2 g of dried samples. After 72 h, the glass containers were covered with airtight lids to avoid moisture loss and weight for moisture absorption content.

9.2.7. *Moisture Content of Pellets*

About two grams of pellets were dried in THELCO laboratory PRECISION oven at 103°C for 24 h. After drying, the containers were covered and placed in a desiccator. The moisture content of pellets was determined by the weight and expressed in moisture content wet basis, percent.

9.2.8. *Higher Heating Value of Pellets*

The higher heating value (HHV) was analyzed with an oxygen bomb calorimeter (Parr Instruments Co., model 6400, Moline, IL) calculating the amount of heat units released in Mega Joules per kilogram (MJ/kg). The HHV of each sample was reported on a dry mass basis, percent (d.b.%). Pellet binders were also analyzed through the same procedure, and a mass and energy balance of biomass binder estimated the final HHV of pellets.

9.2.9. *Ash Content of Pellets*

The ash content was determined based on the National Renewable Energy laboratory standard (Sluiter *et al.*, 2008). 2.0 g ± 0.2 g of oven-dried samples were weighed into tared dried crucibles and placed into a muffle furnace (Model F-A1T30, Thermolyne Sybron Corp., Dubuque, IA) at 575°C ± 25°C for 24 h ± 6 h. The sample was removed and placed inside an oven 105°C for 20–30 min before being

Table 9.1. Experimental design for microwave torrefaction level and pellet variables.

Sample	Die temperature (°C)	Particle size	Binder
Untorrefied	95, 145	Fine, coarse	Sawdust, lignosulfonate
225°C			
255°C			
285°C			

placed in a desiccator to cool. Ash content was calculated by oven dry weight (reported on dry basis, d.b.) and replicated three times.

9.2.10. *Data Analysis*

The objective was to study the effects of microwave torrefaction levels as well as the pelletization process variables: die temperature, particle size, and binder on pellet quality. Table 9.1 presents the experimental design. A 3-way analysis of variance (ANOVA) followed by means comparison was developed by R-Studio statistical software (RStudio, Boston, MA).

9.3. Results and Discussion

9.3.1. *Particle Size*

Table 9.2 presents the d_{gw} and S_{gw} of biomass samples, where a high severity microwave torrefaction significantly corresponded to a decrease in the biomass particle size. This could be explained by a higher degradation of the lignocellulosic structure resulting in better grindability. Arias *et al.* (2008) reported similar results, concluding that torrefaction decomposed the fibrous nature by breaking the particles' links and enhancing handling. Adapa *et al.* (2011) reported that a rupture in the lignocellulosic structure leads to a decrease in the geometric mean particle diameter after grinding.

Figures 9.4(a) and 9.4(b) show the particle size distribution of the samples. Untreated dry oat hull resulted in smaller d_{gw} than untreated oat hull with 10% moisture content, indicating

Table 9.2. Particle size of treated and untreated ground oat hull.

Sample	Grinding size	d_{gw} (mm)	S_{gw} (mm)
Untorrefied	Coarse (knife mill)	0.903	0.476
Untorrefied dry		0.620	0.326
225°C		0.614	0.324
255°C		0.503	0.265
285°C		0.481	0.254
Untorrefied	Fine (precision grinder)	0.565	0.242
Untorrefied dry		0.441	0.189
225°C		0.377	0.162
255°C		0.252	0.108
285°C		0.239	0.102

Note: Geometric mean diameter (d_{gw}) and standard deviation of particle diameter (S_{gw}).

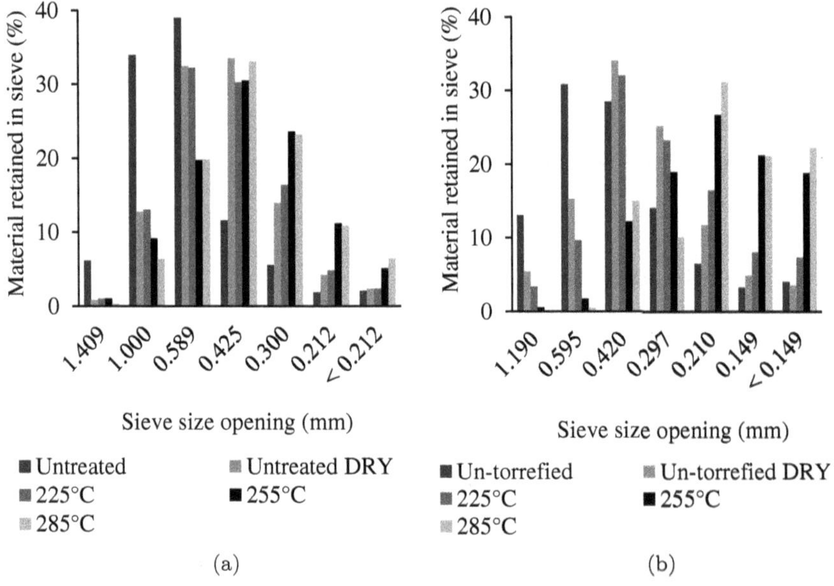

Figure 9.4. (a) Particle size distribution of coarse ground biomass and (b) particle size distribution of fine ground biomass.

that moisture content affected grindability. As the moisture content decreases, fiber cracks due to a structural shrinking produced by dehydration (Repellin *et al.*, 2010). The second factor that improved grindability was the lignocellulosic structure degradation by the removal of hemicellulose through microwave torrefaction. As mentioned by Spiridon and Popa (2008), "the most important biological role of hemicellulose is their contribution to strengthening the cell wall via interactions with cellulose and, in some cases, lignin." Therefore, hemicellulose removal would decrease the strength or grinding resistance of biomass.

The knife mill equipped with a 3.0 mm screen size was defined as "coarse" grinding, and the precision grinder equipped with a 4.6 mm screen size was defined as "fine" grinding, which resulted in smaller particle size due to a higher rpm operation of the equipment (3400 rpm, compared to 1690 rpm of knife mill).

9.3.2. *Physical Quality of Pellets*

Three severity levels (225°C, 255°C, and 285°C) of microwave-torrefied oat hull with different densification formulations, particle size, die temperature, and binder addition, were evaluated for pellet quality. The effects of the densification formulations modified the final pellets' characteristics. Figure 9.5 shows torrefied oat hull samples before and after densification. It was observed that after densification, severely torrefied oat hull pellets presented more superficial fissures or cracks than any other pellet. It was also observed that mild-torrefied oat hull pellets presented a "shiny" and smooth surface. Furthermore, a severe torrefaction treatment reduced the ground biomass particle size.

9.3.3. *Moisture Content of Pellets*

Generally, when compared to untreated pellets, light-torrefied pellets presented a non-significant ($P > 0.05$) difference in the final moisture content. Moreover, moderate and severe torrefied pellets presented a significant reduction in the final moisture content. These results could be attributed to a faster removal of unbonded water from

Figure 9.5. Torrefied biomass before and after densification process. Temperature levels of 225°C, 255°C, and 285°C from left to right.

biomass during the pelletization process, therefore a difference in lower moisture content with the increment in torrefaction level (Peng *et al.*, 2013b).

9.3.4. *Higher Heating Value of Pellets*

Figure 9.6 shows the effect of binder addition on the final HHV of microwave-torrefied pellets. Tukey test performed on data showed that binder addition presented a non-significant effect on HHV of light microwave torrefaction pellets, while 10% lignosulfonate significantly decreased pellet HHV of moderate microwave torrefaction pellets. Moreover, the addition of lignosulfonate or sawdust significantly decreased ($P < 0.001$) the HHV of severe microwave torrefaction pellets. A lower HHV of lignosulfonate and sawdust (17.86 MJ/kg and 19.72 MJ/kg, d.b.) explained the negative effect on HHV of severe microwave torrefaction. In other words, the addition of any of the selected binders will lower the HHV of severe microwave torrefaction pellets. While HHV of light microwave torrefaction pellets was not affected by any type of binders, it was considerably reduced in the case of severe microwave torrefaction pellets. Regarding moderate microwave torrefaction pellets, HHV was only negatively influenced by the use of 10% lignosulfonate.

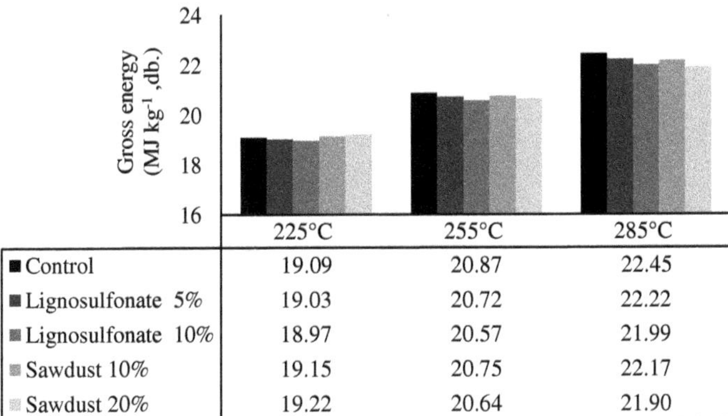

	225°C	255°C	285°C
Control	19.09	20.87	22.45
Lignosulfonate 5%	19.03	20.72	22.22
Lignosulfonate 10%	18.97	20.57	21.99
Sawdust 10%	19.15	20.75	22.17
Sawdust 20%	19.22	20.64	21.90

Figure 9.6. Influence of binders on higher heating value of oat hull pellets.

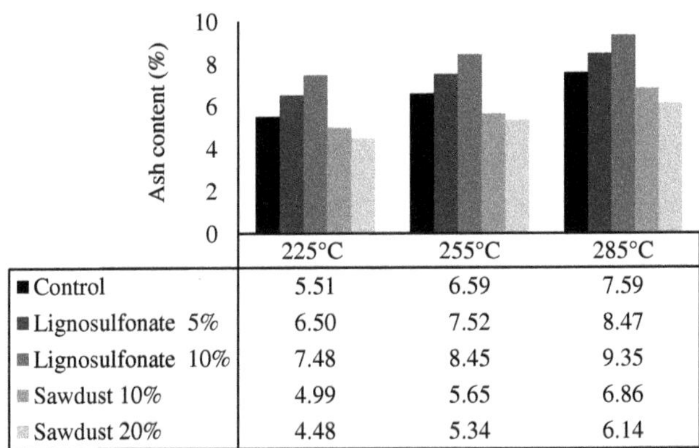

	225°C	255°C	285°C
Control	5.51	6.59	7.59
Lignosulfonate 5%	6.50	7.52	8.47
Lignosulfonate 10%	7.48	8.45	9.35
Sawdust 10%	4.99	5.65	6.86
Sawdust 20%	4.48	5.34	6.14

Figure 9.7. Effects of binders on the pellet total ash content.

9.3.5. *Ash Content of Pellets*

Figure 9.7 shows the total ash content of the different pellets in percentage dry mass basis. Tukey test performed on data shows that pellets with lignosulfonate increased ash content significantly ($P < 0.01$) for all microwave torrefaction levels. Moreover, pellets treated with sawdust decreased ash content significantly for all microwave torrefaction levels. A higher ash content of lignosulfonate (25.25% d.b.) and a much lower ash content of sawdust (0.36% d.b.) binders explained why lignosulfonate increased while sawdust decreased the ash content of microwave torrefaction pellets. According to

McKendry (2002), the ash content decreases the available energy of a fuel and can react forming a liquid phase (slag), therefore, increasing operating cost by reducing plant throughput. Therefore, low ash content is required for a suitable combustion fuel.

9.3.6. *Pellet Unit Density*

Light microwave torrefaction pellets presented the highest unit density, followed by moderate, untorrefied, and severe torrefaction. The low unit density of severe torrefaction pellets may be explained by the removal of low melting point volatiles, therefore, the remaining volatiles needed high die temperatures (25°C below torrefaction level) to melt and increase binding forces (Peng *et al.*, 2013a; Verhoeff *et al.* 2011). Increasing die temperature to 145°C increased pellet density and tensile strength of oat hull samples. As explained by Peng *et al.* (2013a), through higher die temperatures, volatiles, and water, we can improve pellet unit density and strength by increasing binding forces that act between individual particles. Kaliyan and Morey (2009) indicated that high temperatures and pressures form solid bridges, due to diffusion of molecules, and crystallization of components. Verhoeff *et al.* (2011) also reported that high die temperatures resulted in higher pellet density. Moreover, a decrease in particle size also increased pellet unit density and strength of pellets. This could be explained by a higher moisture absorption capability of small particles (Fig. 9.12), leading to a better pre-conditioning of the samples by superior distribution of the free moisture between particles (Adapa *et al.*, 2009). Kaliyan and Morey (2009) mentioned that unbonded water present in the biomass particles activates cohesion forces, enhancing pellet quality. Fine particle size also increased pellet unit density and strength as smaller particles filled more voids, increasing inter-particle bonding (Adapa *et al.*, 2011; Harun and Afzal, 2016; Shaw *et al.*, 2009).

Tables 9.3 and 9.4 show the effect of torrefaction level, die temperature, particle size, and binder use on pellet unit density and tensile strength. Generally, after the 14 days of storage (relaxed density), a loss of 2–5% on unit pellet density was observed. It was also noted that a high die temperature and smaller particle size lead to a minor density change. Similar results were found by Shaw *et al.* (2009), where biomass pellets expanded or contracted after the 14-day storage, and a higher die temperature and smaller particle size increased pellet density.

Furthermore, increasing microwave torrefaction severity generally decreased the density change percentage for control pellets. Theoretically, the decrease in hemicellulose content caused by microwave torrefaction was responsible for a lower elasticity of the biomass pellets. Shaw *et al.* (2009) suggested that an increase in the available lignin lowered the change in pellet density percentage. As explained by different researchers, the unit density change was related to relaxation of the pellet particles after pressure release causing expansion or contraction during the 14-day storage period (Emadi *et al.*, 2017; Iroba *et al.*, 2017; Kashaninejad *et al.*, 2014).

9.3.7. *Tensile Strength of Pellets*

The tensile strength of a pellet was determined by means of a compressive test between two flat surfaces and calculated through Eq. (9.1). Tables 9.3 and 9.4 show that a decrease in the particle size and increase in pellet die temperature leads to an increment in the tensile strength of pellets. Pellets produced with these parameters resulted in higher strength as mentioned in Section 3.6. The test results showed that moderate microwave torrefaction had a strong significant effect on pellet strength. As observed in Fig. 9.8(a), moderate microwave torrefaction control pellets were 4.28 times stronger than the untorrefied and 3.04 times stronger than light microwave torrefaction control pellets. The strength increment could be attributed to rupture of the lignocellulosic structure, where hemicellulose decreased and lignin increased (Valdez, 2021) acting as a natural binder by allowing adhesion and increasing pellet strength (Anglès *et al.*, 2001). Similar results were found by Kashaninejad and Tabil (2011), concluding that microwave chemical treatment ruptured the lignocellulosic structure, dissolving lignin and using it as a binder during pelletization. Verhoeff *et al.* (2011) also found a

Table 9.3. Effect of microwave torrefaction, coarse particle size, and binder on pellet unit density and tensile strength at die temperatures of 95 and 145°C.

Sample	Binder type	Binder (%)	Relaxed pellet unit density (kg/m^3)		Density change (%)		Tensile strength (MPa)	
			A	B	A	B	A	B
Raw	—	0	1116	1055	4.62	6.2	0.5	0.13
225°C	—	0	1162	1058	3.41	5.69	0.46	0.2
255°C	—	0	1128	1065	2.27	4.05	0.88	0.37
285°C	—	0	1074	964	1.46	4.35	0.42	0.19
225°C	Lignosulfonate	5	1168	1125	4.6	1.97	0.71	0.25
255°C	Lignosulfonate	5	1130	1132	2.09	0.43	1.1	0.41
285°C	Lignosulfonate	5	1066	967	0.62	1.56	0.69	0.34
225°C	Lignosulfonate	10	1215	1167	1.66	2.35	1.01	0.47
255°C	Lignosulfonate	10	1215	1166	1.25	0.58	1.23	0.67
285°C	Lignosulfonate	10	1094	1030	0.99	2.39	0.79	0.52
225°C	Sawdust	10	1166	1067	2.35	7.15	0.46	0.2
255°C	Sawdust	10	1142	1084	1.79	5.93	0.91	0.47
285°C	Sawdust	10	1083	975	0.87	7.44	0.54	0.24
225°C	Sawdust	20	1169	1095	0.29	4.5	0.48	0.21
255°C	Sawdust	20	1157	1110	0.24	2.95	0.96	0.53
285°C	Sawdust	20	1080	997	0.25	3.02	0.48	0.27

Note: Column A corresponds to pellets produced under die temperature of 145°C. Column B corresponds to pellets produced under die temperature of 95°C.

significant increment of biomass pellet strength through torrefaction temperatures of 260°C.

Furthermore, pellet tensile strength decreased through severe microwave torrefaction (Figs. 9.8 and 9.9). This behavior could be attributed to a higher degradation of the lignin, therefore decreasing its ability to be used as a binder during pelletization (Peng *et al.*, 2013a; Pirraglia *et al.*, 2013). Meaning that even though more lignin is available in the severely torrefied oat hull sample, a higher die temperature is needed in order to enable it as a natural binder. However, special care must be taken when selecting high die temperatures (e.g., 220°C), due to an extensive or faster wear down of the pellet mill components: roller mill and ring die (Verhoeff

Table 9.4. Effect of microwave torrefaction, fine particle size, and binder on pellet unit density and tensile strength at die temperatures of 95 and 145°C.

Sample	Binder type	Binder (%)	Relaxed pellet unit density (kg/m^3)		Density change (%)		Tensile strength (MPa)	
			A	B	A	B	A	B
Raw	—	0	1140	1059	3.88	6.07	0.51	0.17
225°C		0	1198	1119	1.91	3.15	0.55	0.24
255°C		0	1160	1110	1.21	2.64	1.11	0.73
285°C		0	1099	1015	0.71	0.42	0.54	0.25
225°C	Lignos-ufonate	5	1206	1188	2.63	0.94	0.86	0.39
255°C		5	1176	1104	0.97	2.96	1.37	0.74
285°C		5	1135	1080	0.51	1.44	0.76	0.54
225°C		10	1260	1199	1.17	1.26	1.16	0.49
255°C		10	1226	1170	1.09	0.56	1.69	1.22
285°C		10	1166	1093	0.73	1.54	1.11	0.65
225°C	Sawdust	10	1197	1133	1.67	4.73	0.62	0.29
255°C		10	1189	1117	0.28	4.17	1.33	0.9
285°C		10	1157	1058	0.44	2.5	0.8	0.61
225°C		20	1199	1139	0.28	2.5	0.65	0.34
255°C		20	1165	1132	0.88	2.12	1.14	0.85
285°C		20	1127	1062	0.32	1.71	0.73	0.46

Note: Column A corresponds to pellets produced under die temperature of 145°C. Column B corresponds to pellets produced under die temperature of 95°C.

et al., 2011). For both particle sizes, high die temperature enhances considerably pellet strength. This could be attributed to a better softening and thermosetting behavior of lignin at a temperature of 140°C (van Dam *et al.*, 2004).

Pellets reached the highest strength with the use of 10% lignosulfonate (Figs. 9.8 and 9.9). However, due to the high cost of the binder, its use would not be recommended. Moreover, there are environmental consequences by a significant increment of sulfur content which increases sulfur oxides emissions (Tarasov *et al.*, 2013). Additionally, lignosulfonate contains a high percentage of ash content (Section 3.5) making it unattractive as a binder.

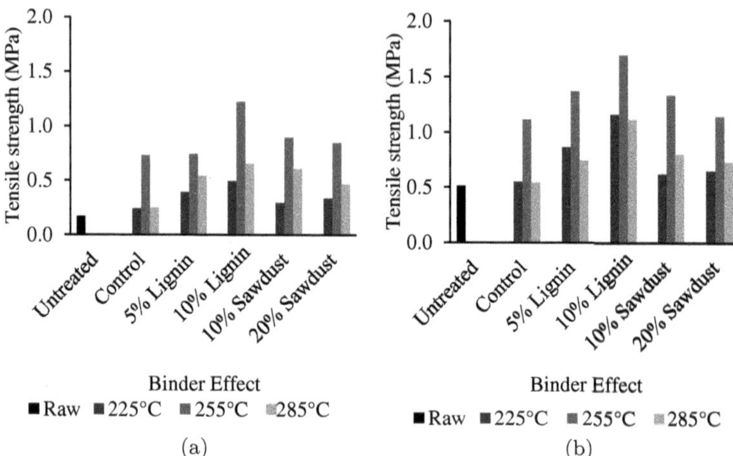

Figure 9.8. (a) Tensile strength of pellets with fine particle size and pellet die temperature of 95°C. (b) Tensile strength of pellets with fine particle size and pellet die temperature of 145°C.

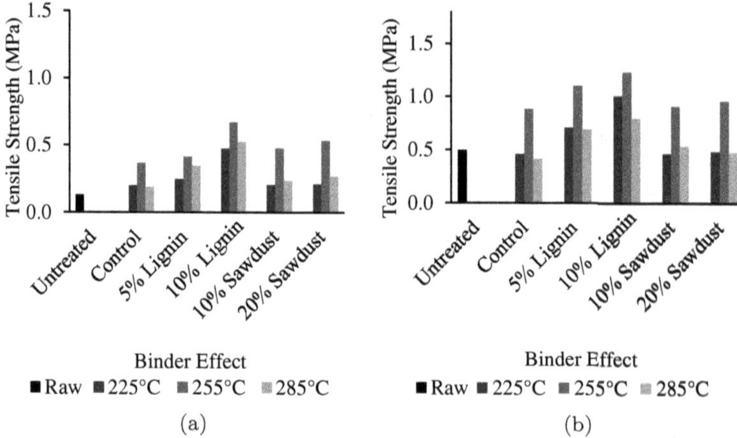

Figure 9.9. (a) Tensile strength of pellets with coarse particle size and pellet die temperature of 95°C. (b) Tensile strength of pellets with coarse particle size and pellet die temperature of 145°C.

9.3.8. *Effect of Material and Pelleting Variables on Pellet Unit Density and Tensile Strength*

The experimental data were analyzed to understand the significance of microwave torrefaction levels as well as the pelletization process

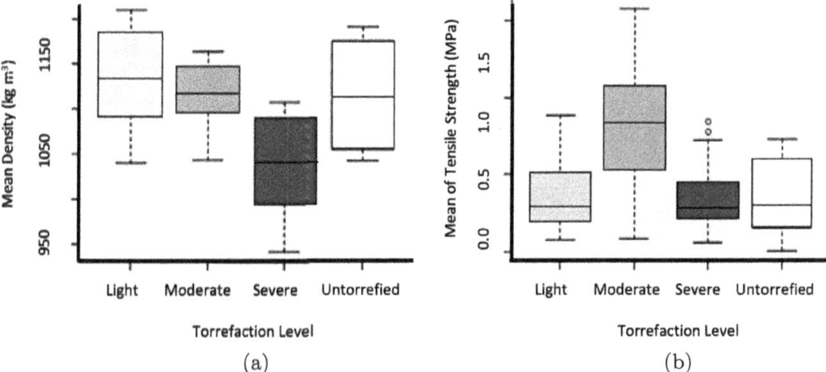

Figure 9.10. Effect of torrefaction level on (a) pellet density and (b) pellet tensile strength.

variables on pellet quality. Figure 9.10 shows the effect of microwave torrefaction level on pellet's unit density and tensile strength. Average unit density significantly dropped at severe microwave torrefaction while the parameter was roughly identical between the remaining microwave torrefaction levels (Fig. 9.10(a)). Meanwhile, pellet's tensile strength remarkably surged at moderate torrefaction conditions and remained stable at the other microwave torrefaction levels (Fig. 9.10(b)). Intensified microwave torrefaction levels generally increased calorific value and ash content but significantly decreased the pellet's final moisture content.

Unit density and tensile strength of pellets were significantly influenced by all process variables, while moisture absorption was mostly controlled by particle size and binder (Table 9.5). Fine particles were shown to enhance unit density, tensile strength, moisture absorption, and moisture content in comparison with coarse particles. Furthermore, unit density and tensile strength were also improved by increasing the die temperature during the pellet process.

Figure 9.11 shows the effect of binder on pellet's unit density and tensile strength for the three microwave torrefaction treatments. In most cases, additional binders were found to improve the pellet unit density except for the case of 10% sawdust (SD10) in light microwave torrefaction. In all microwave torrefaction treatments, increasing the

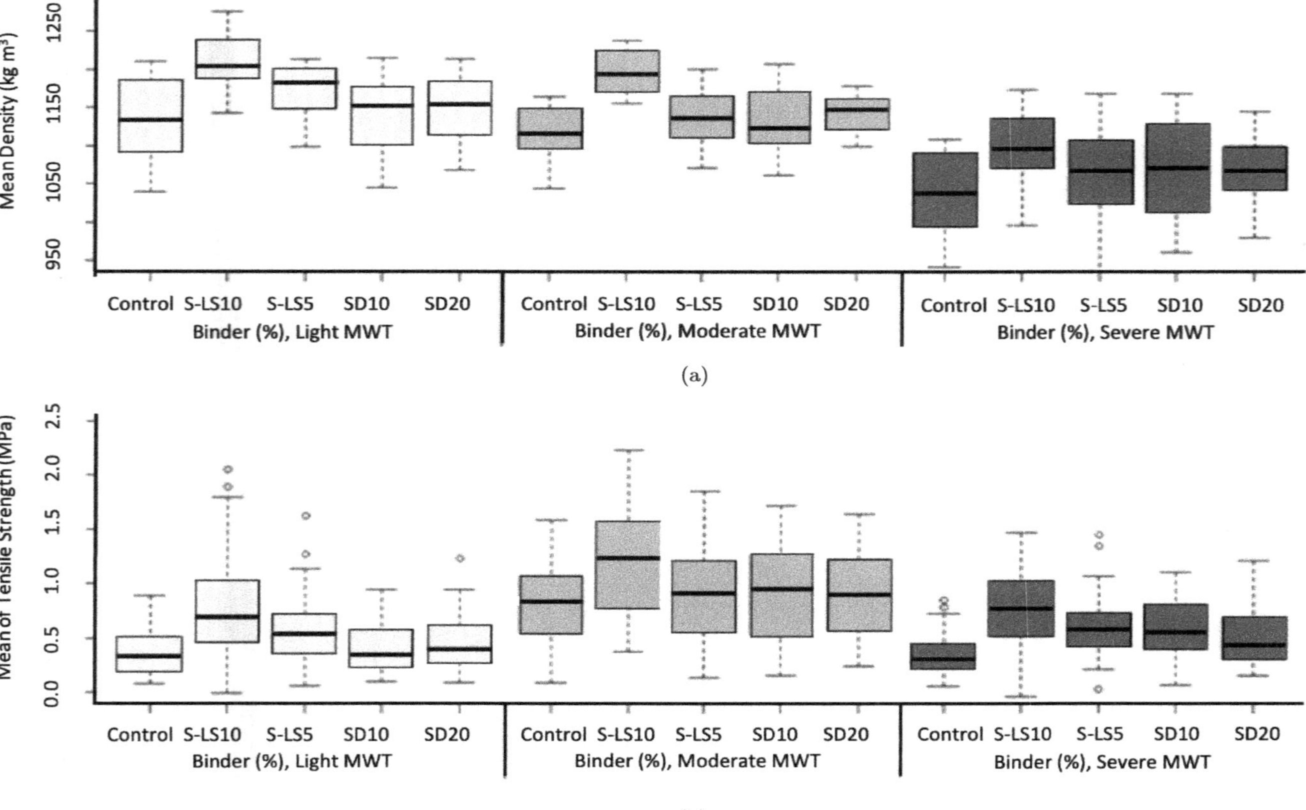

Figure 9.11. (a) Effect of binder use on pellet unit density for different microwave torrefaction levels and (b) effect of binder use on pellet tensile strength for different microwave torrefaction levels.

Table 9.5. Level of significance of process variables on pellet properties.

	Process variables		
Pellet quality	Particle size (fine)	Die temperature (high)	Binder use
Density	$(+)^{***}$	$(+)^{***}$	$(+)^{***}$
Tensile strength	$(+)^{***}$	$(+)^{***}$	$(+)^{***}$
Moisture absorption	$(+)^{**}$	ns	$(+)^{***}$
Moisture content	$(+)^{**}$	$(-)^{***}$	$(+)^{***}$
Higher heating value	ns	ns	$(-)^{***}$
Ash content	ns	ns	$(+)^{***}$

Note: $^{**}P < 0.01$, $^{***}P < 0.001$, ns (non-significant).

sawdust content from 10 to 20% did not statistically affect the unit density. In the case of moderate and severe treatments, there were no significant changes in unit density when switching binders between 5% lignosulfonate, 10% sawdust, and 20% sawdust. While tensile strengths in light microwave torrefaction pellets increased significantly ($P < 0.05$) with the use of 5% lignosulfonate and 10% lignosulfonate, the parameter on moderate microwave torrefaction pellets was enhanced only by lignosulfonate 10 at $P < 0.001$. Additionally, all binders enhanced the tensile strength of severe microwave torrefaction pellets at $P < 0.05$. Interestingly, in all microwave torrefaction treatments, there were no significant changes in pellet's tensile strength between 5% lignosulfonate, 10% sawdust, and 20% sawdust.

9.3.9. *Moisture Absorption of Pellets*

The moisture absorption of microwave-torrefied pellets was significantly lower than that of untreated pellets. At full saturation (after 72 h at 30°C and 90% relative humidity), the coarse samples with light, moderate, and severe microwave torrefaction presented a respective moisture absorption of 17.32 wt%, 12.29 wt%, and 11.39 wt% vs. 22.10% for untreated pellets (Fig. 9.12). Pellets made with fine particle sizes presented higher moisture absorption than unground or coarse particle pellets. A larger surface area of

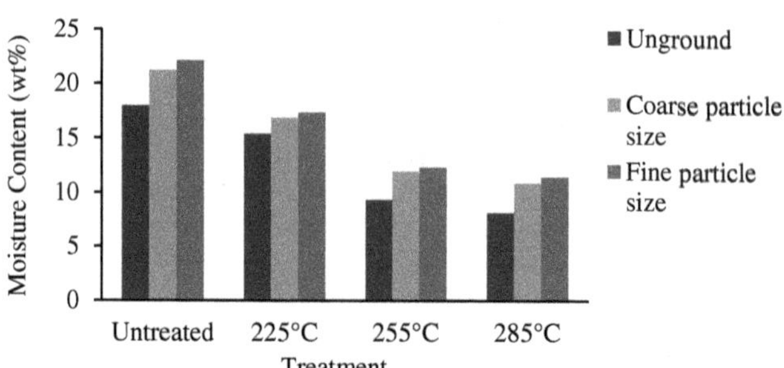

Figure 9.12. Moisture absorption of different particle sizes and pre-treatment pellets after full saturation.

small particles leading to higher contact with the relative humidity resulting in higher moisture absorption could explain the difference. Figure 9.13(a) shows the measured moisture absorption rate of untreated and treated samples. It can be seen that pellets reached full saturation after 12 h, where the saturated moisture absorption was mainly dependent on microwave torrefaction conditions.

A higher torrefaction level reduced the total moisture absorption (Fig. 9.10(a)), which according to Verhoeff *et al.* (2011) was produced by depolymerization of polymers and removal of oxygen groups from the cell wall replacing hydrophilic for hydrophobic groups. Peng *et al.* (2013a) contributed the decrease in moisture absorption of torrefied pellets to decremented hydrophilic hydroxyl (OH) groups and incremented hydrophobic carbon content. Moreover, Anglès *et al.* (2001) explained that hemicelluloses were responsible for moisture absorption and biological degradation, therefore a decreased hemicellulose content reduced the total moisture absorption of the biomass. Similar results were observed by Valdez (2021), where a decrease in hemicellulose content produced by microwave torrefaction treatment lowered the total moisture absorption of the biomass samples.

Figure 9.13(b) shows the measured moisture absorption rate of 255°C microwave torrefaction of hull pellets with different binders. Pellets formulated with lignosulfonate as a binder presented the highest moisture absorption saturation. This could be explained by the availability of OH groups and the highly amorphous structure

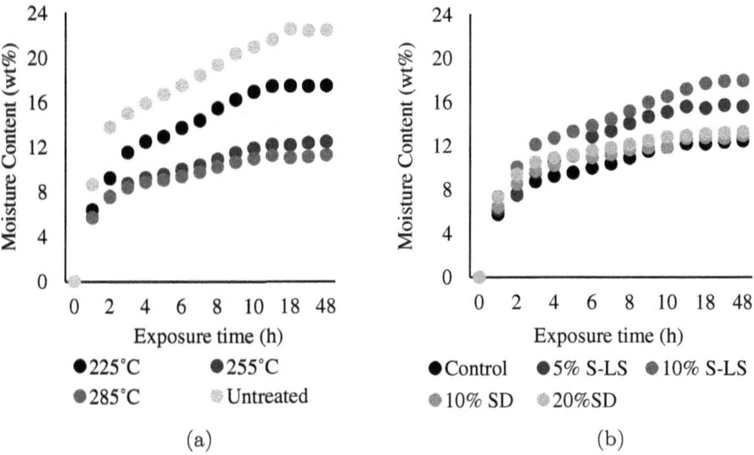

Figure 9.13. (a) Moisture content of torrefied and untorrefied oat hull and (b) moisture content of moderate microwave-torrefied oat hull pellets with different binders.

of lignosulfonates, which increases their sorption ability (Hemmilä *et al.*, 2019). Pellets formulated with sawdust as a binder presented higher moisture absorption than control pellets due to a higher absorption of raw sawdust.

Figure 9.14 shows the pellet deformation after the moisture absorption test, where untorrefied pellets are completely decompressed. Moreover, pellets formulated by coarse particle size presented a larger deformation compared to pellets with fine particle size (Figs. 9.14(b)–9.14(d)). Furthermore, deformation was minor for fine particle size pellets made by moderate and severe microwave torrefaction. Those results are consistent with experiments by Peng *et al.* (2013a), where untorrefied biomass pellets presented considerable decompression, light torrefaction presented a large deformation, while moderate and severe torrefaction remained in good condition.

Moisture absorption of light microwave torrefaction pellets was significantly increased ($P < 0.001$) by 5 and 10% lignosulfonate. Whereas for moderate and severe microwave torrefaction, all binders significantly increased ($P < 0.05$) pellet moisture absorption except 10% sawdust. Binder use was non-significant at $P < 0.05$ for moisture content for any of the microwave torrefaction treatment levels.

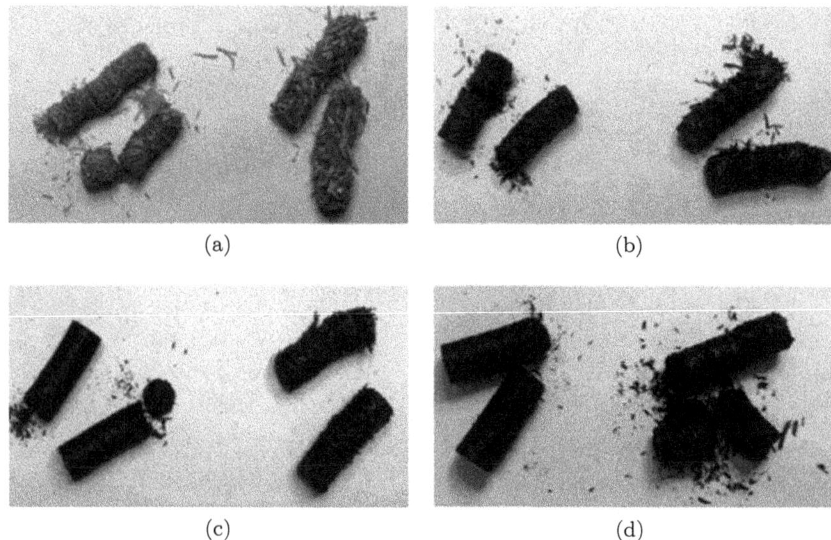

Figure 9.14. Control pellet deformation, fine (left-hand side) and coarse (right-hand side) particle size after moisture absorption test: (a) untorrefied, (b) light — 225°C, (c) moderate — 255°C, and (d) severe — 285°C.

Furthermore, microwave torrefaction level was found to positively affect the pellet's water uptake.

9.4. Conclusions

Microwave torrefaction severity, die temperature, particle size, and binder use generally presented a significant effect on pellet quality. Based on this work, the following conclusions are drawn:

(1) There was a significant influence of microwave torrefaction level on oat hull pellet quality. Moderate torrefaction showed advantage over severe torrefaction in terms of pellet's unit density, while it was observed to improve tensile strength, HHV, moisture absorption, and reduce the moisture content of pellets in comparison with untreated and light torrefaction pellets.

(2) There was a significant influence of particle size on most of the pellet quality parameters. Fine particles of feedstock resulted in higher pellet's tensile strength, unit density, moisture absorption,

and moisture content as compared to those in pellets made from coarse feedstock particles.

(3) There was a significant influence of die temperature on some of the pellet quality parameters. Pellets made with high die temperature increased pellet density and tensile strength.

(4) There was a significant influence of binder on most of the pellet quality parameters. Lignosulfonate was observed to improve unit density and tensile strength while resulting in the highest moisture absorption, ash content, and lowest heating value.

(5) The use of sawdust binder generally had a significant positive impact on pellet density in all types of microwave torrefaction treatments, whereas it was only significant on tensile strength for severe microwave torrefaction treatment. Moreover, it also yielded lower moisture absorption and ash content pellets when compared to those with lignosulfonate. However, both sawdust and lignosulfonate significantly lowered the HHV of severe microwave torrefaction pellets.

Acknowledgments

The authors would like to acknowledge the funding support from the Natural Sciences and Engineering Research Council of Canada (NSERC) and BioFuelNet Canada. We also acknowledge the support of our industry partners, particularly Richardson Milling.

References

Abedi, A. & Dalai, A. K. (2017). Study on the quality of oat hull fuel pellets using bio-additives. *Biomass Bioenergy*, *106*, 166–175. https://doi.org/10.1016/j.biombioe.2017.08.024.

Acharjee, T. C., Coronella, C. J., & Vasquez, V. R. (2011). Effect of thermal pretreatment on equilibrium moisture content of lignocellulosic biomass. *Bioresour. Technol.*, *102*(7), 4849–4854. https://doi.org/10.1016/j.biortech.2011.01.018.

Adapa, P. K., Tabil, L. G., & Schoenau, G. J. (2010). Compression characteristics of non-treated and steam-exploded barley, canola, oat, and wheat straw grinds. *Appl. Eng. Agric.*, *26*(4), 617–632.

Adapa, P., Tabil, L., & Schoenau, G. (2009). Compaction characteristics of barley, canola, oat and wheat straw. *Biosyst. Eng.*, *104*(3), 335–344. https://doi.org/10.1016/j.biosystemseng.2009.06.022.

Adapa, P., Tabil, L., & Schoenau, G. (2011). Grinding performance and physical properties of non-treated and steam exploded barley, canola, oat and wheat straw. *Biomass Bioenergy*, *35*(1), 549–561. https://doi.org/10.1016/j.biombioe.2010.10.004.

Anglès, M. N., Ferrando, F., Farriol, X., & Salvadó, J. (2001). Suitability of steam exploded residual softwood for the production of binderless panels. Effect of the pre-treatment severity and lignin addition. *Biomass Bioenergy*, *21*(3), 211–224. https://doi.org/10.1016/S0961-9534(01)00031-9.

Arias, B., Pevida, C., Fermoso, J., Plaza, M. G., Rubiera, F., & Pis, J. J. (2008). Influence of torrefaction on the grindability and reactivity of woody biomass. *Fuel Process. Technol.*, *89*, 169–175. https://doi.org/10.1016/j.fuproc.2007.09.002.

Castellano, J. M., Gómez, M., Fernández, M., Esteban, L. S., & Carrasco, J. E. (2015). Study on the effects of raw materials composition and pelletization conditions on the quality and properties of pellets obtained from different woody and non woody biomasses. *Fuel*, *139*, 629–636. https://doi.org/10.1016/j.fuel.2014.09.033.

Emadi, B., Iroba, K. L., & Tabil, L. G. (2017). Effect of polymer plastic binder on mechanical, storage and combustion characteristics of torrefied and pelletized herbaceous biomass. *Appl. Energy*, *198*, 312–319. https://doi.org/10.1016/j.apenergy.2016.12.027.

Gilbert, P., Ryu, C., Sharifi, V., & Swithenbank, J. (2009). Effect of process parameters on pelletisation of herbaceous crops. *Fuel*, *88*(8), 1491–1497. https://doi.org/10.1016/j.fuel.2009.03.015.

Harun, N. Y., & Afzal, M. T. (2016). Effect of particle size on mechanical properties of pellets made from biomass blends. *Procedia Eng.*, *148*, 93–99. https://doi.org/10.1016/j.proeng.2016.06.445.

Hemmilä, V., Hosseinpourpia, R., Adamopoulos, S., & Eceiza, A. (2019). Characterization of wood-based industrial biorefinery lignosulfonates and supercritical water hydrolysis lignin. *Waste Biomass Valorization*, 1–11. https://doi.org/10.1007/s12649-019-00878-5.

Iroba, K. L., Baik, O.-D., & Tabil, L. G. (2017). Torrefaction of biomass from municipal solid waste fractions II: Grindability characteristics, higher heating value, pelletability and moisture adsorption. *Biomass Bioenergy*, *106*, 8–20. https://doi.org/10.1016/j.biombioe.2017.08.008.

Kaliyan, N. & Morey, V. (2009). Factors affecting strength and durability of densified biomass products. *Biomass Bioenergy*, *33*(3), 337–359. https://doi.org/10.1016/j.biombioe.2008.08.005.

Kashaninejad, M., Tabil, L. G., & Knox, R. (2014). Effect of compressive load and particle size on compression characteristics of selected varieties of wheat straw grinds. *Biomass Bioenergy*, *60*, 1–7. https://doi.org/10.1016/j.biombioe.2013.11.017.

Kashaninejad, T. & Tabil, L. G. (2011). Effect of microwave–chemical pre-treatment on compression characteristics of biomass grinds. *Biosyst. Eng.*, *108*, 36–45.

Li, H., Liu, X., Legros, R., Bi, X. T., Jim Lim, C., & Sokhansanj, S. (2012). Pelletization of torrefied sawdust and properties of torrefied pellets. *Appl. Energy*, *93*, 680–685. https://doi.org/10.1016/j.apenergy.2012.01.002.

Lu, D., Tabil, L. G., Wang, D., Wang, G., & Emami, S. (2014). Experimental trials to make wheat straw pellets with wood residue and binders. *Biomass Bioenergy*, *69*, 287–296. https://doi.org/10.1016/j.biombioe.2014.07.029.

Mani, S., Tabil, L. G., & Sokhansanj, S. (2006). Effects of compressive force, particle size and moisture content on mechanical properties of biomass pellets from grasses. *Biomass Bioenergy*, *30*(7), 648–654. https://doi.org/10.1016/j.biombioe.2005.01.004.

McKendry, P. (2002). Energy production from biomass (Part 1): Overview of biomass. Bioresour. Technol. 83, 37–46. *Bioresour. Technol.*, *83*, 37–46. https://doi.org/10.1016/S0960-8524(01)00118-3.

Peng, J. H., Bi, X. T., Lim, C. J., Peng, H., Kim, C. S., Jia, D., & Zuo, H. (2015). Sawdust as an effective binder for making torrefied pellets. *Appl. Energy*, *157*, 491–498. https://doi.org/10.1016/j.apenergy.2015.06.024.

Peng, J. H., Bi, H. T., Lim, C. J., & Sokhansanj, S. (2013a). Study on density, hardness, and moisture uptake of torrefied wood pellets. *Energy Fuels*, *27*(2), 967–974. https://doi.org/10.1021/ef301928q.

Peng, J. H., Bi, X. T., Sokhansanj, S., & Lim, C. J. (2013b). Torrefaction and densification of different species of softwood residues. *Fuel*, *111*, 411–421. https://doi.org/10.1016/j.fuel.2013.04.048.

Pirraglia, A., Gonzalez, R., Saloni, D., & Denig, J. (2013). Technical and economic assessment for the production of torrefied ligno-cellulosic biomass pellets in the US. *Energy Convers. Manag.*, *66*, 153–164. https://doi.org/10.1016/j.enconman.2012.09.024.

Repellin, V., Govin, A., Rolland, M., & Guyonnet, R. (2010). Energy requirement for fine grinding of torrefied wood. *Biomass Bioenergy*, *34*(7), 923–930. https://doi.org/10.1016/j.biombioe.2010.01.039.

Satpathy, S. K., Tabil, L. G., Meda, V., Narayana, S., & Rajendra, P. (2013). Torrefaction and grinding performance of wheat and barley straw after microwave heating. 17. http://www.csbe-scgab.ca/docs/meetings/2013/CSBE13065.pdf.

Serrano, C., Monedero, E., Lapuerta, M., & Portero, H. (2011). Effect of moisture content, particle size and pine addition on quality parameters of barley straw pellets. *Fuel Process. Technol.*, *92*(3), 699–706. https://doi.org/10.1016/j.fuproc.2010.11.031.

Shaw, M. D., Karunakaran, C., & Tabil, L. G. (2009). Physicochemical characteristics of densified untreated and steam exploded poplar wood and wheat straw grinds. *Biosyst. Eng.*, *103*(2), 198–207. https://doi.org/10.1016/j.biosystemseng.2009.02.012.

Sluiter, A., Hames, B., Ruiz, R., Scarlata, C., Sluiter, J., Templeton, D., & Crocker, D. (2008). Determination of structural carbohydrates and lignin in biomass (Technical Report NREL/TP-510-42618; NREL Laboratory

Analytical Procedures for Standard Biomass Analysis, pp. 1–15). U.S. Department of Energy, https://www.nrel.gov/docs/gen/fy13/42618.pdf.

Soleimani, M., Tabil, X. L., Grewal, R., & Tabil, L. G. (2017). Carbohydrates as binders in biomass densification for biochemical and thermochemical processes. *Fuel*, *193*, 134–141. https://doi.org/10.1016/j.fuel.2016.12.053.

Spiridon, I. & Popa, V. (2008). Chapter 13 — Hemicelluloses: Major sources, properties and applications. In *Monomers, Polymers and Composites from Renewable Resources* (pp. 289–304). https://doi.org/10.1016/B978-0-08-045316-3.00013-2.

Tabil, L. G. & Sokhansanj, S. (1996). Compression and compaction behavior of alfalfa grind: Part 2: Compaction behavior. *Power Handling Proces.*, *8*(2), 177–122.

Tarasov, D., Shahi, C., & Leitch, M. (2013). Effect of additives on wood pellet physical and thermal characteristics: A review. *Int. Scholarly Res. Not.*, *2013*, 6. https://doi.org/10.1155/2013/876939.

Tumuluru, J., Sokhansanj, S., Hess, J. R., Wright, C. T., & Boardman, R. D. (2011). A review on biomass torrefaction process and product properties for energy applications. *Ind. Biotechnol.*, *7*(5), 384–401. https://doi.org/10.1089/ind.2011.7.384.

Valdez, E. (2021). Microwave torrefaction and densification of oat hulls for heat and power production [Master dissertation, University of Saskatchewan; Electronic resource]. Retrieved from http://hdl.handle.net/10388/13245.

van Dam, J. E. G., van den Oever, M. J. A., Teunissen, W., Keijsers, E. R. P., & Peralta, A. G. (2004). Process for production of high density/high performance binderless boards from whole coconut husk: Part 1: Lignin as intrinsic thermosetting binder resin. *Ind. Crops Prod.*, *19*(3), 207–216. https://doi.org/10.1016/j.indcrop.2003.10.003.

Verhoeff, F., Adell i Arnuelos, A., Boersma, A. R., Pels, J. R., Lensselink, J., Kiel, J. H. A., & Schukken, H. (2011). TorTech: Torrefaction technology for the production of solid bioenergy carriers from biomass and waste | Titel (ECN-E–11-039; p. 82). Energy research Centre of the Netherlands. Retrieved from https://library.wur.nl/WebQuery/titel/2012815.

https://doi.org/10.1142/9781800613799_0010

Chapter 10

Densification Characteristics of Ammonia Fiber Explosion Pretreated Biomass and Applications

Venkatesh Balan[*,‡], Bryan Ubanwa[*,§], and Bruce Dale[†,¶]
[*]*Department of Engineering Technology, College of Technology University of Houston, Sugar Land, Texas, USA*
[†]*Chemical Engineering and Material Science Great Lakes Bioenergy Center, Michigan State University East Lansing, Michigan, USA*
[‡]*vbalan@uh.edu*
[§]*bryan.ubanwa@yahoo.com*
[¶]*bdale@egr.msu.edu*

Abstract

Lignocellulosic biomass is the most abundant cheap resource produced from agriculture and forest residues. Among several thermochemical pretreatments reported in the literature, Ammonia Fiber Expansion (AFEX) pretreatment uses ammonia that can be recovered and re-used benefiting the environment. Ammonia helps open up the complex cell wall by cleaving the ester linkages and increasing the digestibility during enzyme hydrolysis and rumen in the animal gut. The lignin solubilized by ammonia during AFEX pretreatment is partially relocated to the surface of the biomass, while it is released to act as a natural binding when densifying the biomass. The AFEX pellets have a bulk density comparable to corn, can be stored in grain elevators, and can be used for diverse applications, such as feedstock for biofuels, animal feed, anaerobic digestion, and producing biocomposite materials. A local biomass pretreatment depot (LBPD) will help produce the AFEX pellets near the location where the feedstocks are produced and transported using trucks or rail carts. This approach will help overcome the logistic issues of transporting and storing biomass and sold as a commodity product that will enable the biochemical platform and biobased economy.

Keywords: Ammonia fiber explosion, lignocellulosic biomass, densification, pretreatment, animal feed

10.1. Introduction

Liquid transportation fuels and chemicals are important components for a prosperous modern society. At present, 96% of these components are produced using crude oil while only 4% are derived from bio-based feedstocks (Stanchin *et al.*, 2020). Crude oil has seen volatile swings in price over the period from 2008 to 2015 and some have predicted that we have reached the peak in new oil discovery. In addition, burning fuel and producing chemicals from crude oil produces a large amount of greenhouse gases (GHGs) which are responsible for environmental pollution and are linked to global warming and climate change (Gaulin and Le Billion, 2020). Several policy changes are underway to switch from a crude oil-based economy to a bio-based economy. Though it is challenging and requires significant investment from industry government, the transition will result in greater energy security for a given country, a cleaner environment, and rural job creation (Rafiaani *et al.*, 2018). In the past few years, governments around the world have been committed to invest in the creation of the bio-based economy by funding the development of emerging technologies for making this transition a reality. A bio-based economy can be realized by producing renewable biofuels and biochemicals using biochemical (extraction, hydrolysis, and fermentation), thermochemical (pyrolysis/liquefaction, gasification, and combustion), biological, and hybrid conversion platforms in a biorefinery, which will have similarities to a petroleum refinery (Kumar *et al.*, 2019). Since the carbon generated during combustion of fuel is fixed by plants and trees, they are considered carbon neutral and do not put net GHG in the atmosphere.

Lignocellulosic biomass produced using agriculture operations (e.g., corn stover, wheat straw, sorghum, grain hulls, perennial grass such as switch grass, miscanthus) and forest residues (e.g., hard and soft wood) is comprised of sugar polymers, such as cellulose and hemicellulose, and aromatic polymers, such as lignin, ash, and extractives. In a lignocellulosic biorefinery using a biochemical route, the sugar polymers are depolymerized using commercially available carbohydrate-active enzymes (CAZymes), followed by microbial fermentation using different native and genetically organisms (e.g., yeast or bacteria) to produce microbial byproducts such as fuels and chemicals (Chandel *et al.*, 2019). These byproducts are then separated by distillation or fractionated and are sustainable alternatives to fossil-derived fuels and chemicals. Lignocellulosic biomass is high recalcitrant and requires some pretreatment to open up the cell wall for improving the accessibility of sugar polymers by CAZymes (Kubicek *et al.*, 2014). Several pretreatment processes have been developed and can be broadly classified as acids (e.g., dilute sulfuric acid and phosphoric acid), bases (e.g., $NaOH/H_2O_2$, KOH, and NH_3), solvents (e.g., ethanol, ionic liquid, and tetra hydro furan), steam (e.g., acid- or base-catalyzed), and biological (e.g., fungi and bacteria) (Chen *et al.*, 2017). Each pretreatment has its own advantages/disadvantages and uses different mechanisms to open up the cell wall (Sousa *et al.*, 2009; Zhao *et al.*, 2020). Several new environmentally friendly pretreatment techniques have been developed (Haldar and Purkait, 2021). Acid pretreatment usually solubilizes hemicellulose to monomeric and oligomeric sugars and leaves behind cellulose and lignin. On the other hand, base pretreatment solubilizes lignin leaving behind cellulose and hemicellulose intact. Solvent-based extraction of biomass (commonly called organosolv pretreatment) extracts lignin and some hemicellulose. Steam pretreatments are carried out using super-critical steam that opens up lignin–carbohydrate complexes (LCCs) and solubilizes some of the hemicelluloses. Biological pretreatment uses microorganisms that secrete oxidative and hydrolytic enzymes that de-construct lignocellulosic biomass (Paritosh *et al.*, 2021).

Lignocellulosic biomass is essential for making this bio-based economy possible. However, there are logistical challenges in sustainably transporting the biomass from the field to biorefinery where they will be processed since they are low density and often unstable toward biological and thermal degradation. In this chapter, we will summarize a method of pretreating the biomass using Ammonia Fiber Expansion (AFEX) followed by densification to produce stable biomass pellets that could be easily shipped as we currently do with corn. The lignin and other ammonia soluble products produced during AFEX pretreatment act as a natural binder which helps 'hold the biomass together. We will discuss the method of pretreating the biomass and AFEX pellet's characteristics and its use in different processes such as carbon sources.

10.2. AFEX Pretreatment Process

In the AFEX process, hot concentrated ammonia (temperatures of 70–200°C) is mixed with biomass under pressure (20–30 bar) either in a liquid or gaseous state for residence times of 15–30 min in a small stainless steel reactor (25–1000 mL, Balan *et al.*, 2009a). Both ammonia and water have a loading range of 0.4–2.0 g/g dry biomass feedstock (Balan *et al.*, 2009a; Garlock *et al.*, 2009). In some cases, the reactor is externally heated to raise the temperature. After the desired temperature is reached, the reaction is continued for a few more minutes and the pressure is rapidly released, causing the system to cool and the ammonia is recovered (Fig. 10.1(a)). Hemicellulose gets cleaved, and the internal structures of the biomass are swollen and expanded, thereby making the sugar polymers in the biomass more accessible to CAZymes' breakdown or ruminant digestion (Balan *et al.*, 2011; Chundawat *et al.*, 2013). The AFEX process is performed under high solids loading (30–75% solids on a total weight basis) (Jin *et al.*, 2016).

During the pretreatment process, both ammonolysis and hydrolysis reactions take place to cleave the ester linkages and LCC complex (Fig. 10.1(b)). Several degradation products are produced during AFEX pretreatment; the most noticeable ones include ferulic acid/feruloyl amides, acetic acid/acetamide, coumaric acid/coumaroyl amide, oligosaccharides, and lignins which are soluble in ammonia (Chundawat *et al.*, 2010, 2013). Once the pretreatment is complete, the valve is open to release the pressure and during this process the soluble lignin and degraded compounds move to

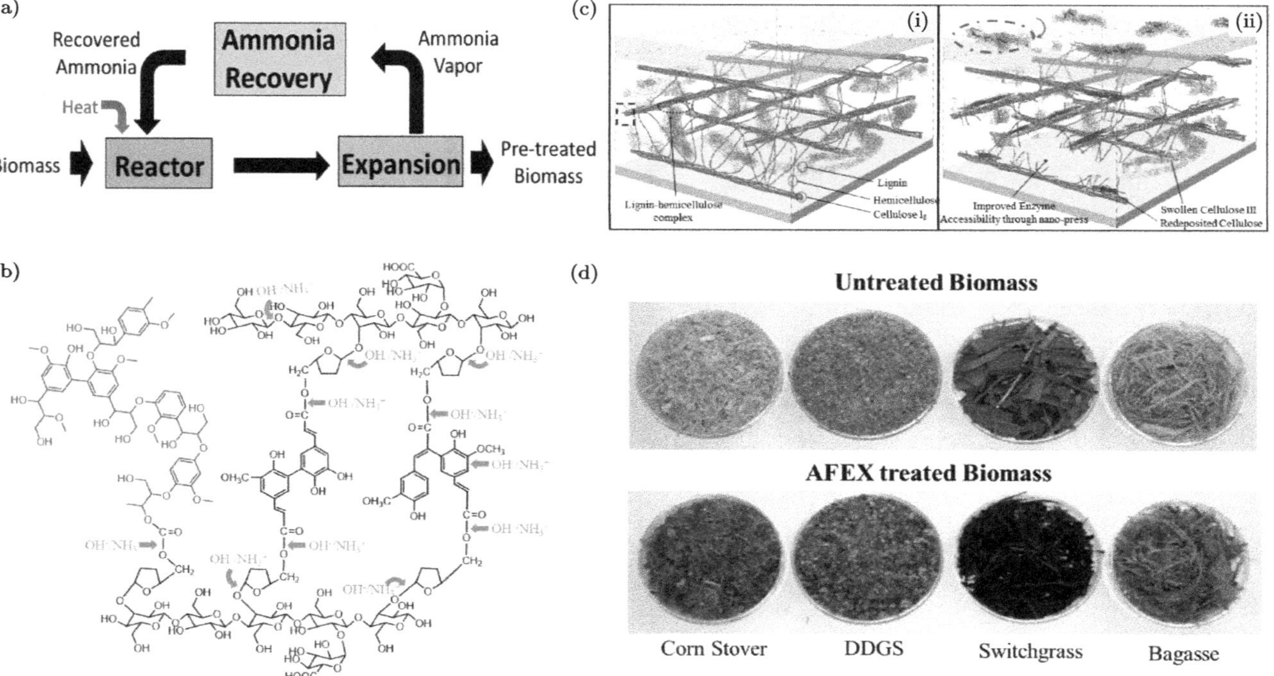

Figure 10.1. Ammonia pretreatment of lignocellulosic biomass: (a) process diagram of the AFEX process where ammonia is being recovered and re-used for pretreating the biomass; (b) lignin–carbohydrate complex comprise of ester linkages which are cleaved by ammonium and hydroxyl ion during AFEX process; (c) plant cell wall structure before (i) and after (ii) AFEX pretreatment solubilizing some of the cell wall component and re-depositing on the surface when ammonia is released after pretreatment; (d) appearance of lignocellulosic biomass such as corn stover, dry distillers grains and soluble (DDGS), switchgrass, and sugarcane bagasse before and after AFEX pretreatment.

the surface of the biomass (Fig. 10.1(c)) (Chundawat *et al.*, 2011a). The AFEX pretreated biomass appears to be slightly dark brown in color due to re-localization of lignin and degradation products to the surface (Fig. 10.1(d)). In a different improvised Extractive Ammonia (EA) pretreatment process, ammonia-soluble degradation products including lignin have been separated into a different vessel that could be further fractionated (Sousa *et al.*, 2016a). Removing lignin up to 45% in lignocellulosic biomass using EA pretreatment makes the biomass less recalcitrant and highly digestible using commercial enzymes during hydrolysis (Chundawat *et al.*, 2011b).

10.3. Advantages of Using AFEX Pretreated Biomass

Among the various available pretreatments, AFEX technology has several advantages. AFEX is a dry-to-dry process where a single solid stream is generated. Unlike other pretreatment technologies that produce liquid streams/slurries, the used catalyst is retained in pretreated biomass. AFEX uses volatile ammonia that can be recovered and re-used (up to 97%). AFEX pretreatment can also improve the densification process due to the relocation of lignin, which acts as a natural binding agent (Dale *et al.*, 2013). Generation of denser briquettes is also possible using AFEX-pretreated biomass that can further decrease energy costs associated with the densification process. AFEX pretreated agricultural residues were found to be highly digestible (an increase of 3- to 4-fold in digestibility when compared to untreated feedstock) by commercial CAZymes. Finally, AFEX-treated biomass can be used as animal feed, allowing a second market for the biomass processor, and used for producing biogas and biomaterials. As mentioned, it is critically important to understand that the AFEX process produces no separate liquid phase of waste biomass or spent catalyst, as do nearly all the competing thermochemical pretreatment processes. Also, minimal degradation of sugars occurs during AFEX, avoiding the loss of yield and inhibition of fermentation that many pretreatments cause. After the ammonia is removed and recovered, the treated biomass gives much higher biofuel yields and becomes a much better cattle feed (Blummel *et al.*, 2014).

10.4. AFEX Pretreatment in Pilot Scale

A simplified diagram of the actual AFEX pilot plant is designed, built, and operated by MBI, International, Lansing, MI, which is shown in Fig. 10.2. MBI conducted over 1500 separate runs on this pilot plant, producing over 36 t of treated biomass with no safety problems or ammonia releases ever occurring (Campbell *et al.*, 2018, 2019). A large pilot scale AFEX process was possible with the development of gaseous ammonia-based pretreatment developed at Michigan State University (MSU) (Balan *et al.*, 2015). Gaseous ammonia reactor with water in the biomass producing exothermic reaction instantaneously heats the biomass to 140°C. The pilot scale AFEX system comprised of a twin high-pressure reactor which could be operated in a semi-continuous manner to produce AFEX pretreated biomass up to 1 t/day followed by producing AFEX pellets.

First biomass is pre-processed before subjecting to AFEX pretreatment (Fig. 10.2(a)). Biomass bales (square or round bales) are ground to smaller particle sizes (roughly 4–6 cm), filled in the metal wire mesh baskets, and stacked in the reactor on the top of each other (Fig. 10.2(b)). Each packed bed of biomass is treated in five steps in a twin reactor: pre-steam, ammonia charge, soak, depressurize, and steam strip (Fig. 10.2(c)) (Campbell *et al.*, 2013, 2019). After the pretreatment, the biomass baskets are removed (Fig. 10.2(d)) and finally, the AFEX-treated biomass is subjected to densification after adjusting the moisture, followed by drying in a 50°C oven to produce densified AFEX pellets with moisture content <8% and stored dry (Fig. 10.2(e)). The details about the AFEX pretreatment operational details of these five steps in the pilot-scale reactors are given in the following:

(1) **Pre-steam:** Each bed was first pre-steamed by admitting saturated steam at 2.5 kg/min mass flow rate through the reactor vessel's inlet port. During pre-steaming, a vent valve in the exit pipe at the bottom of the reactor vessel was kept open to allow

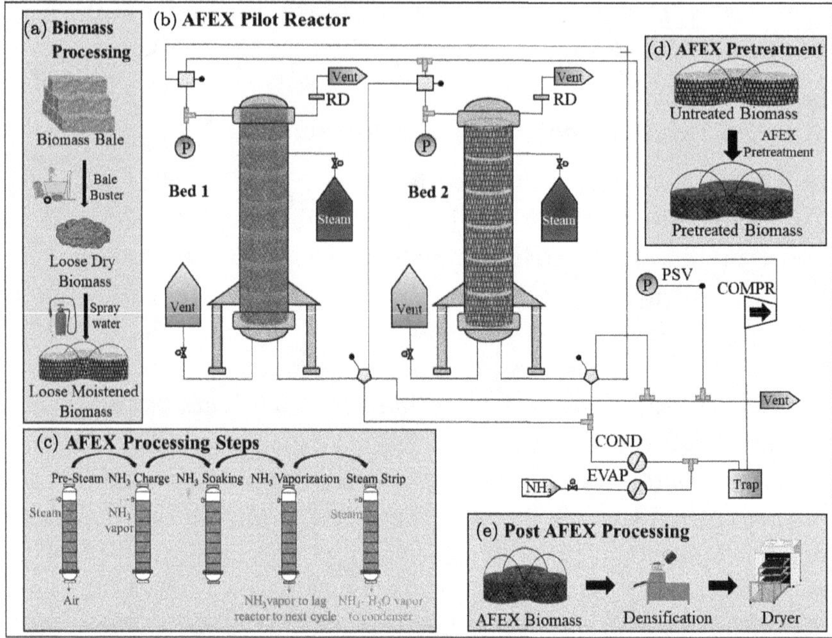

Figure 10.2. Semicontinuous AFEX pretreatment process using twin reactor system: (a) biomass pre-processing is carried out, where bales are busted to lose biomass, sprayed with water, and transferred to baskets; (b) the biomass baskets are stacked in the AFEX reactor connected to a ammonia tank (NH_3), condenser (COND), pressure safety valve (PSV), compressor (COMPR), rotating disc gate valve (RD), pressure gauge (P), and steam and venting system; (c) step-by-step AFEX processing where the biomass is loaded in the reactor first, reactor closed, pre-steamed followed by charging with ammonia and holding for 30 min residence time before the valve is open to release ammonia in vapor phase. Subsequently, steam is used to strip residual ammonia in the biomass. The released ammonia is compressed and re-used for subsequent biomass pretreatment in the second reactor that is stacked with biomass. About 95–97% of ammonia could be recovered and re-used in the process developed at MBI. (d) Picture showing untreated and pretreated biomass in basket and (e) post process of AFEX biomass to biomass pellets and drying.

air to escape from the vessel to the scrubber. Pre-steaming was stopped when the temperature at the bottom of the bed reached 80°C (~5 min); then the bottom valve was closed.

(2) **Ammonia charge:** The first bed in a series of two reactor beds was charged with fresh anhydrous ammonia vapor

from the vaporizer, while each subsequent bed in the series was first charged with ammonia from the opposite reactor's depressurization and steam stripping steps and then with fresh makeup ammonia. The target ammonia load for each bed was 0.7 kg ammonia per kg dry biomass. As ammonia is added, the temperature at the top of the reactor cools to 60°C (the temperature of the ammonia), while the temperature at the bottom increases to ~140°C due to the exothermic reaction of ammonia vapor and water. Bed pressure at the end of ammonia charging was typically 2.04 MPa ±0.03 MPa.

(3) **Soak:** Once charged with ammonia, each bed was allowed to stand for at least 20 min of soak time before starting depressurization. No upper limit was assigned to the soak time to allow for time differences in transferring biomass in the other reactor. The reactors do not have insulation and thus naturally cool down during this time, and the pressure correspondingly decreases to ~1.4 MPa.

(4) **Depressurize:** After soaking, each bed was depressurized by gradually opening a flow control valve, releasing vapor from the bottom of the bed. The vapor released could be sent to the opposite bed; in this case, pressures in the two beds were first equalized and then vapor from the bed being depressurized was sent through the condenser and trap, recompressed by the compressor, and charged to the opposite reactor. Alternatively, the ammonia vapor released by depressurization could be sent to the scrubber.

(5) **Steam strip:** Following depressurization, each bed still contained 5–10 kg of residual ammonia bound to lignocellulosic biomass, which was removed by steam stripping. Steam was admitted to the top of the reactor vessel at 2.5 kg/min. As with depressurization, gaseous ammonia vapor removed during steam stripping could be recompressed in the compressor and sent to the bed in the opposite reactor or sent to the scrubber. Steam stripping was stopped when the temperature of the vapor exiting the bed reached 100°C. After steam stripping was complete, the bed was allowed to cool with air flowing to sweep any

residual ammonia vapor into the scrubber. The moisture of the treated biomass removed from the reactors was typically 45 wt% ±5 wt%.

10.5. AFEX Biomass Pellets and Bulk Densities

This AFEX-treated biomass was then partially dried and pelleted into easily flowing pellets using a densification machine with a moisture of 20–30% (Campbell and Bals, 2018). The untreated loose biomass has a bulk density of 50–100 kg/m^3, and after AFEX pretreatment, densification, and drying, the bulk density increase to 300–575 kg/m^3. The solubilized lignin deposited on the surface of the biomass after AFEX pretreatment helps bind biomass together (Karki *et al.*, 2015). The AFEX-pretreated biomass bulk density is higher than the untreated biomass bulk density of 250–470 kg/m^3 which requires some binding agents like starch or lignin to hold the biomass together (Fig. 10.3). The AFEX pellets' bulk density is closer to corn grains (760 kg/m^3) and enabled them to store and transport more efficiently. These pellets are stable (durability of 99.2) indefinitely about their increased reactivity and do not degrade unless they are exposed to water. A commercial-scale depot based on the AFEX process is envisioned to operate in much the same way as this pilot plant.

10.6. Necessity of Biomass Densification

The dominant existing conceptual model for cellulosic biofuel production envisions biorefineries processing a few thousand (2000–3000) tons per day of low bulk density, heterogeneous, essentially unprocessed biomass (Kim and Dale, 2015b, 2015c). These fully integrated, centralized facilities contain all biofuel conversion processing operations in a single location. However, a system such as this is burdened with severe logistical constraints such as high biomass transportation and storage costs and the need to contract with literally thousands of farmers to obtain adequate feedstock.

Oil giant BP identified this problem explicitly as they entered and then quickly left the nascent cellulosic ethanol business about

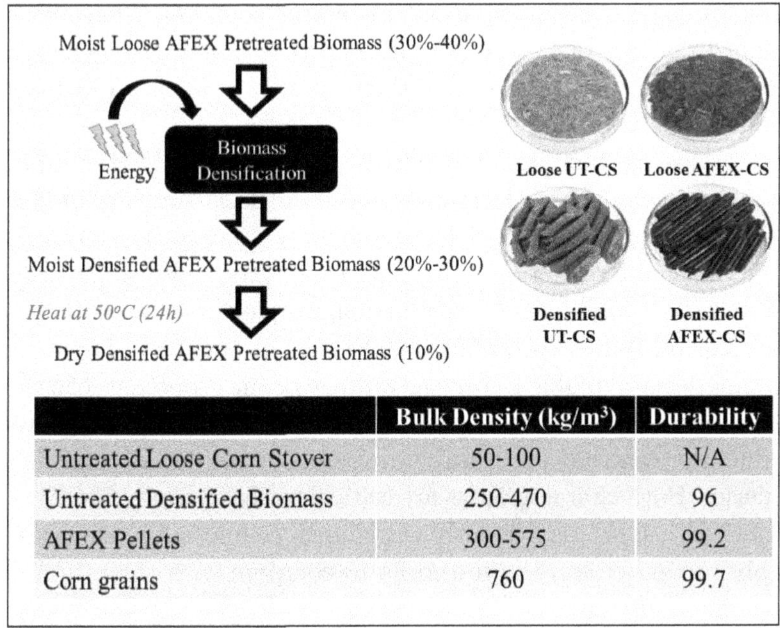

	Bulk Density (kg/m³)	Durability
Untreated Loose Corn Stover	50-100	N/A
Untreated Densified Biomass	250-470	96
AFEX Pellets	300-575	99.2
Corn grains	760	99.7

Figure 10.3. Processing of AFEX-pretreated biomass to densified pellets. Loose AFEX-pretreated biomass with 30–40% moisture is passed to a densification mill producing wet pellet, which is then dried in an oven at 50°C to bring the moisture content to less than 10% for long shelf stability. The images of untreated corn stover (UT-CS) and AFEX-treated corn stover (AFEX-CS) before and after densification. The biomass bulk density of untreated loose, untreated densified, AFEX pellets, and corn grains and their durability are given in the table.

a decade ago. BP could not see a way to scale the supply chains and therefore they could not see how to scale the biorefineries to a size that was commercially interesting to them. The oil companies deal in very large facilities — small fuel-processing systems are simply not attractive to such firms. Similarly, DuPont learned to their surprise that the cost of establishing the supply chains to supply their Nevada, Iowa cellulosic ethanol biorefinery was roughly twice the capital cost of the biorefinery itself, demonstrating once more the critical importance of the supply chain. DuPont and other such companies also found that their biomass storage systems were subject to complete loss by fire due to spontaneous combustion, lightning, arson, and machinery-generated sparks. Other companies, for example, Amyris, started out as biofuel companies, encountered the hard realities of scale in an industry that must produce low-cost fuels, and were forced to change their mission to focus on high-value products (Dale, 2017).

The cost of biomass transportation is the chief factor limiting the scale of cellulosic biorefineries, and it is the reason that modeling of such systems by the US Department of Energy has focused on relatively small refineries (Kim and Dale, 2015b, c). This scale of operation is roughly one order of magnitude lower than the scale of typical oil refineries (about 30,000 t of oil per day). These small biorefineries are unable to take advantage of the economies of scale of oil refineries. Small biorefineries are also unable to make a high-value product nor can they profitably process the many side streams like oil refineries. Both characteristics are critical to the success of modern oil refineries. In short, as currently envisioned, cellulosic biorefineries are simply too small and too inflexible to compete with the established liquid fuel industry.

10.7. AFEX-Pretreated Biomass for Biorefinery Process

(1) Producing fermentable sugars in a bioreactor: The AFEX-pelleted biomass gave higher sugar conversion than loose-pretreated biomass in a shake flask (Bals *et al.*, 2013). At 18% solids loading hydrolysis using fed-batch addition, AFEX corn stover pellet gave 68% and 65% of glucose and xylose conversion respectively when using 20 mg of commercial enzymes (cellulase and hemicellulose)/g of glucan after 72 h. Under similar conditions, AFEX-unpelletized corn stover gave 61% and 59% of glucose and xylose conversion, respectively. It was also reported that AFEX pellet biomass resulted in a faster initial hydrolysis rate when compared to using loose AFEX biomass. The reason for such an increase in the rate of sugar conversion was due to better water absorption and retention which facilitate mixing (rheological properties) when using AFEX pellet biomass compared to lose biomass in a bioreactor. Loading AFEX pelleted to bioreactor releases few dust particulates when compared to using loosed biomass prior to hydrolysis which facilitates cleaner processing conditions.

For large-scale enzyme hydrolysis, the experiment was carried out using AFEX corn stover pellet in a 2500 L bioreactor using the combination of commercial enzymes such as cellulase, and hemicellulase at a 1:1 ratio was added at a 7 g protein/kg biomass total in a bioreactor (Sarks *et al.*, 2016; Campbell and Bals, 2018). AFEX pellets were added into the feed batch to overcome a significant drop in sugar conversion due to the non-availability of water for the enzymes to act efficiently. About one-third of AFEX pellets (with and without dry sterilization) were added to autoclaved water which were added at the beginning of hydrolysis and two-thirds were added after 3 h of hydrolysis. Enzymes were diluted in distilled water which was also added in two batches after passing through a 22-μm filter. Half of the enzymes were added at the beginning of hydrolysis and the rest after 3 h of hydrolysis after adding the second batch of AFEX pellets. The pH of the hydrolysis was maintained at 5.0 by using 4 M H_2SO_4 and the temperature was maintained at 50°C. The impeller speed was maintained at 900 rpm when the biomass slurry is viscous in the beginning and reduced to 400 rpm after liquefaction. Some of the fed-batch hydrolysis experiments were carried out at industrial-relevant high solids loading conditions (22% solids loading) and achieved 85% of sugar conversion with a total sugar yield of 121 g/L (81 g/L of glucose and 40 g/L of xylose) within 48 h. When compared to using AFEX biomass pellet produced in the lab reactor (1 kg/batch), AFEX biomass pellet produced using a pilot plant reactor (300 kg/batch) resulted in higher sugar release (19% glucose and 15% xylose) after 48 h of hydrolysis (Sarks *et al.*, 2016).

(2) Fermentation to produce fuels and chemicals: Subsequently, these sugars were fermented by yeast and bacteria to produce ethanol (Sarks *et al.*, 2016) or other organic acids, such as gluconic or xylonic acid (Zhou *et al.*, 2018). It is important to note that monocot biomass (perennial grasses, corn stover, and wheat straw) pretreated using AFEX is more digestible than dicot biomass (hard or soft wood). This is due to the presence of arabinoxylan linkages in monocot compared to dicot biomass (Balan *et al.*, 2009b, 2011). For ethanol fermentation using bacteria *Zymomonas mobilis* is capable of fermenting both glucose and xylose. The hydrolysate was cooled to 30°C and pH-adjusted to 6.0 using 4 M KOH. For *Z. mobilis* seed culture preparation, sugar-rich media (105 g/L glucose, 20 g/L xylose, 10 g/L Bacto yeast extract, and 2 g/L monopotassium phosphate) was used to prepare in 15 mL for 8 h until the optical density reaches 0.7 at 600 nm. After this, the *Z. mobilis* was centrifuged and transferred to 500 mL and then to 5 L media. The *Z. mobilis* inoculum used for fermentation was 10% of the total weight of AFEX biomass used for hydrolysis. About 2.5 g/L of sterilized corn steep liquor (0.5% of the total weight of AFEX biomass) was used as the nutrient source for fermentation. Total ethanol produced after 48 h of fermentation was 58 g/L. Almost all the glucose and xylose were consumed within the first 14 h and 36 h of fermentation, respectively. It is important to note that scaling the process from shake flask (500 mL) to 2500 L gave a similar sugar and ethanol yield during hydrolysis and fermentation, respectively.

10.8. AFEX-Pretreated Biomass Pellet As Animal Feed

Already the world population is 7.6 billion and is predicted to reach 9 billion in 2050. With increasing population, the demand for food grain will increase and that will add more pressure on the agricultural systems. About half of the grains produced in the world are fed to domestic animals to produce useful food for human society (Bernes-Lee *et al.*, 2018). There is concern that the growing demand for animal feed will drive up the price of food (Berners-Lee *et al.*, 2018). One of the main reasons for such an increase in price is due to transportation fuel cost and growing population demand. One solution to overcome this problem is to convert unconventional feedstocks, like agricultural residues, into more digestible animal feeds (Antar *et al.*, 2021).

AFEX increases the ruminant digestibility of biomass in large part by breaking down the LCC complex and cleaving the ester linkages present in hemicelluloses. This chemical cleavage helps them to be available for rapid digestion in the rumen by simultaneously adding non-protein nitrogen. For example, AFEX-treated corn stover and late-harvest switchgrass showed improved rumen digestibility by 52% and 128%, respectively, over untreated material and produced up to 27% more milk in dairy cow and buffalos during feed trials carried out at the National Dairy Research Institute, India (Mokomele *et al.*, 2018; Mor *et al.*, 2018; Bals *et al.*, 2019). Extensive economic and digestibility analyses suggest that a very large fraction of both beef cattle and dairy cattle nutritional requirements can be met with AFEX-treated animal feed depending on the age of the cattle (Mor *et al.*, 2018). Thus, AFEX-treated feedstocks might be viable alternatives to traditional forages and, with pelleting and subsequent blending, could help convert diverse biomass feedstocks into more uniform and salable commodities (Mor *et al.*, 2019).

By mass, the ruminant animal (e.g., dairy cow, beef cow, goat, sheep, and horse) feed market is approximately six times larger than the human food market in the US (Dale, 2017). Therefore, there is a large existing market for animal feeds that can undergird and help "jump-start" the development of a cellulosic bioenergy/bioproducts industry in the same way that the large established supply of corn grain helped in the rapid scale-up of the corn ethanol industry in the 20th century. Importantly, the ability to coproduce food (animal feed), fuel, and other products using AFEX-pretreated biomass also addresses the "food vs. fuel" controversy for related bioenergy systems (Muscat *et al.*, 2020).

Because of its relatively low capital investment costs, it is likely that AFEX can be practiced in widely distributed depots, located close to areas where the biomass is produced. These depots could be owned wholly or in part by farmers or other rural interests and thereby provide opportunities for rural economic growth. Because of physical and chemical changes in the biomass caused by AFEX, it is easy and inexpensive to produce stable, dense biomass pellets from AFEX-treated biomass (Dale, 2017). These dense pellets can be handled and stored using conventional grain handling equipment, thereby enabling supply chain development for cellulosic bioenergy (Kim *et al.*, 2019). Thus, AFEX does, in fact, produce a dense, stable, easily handled and stored, conversion-ready material like corn. AFEX was scaled up and studied thoroughly by MBI (Lansing, Michigan,

USA) to approximately a Technology Readiness Level (TRL) of 7, the stage just prior to full commercialization. However, it remains for another organization to fully commercialize AFEX.

The AFEX process has much to offer the cellulosic biofuels industry, including the potential to coproduce cellulosic ethanol and purified lignin streams as potential feedstocks for bio-based aromatics. Lignin may be the only means of providing renewable aromatics at scale. If so, it will be critical to generate these aromatic streams in a biorefinery where they are available at scale, relatively reactive, and "clean," i.e., suitable for further conversion and processing without extensive clean-up. The AFEX process does produce a clean lignin stream that can be separated, fractionated using a different EA process, and sold for different applications. Alternatively, lignin can be or catalytically converted to various aromatic compounds and displace fossil-derived aromatic molecules (Sousa *et al.*, 2016b).

10.9. Preliminary Depot Design Using the AFEX Process

As a possible solution to these logistical and scale challenges, we propose an alternative model incorporating Local Biomass Processing Depots (LBPDs but referred to simply as "depots" in this chapter) (Fig. 10.4) (Carolan *et al.*, 2007; Bals and Dale, 2012; Eranki *et al.*, 2013). The chief purpose of the depot is to process low bulk density, difficult-to-handle biomass materials into energy-dense, stable, shippable, storable intermediate products. Biomass is also subject to degradation and loss due to adverse weather events, both of which problems would be substantially lessened by depot-level processing to more stable and storable materials (Dale, 2017).

Depots are envisioned to be relatively small (100–1000 t/day) and are supplied by local biomass feedstocks but could have multiple markets for their products. Depots could market products to co-fired power plants, biorefineries, oil refineries, and animal feeding operations. For the purposes of this chapter, depots are decoupled biomass pretreatment, storage, and formatting facilities. Pretreatment is a

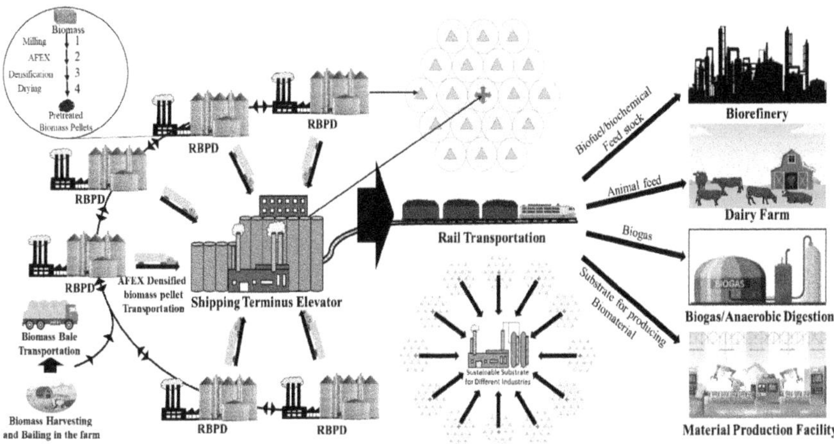

Figure 10.4. Local Biomass Processing Depot (LBPD) concept. Here, biomass that is locally produced is baled and transported to RBPD using trucks through roadways. Each LBPD can process biomass covering 10,000 acres of farmland. Four different unit operations are carried out in LBPD (1: milling, 2: AFEX pretreatment, 3: densification, and 4: drying). The densified AFEX-pretreated biomass is transported using shipping containers through roadways to the shipping terminus where the pellets are stored in a grain elevator. The densified AFEX-pretreated biomass is transported by railroad carts to different end users, such as biorefinery (producing fuels/chemicals), dairy farm (animal feed), anaerobic digestion (biogas), and material production facility (paper boards and composites). This LBPD concept can sustainably produce densified biomass and help solve the biomass logistic issues for end users.

process required for the biochemical conversion of biomass to sugars for biofuel production. In some cases, depots may also contain other technologies (e.g., protein recovery) that produce additional valuable products (Dale *et al.*, 2009; Eranki *et al.*, 2011; Bals and Dale, 2010) or produce biodiesel using oil seed or animal fat.

In this chapter, we emphasize depots employing four steps (biomass milling, AFEX pretreatment, subsequent densification, and densified biomass drying). The produced AFEX-densified biomass will be ready to be shipped for biological conversion of biomass sugars to fuels and animal feed (Fig. 10.4). However, other processing technologies might also serve as the basis for depots, including anaerobic digestion or pyrolysis or producing biomaterials (Eranki *et al.*, 2011). Anaerobic digestion produces a mixture of methane and carbon dioxide, which can be burned to produce biopower or processed to pipeline quality biomethane by removing the carbon dioxide (Rojas-Sossa *et al.*, 2019). Pyrolysis produces a liquid "bio-oil" as well as some gases and a solid called "biochar" (Sundaram *et al.*, 2017). The bio-oil could be stabilized and shipped as a liquid to existing oil refineries for subsequent upgrading to hydrocarbon fuels. Biocomposites could be produced by combining AFEX pretreated biomass with different polymers (Huda *et al.*, 2007). All depots, whether producing energy-dense, shippable solids, liquids, and/or gases, are likely to vary in capacity, operational characteristics, the feedstocks processed, and co-products produced, as dictated by the characteristics of the landscapes in which they are embedded (e.g., largely agricultural landscapes vs. heavily forested landscapes). The AFEX produced in different LBPDs will be transported by trucks to a large central shipping terminus where it will be stored and transported to biorefinery, dairy farm, AD processing, and biomaterial producing facility using rail transportation. Developing this biomass distribution system will also enable us to transport AFEX-treated biomass by large cargo ships to foreign countries.

Another significant problem with the current approach to supplying a large-scale biomass conversion industry is that those who manage the land (the farmers and forest owners) are not currently organized to supply cellulosic feedstocks to biorefineries. Without the active participation of farmers and foresters, we will not get the scale we need for a biomass processing industry to thrive. Without scale, the cost of biofuel will not decrease as it must. For example, the cost of cellulosic ethanol produced in a large quantity (30,000 t/day) based on pellets is at least $1.00 per gallon less than ethanol produced in a 3000 t/day biorefinery, all other things being equal (Kim and Dale, 2015b). As noted above, without scale and scalability of the system, the existing fuel refining industry is unlikely to bring its deep expertise and capital resources to bear on the problem of large-scale biofuels.

Interestingly, increased farmer participation is so important, so central to the success of this industry, that if farmers are paid more for their biomass, there may be little or no impact on the resulting cellulosic biofuel cost (Kim and Dale, 2015a, b). This is because increased farmer participation due to higher prices leads to shortened supply chains, with reduced transportation costs, and to larger biorefineries, with better economies of scale. The result is little or no effect on the overall cost to manufacture ethanol.

This surprising result reminds us that not all aspects of renewable bioenergy and bioproducts will be like their petroleum counterparts, particularly in the feedstock supply chains. The reality is that, unlike oil, cellulosic biomass does not handle, travel, or store well. It is bulky, combustible, and degradable. Also, like other solar-based resources, biomass is produced over extensive land areas. Biomass must therefore be aggregated at some significant scale before it can be used as a biofuel feedstock. The aggregation of cellulosic biomass at scale and all the associated issues have yet received comparatively little research and development attention. This is a critical failing of the current biofuels policy (Saravanan *et al.*, 2020; Khan *et al.*, 2020).

Raw cellulosic biomass has no large, well-established markets to help provide feedstocks for bioenergy production in the way that the large existing corn market enabled the rapid growth of the corn ethanol industry. In fact, an "ideal" cellulosic biomass would have properties like those of corn. An ideal cellulosic biomass would be dense, stable, storable, shippable, ready for conversion to fuels and other products and would also have alternative uses to help create scale. Can such a "corn-like" cellulosic biomass feedstock be produced? Yes, the AFEX process, does indeed produce such a dense, stable, storable, shippable, conversion-ready cellulosic material (Karki *et al.*, 2015).

10.10. Conclusion

If the cellulosic biofuel industry is to have a large, positive impact, there are critical issues of biomass aggregation, storage, transport, and handling that must be addressed and solved. However, comparatively little attention has been paid to such issues thus far. The AFEX process, practiced in widely dispersed local depots, offers real potential to address these issues because it (1) produces a dense, stable, storable, conversion-ready product, (2) does not generate waste streams, (3) requires relatively little capital investment, (4) generates a valuable animal feed useful to farmers and which addresses the "food versus fuel" issue, and therefore, (5) overall, gives farmers the opportunity and incentive to participate directly and profitably in the biofuel system so that the necessary scale and logistics can be achieved.

Acknowledgments

The research was also supported by the Office of Biological and Environmental Research, Office of Science, United States, Department of Energy through DOE Great Lakes Bioenergy Research Center (GLBRC) Grant DE-FC02-07ER64494. B.V. acknowledges the University of Houston High Priority Area Research Seed Grants and the State of Texas for startup funds.

References

Antar, M., Lyu, D., Nazari, M., Shah, A., Zhou, X., & Smith, D. L. (2021). Biomass for a sustainable bioeconomy: An overview of world biomass production and utilization. *Renewable Sustainable Energy Rev., 139*, 110691.

Balan, V., Bals, B., Chundawat, S. P. S., Marshall, D., & Dale, B. E. (2009a). Lignocellulosic biomass pretreatment using AFEX. In J. R. Mielenz (Ed.), *Biofuels: Methods and Protocols, Methods in Molecular Biology* (Vol. 581, pp. 61–77). Humana Press, a part of Springer Science + Business Media, LLC.

Balan, V., Sousa, L. D. C., Chundawat, S. P. S., Marshall, D., Sharma, L. N., Chambliss, K., & Dale, B. E. (2009b). Enzymatic digestibility and pretreatment degradation products for AFEX treated hardwoods (Populus nigra). *Biotechnol. Prog., 25*(2):365–375.

Balan, V., Bals, B., Souse, L. D. C., Garlock, R., & Dale, B. E. (2011). A short review on ammonia based lignocelluloses' biomass pretreatment. In Blake Simmons Royal Society of Chemistry (Ed.), *RSC Energy and Environment,* Series No. 4, Chemical and Biochemical Catalysis for Next Generation Biofuel (Chapter 5, pp. 89–114). Published by the Royal Society of Chemistry. www.rsc.org.

Balan, V., Dale, B. E., Chundawat, S. P. S., & Sousa, L. D. C. (2015). Method of pretreating biomass. United States Patent US8,968,515 B2.

Bals, B. & Dale, B. E. (2010). Economic comparison of multiple techniques for recovering leaf protein in biomass processing. *Biotechnol. Bioeng.*, *108*, 530–537.

Bals, B. D. & Dale, B. E. (2012). Developing a model for assessing biomass processing technologies within a local biomass processing depot. *Bioresour. Technol.*, *106*, 161–169.

Bals, B. D., Gunawan, C., Moore, J., Teymouri, F., & Dale, B. E. (2013). Enzymatic hydrolysis of pelletized AFEX-treated corn stover at high solid loadings. *Biotechnol. Bioeng.*, *111*, 264–271.

Bals, B., Teymouri, F., Haddad, D., Julian, W. A., Vismeh, R., Jones, A. D., Mor, P., Soest, B. V., Tyagi, A., VandeHaar, M., & Bringi, V. (2019). Presence of acetamide in milk and beef from cattle consuming AFEX-treated crop residues. *J. Agric. Food Chem.*, *67*, 10756–10763.

Berners-Lee, M., Kennelly. C., Watson, R., & Hewitt, C. B. Current global food production is sufficient to meet human nutritional needs in 2050 provided there is radical societal adaptation. *Elementa Sci. Anthropocene*, *6*, 52.

Blummel, M., Steele, B., & Dale, B. E. (2014). Opportunities from second-generation biofuel technologies for upgrading lignocellulosic biomass for livestock feed. *CAB Rev.*, *9*, 1–8.

Chundawat, S. P. S., Vismeh, R., Sharma, L. N., Humpula, J. F., Sousa, L. D., Chambliss, C. K., Jones. A. D., Balan, V., & Dale. B. E. (2010). Multifaceted characterization of cell wall decomposition products formed during ammonia fiber expansion (AFEX) and dilute acid-based pretreatments. *Bioresr. Technol.*, *101*, 8429–8438.

Chundawat, S. P. S., Donohoe, B. S., Sousa, L. D., Elder, T., Agarwal, U. P., Lu, F., Ralph, J., Himmel, M. E., Balan, V., & Dale, B. E. (2011a). Multi-scale visualization and characterization of lignocellulosic plant cell wall deconstruction during thermochemical pretreatment. *Energy Environ. Sci.*, *4*, 973-984.

Chundawat, S. P. S, Bellesia, G., Uppugundla, N., Sousa, L. D., Gao, D., Cheh, A., Agarwal, U. P., Bianchetti, C. M., Phillips, Jr. G. N., Langan, P., Balan, V., Gnanakaran, S., & Dale, B. E. (2011b). Restructuring crystalline cellulose hydrogen bond network enhances its de-polymerization rate. *J. Am. Chem. Soc.*, *133*, 11163–11174.

Chundawat, S. P. S., Bals, B., Campbell, T., Sousa, L., Gao, D., Jin, M., Eranki, P., Garlock, R., Teymouri, F., Balan, V., & Dale, B. E. (2013). Primer on ammonia fiber expansion pretreatment. C. Wyman (Ed.), Chapter 9 for the book titled *Aqueous Pretreatment of Plant Biomass for Biological and Chemical Conversion to Fuels and Chemicals, First Edition* (pp. 169–195). Wiley Series in Renewable Resources.

Campbell, T. J., Teymouri, F., Bals, B., Glassbrook, J., Nielson, C. D., and Videto, J. A. (2013). A packed bed Ammonia Fiber Expansion reactor system for pretreatment of agricultural residues at regional depots. *Biofuels*, *4*, 23–34.

Campbell, T. J. & Bals, B. (2018). Ammonia fiber expansion and its impact on subsequent densification and enzymatic conversion. In J. S. Tumuluru (Ed.),

Biomass Preprocessing and Pretreatments for Production of Biofuels — Mechanical, Chemical, and Thermal Methods (Chapter 16, pp. 437–456). CRC Press, Taylor and Francis Group.

Campbell, T.J., Bals, B., Teymouri, F., Glassbrook, J., Nielson, C., Videto, J., Rinard, A., Moore, J., Julian, A., & Bringi, V. (2019). Scale-up and operation of a pilot-scale ammonia fiber expansion reactor. *Biotechnol. Bioeng.*, *117*, 1241–1246.

Carolan, J. E., Joshi, S. V., & Dale, B. E. (2007). Technical and financial feasibility analysis of distributed bioprocessing using regional biomass pre-processing centers. *J. Agric. Food Ind. Organ.*, *5*, 1–27.

Chandel, A. K., Garlapati, V. K., Singh, A. K., Antunes, F. A. F., & da Silva, S. L. (2019). The path forward for lignocellulose biorefineries: Bottlenecks, solutions, and perspective on commercialization. *Bioresource Technol.*, *264*, 370–381.

Chen, H., Liu, J., Chang, X., Chen, D., Xue, Y., Liu, P., Lin, H., & Han, S. (2017). A review on the pretreatment of lignocellulose for high-value chemicals. *Fuel Process. Technol.*, *160*, 196–206.

Dale, B. E., Allen, M. S., Laser, M., & Lynd, L. R. (2009). Protein feeds coproduction in biomass conversion to fuels and chemicals. *Biofuels, Bioprod. Biorefin.*, *3*, 219–230.

Dale, B.E., Ritchie, B., & Marshall, D. (2013). Methods for producing and using densified bomass products containing pretreated biomass fibers. United States Patent Application Pub. No. US20130280762A1.

Dale, B. E. (2017). Feeding a sustainable chemical industry: do we have the bioproducts cart before the feedstocks horse? *Faraday Discuss.* doi: 10.1039/c7fd00173h.

Eranki, P. L., Bals, B. D., & Dale, B. E. (2011). Advanced regional biomass processing depots: A key to the logistical challenges of the cellulosic biofuel industry. *Biofuels, Bioprod. Biorefin.*, *5*, 621–630.

Eranki, P. L., Manowitz, D. H., Bals, B. D., Izaurralde, R. C., Kim, S., & Dale, B. E. (2013). The watershed-scale optimized and rearranged landscape design (WORLD) model and local biomass processing depots for sustainable biofuel production: Integrated life cycle assessments. *Biofuels, Bioprod. Biorefin.*, *7*, 537–550.

Garlock, R. J., Chundwat, S. P. S., Balan, V., & Dale, B. E. (2009). Optimizing harvest of corn stover fractions based on overall sugar yields following ammonia fiber expansion pretreatment and enzymatic hydrolysis. *Biotechnol. Biofuels*, *2*, 29.

Gaulin, N. & Le Billion, P. (2020). Climate change and fossil fuel production cuts: Assessing global supply-side constraints and policy implications. *Climate Policy*, *20*, 888–901.

Haldar, D. & Purkait, M. K. (2021). A review on the environment-friendly emerging techniques for pretreatment of lignocellulosic biomass: Mechanistic insight and advancements. *Chemosphere*, *264*, 128523.

Huda, M. S., Balan, V., Chundawat, S. P. S., Dale, B. E., Drzal, L. T., & Misra, M. (2007). Effect of Ammonia Fiber Expansion (AFEX) and silane treatments of corncob granules on the properties of renewable resource based bio composites. *J. Bio-based Mat. Bioenergy*, *1*, 127–136.

Jin, M., Sarks, C., Bals, B. D., Posawatz, N., Gunawan, C., Dale, B. E., & Balan, V. (2016). Toward high solids loading process for lignocellulosic biofuel production at a low cost. *Biotechnol. Bioeng.*, *114*(5), 980–989.

Khan S. A. R., Zhang, Y., Kumar, A., Zavadskas, E., & Stremikiene, D. (2020). Measuring the impact of renewable energy, public health expenditure, logistics, and environmental performance on sustainable economic growth. *Sustainable Dev.*, *28*, 833-843.

Karki, B., Muthukumarrappan, K., Wang, Y., Dale, B. E., Balan, V., Gibbson, W. R., & Karunanithy, C. (2015). Physical characteristics of AFEX-pretreated and densified switchgrass, prairie cord grass and corn stover. *Biomass Bioenergy*, *78*, 164–174.

Kim, S. & Dale, B. E. (2015a). Potential job creation in the cellulosic biofuel industry: The effect of feedstock price. *Biofuels, Bioprod. Biorefin.*, 9:639–647.

Kim, S. & Dale, B. E. (2015b). Comparing alternative cellulosic biomass biorefining systems: Centralized versus distributed processing systems. *Biomass Bioenergy*, *74*, 135–147.

Kim, S. & Dale, B. E. (2015c). All biomass is local: The cost, volume produced, and global warming impact of cellulosic biofuels depend strongly on logistics and local conditions. *Biofuels, Bioprod. Biorefin.*, *9*, 422–434.

Kim, S., Dale, B. E., Jin, M., Thelen, K. D., Zhang, X., Meier, P., Reddy, A. D., Jones, C. D., Izaurralde, R. C., Balan, V., Runge, T., & Sharara, M. (2019). Integration in depot-based decentralized biorefinery system: I corn stover cellulosic biofuel. *GCB Bioenergy*, *11*, 871–882.

Kubicek, C. P. Starr, T. L., & Glass, N. L. (2014). Plant cell wall–degrading enzymes and their secretion in plant-pathogenic fungi. *Annu. Rev. Phytopathol.*, *52*, 427–51.

Kumar, G., Dharmaraja, J., Arvinnarayan, S., Shoban, S., Bakoniyi, P., Saratale, G. D., Nemestothy, N., Belafi-Bako, K., Yoon, J.-J., & Kim, S.-H. (2019). A comprehensive review on thermochemical, biological, biochemical and hybrid conversion methods of bio-derived lignocellulosic molecules into renewable fuels. *Fuel*, *251*, 352–367.

Mor, P., Bals, B., Tyagi, A. K., Teymouri, F., Tyag, N., Kumar, S., Bringi, V., & VanderHaar, M. (2018). Effect of ammonia fiber expansion on the available energy content of wheat straw fed to lactating cattle and buffalo in India. *J. Dairy Sci.*, *101*, 7990–8003.

Mor, P., Bals, B., Kumar, S., Tyagi, N., Reen, J. K., Tyagi, B., Choudhury, P. K., & Tyagi, A. L. (2019). Influence of replacing concentrate mixture with AFEX pellets on rumen fermentation, blood profile and acetamide content in the rumen of crossbred (Alpine × Beetle) female goats. *Small Ruminant Res.*, *170*, 109–115.

Mokomele, T., Sousa, L. D. C., Bals, B., Balan V., Goosen, B., Dale, B. E., & Gorgens, J. F. (2018). Using steam explosion or AFEX to produce animal feeds and biofuel feedstocks in a biorefinery based on sugarcane residues. *Biofuels, Bioprod. Biorefin.*, *12*, 978–996.

Muscat, A., de Olde, E. M., de Boer, I. J. M., & Ripoll-Bosch, R. (2020). The battle for biomass: A systematic review of food-feed-fuel competition. *Global Food Secur.*, *25*, 100330.

Paritosh, K., Yadav, M., Kesharwani, N., Pareek, N., Karthikeyan, O. P., Balan, V., & Vivekanand, V. (2021). Strategies to improve solid state anaerobic bioconversion of lignocellulosic biomass: an overview. *Bioresour. Technol.*, *331*, 125036.

Rafiaani, P., Kuppens, T., Van Dael, M., Azadi, H., Lebailly, P., & Van Passel, S. (2018). Social sustainability assessments in the biobased economy: Towards a systemic approach. *Renewable Sustainable Energy Rev.*, 82, 1839–1853.

Rojas-Sossa, J. P., Zhong, Y., Valenti, F., Blackhurst, J., Marsh, T., Kirk, D., Fang, D., Dale, B. E., & Liao, W. (2019). Effects of ammonia fiber expansion (AFEX) treated corn stover on anaerobic microbes and corresponding digestion performance. *Biomass Bioenergy*, *127*, 105263.

Saravanan, A. P., Pugzahendhi, A., & Mathimani, T. (2020). A comprehensive assessment of biofuel policies in the BRICS nations: Implementation, blending target and gaps. *Fuel*, *272*, 117635.

Sarks, C., Bals, B. D., Wynn, J., Teymouri, F., Schwegmann, S., Sanders, K., Jin, M., Balan, V., & Dale, B. E. (2016). Scaling up and benchmarking of ethanol production from pelletized pilot scale AFEX treated corn stover using *Zymomonas mobilis* 8b. *Biofuels*, *7*, 253–262.

Sousa, L. D., Chundawat, S. P. S., Balan, V., & Dale, B. E. (2009). Cradle-to-grave - assessment of existing lignocellulose pretreatment technologies. *Curr. Opin. Biotechnol.*, *20*, 339–347.

Sousa, L. D. C., Jin, M., Chundawat, S., Bokade, V., Tang, X., Azarpira, A., Lu, F., Avci, U., Humpula, J., Uppugundla, N., Cheh. A., Kothari, N., Kumar, R., Ralph, J., Hahn, M. G., Wyman, C. E., Singh, S., Simmons, B. A., Dale, B. E., & Balan. V. (2016a). Next-generation ammonia pretreatment enhances biofuel production from biomass via simultaneous cellulose de-crystallization and lignin extraction. *Energy Environ. Sci.*, *9*, 1215–1223.

Sousa, L. D. C., Foston, M., Bokade, V., Azarpira A, Lu, F., Ragauskas, A. J., Ralph, J., Dale, B., & Balan, V. (2016b). Isolation and characterization of new lignin streams derived from Extractive Ammonia (EA) pretreatment. *Green Chem.*, *18*, 4205–4215.

Stanchin, H., Mikulcic, H., Wang, X., & Duic, N. (2020). A review on alternative fuels in future energy system. *Renewable Sustainable Energy Rev.*, *128*, 109927.

Sundaram, V., Muthukumarappa, K., & Gent, S. (2017). Understanding the impacts of AFEX pretreatment and densification on the fast pyrolysis of corn stover, prairie cord grass, and switchgrass. *Appl. Biochem. Biotechnol.*, *181*, 1060–1079.

Zhao, C., Shao, Q., & Chundawat, S. P. S. (2020). Recent advances on ammonia-based pretreatments of lignocellulosic biomass. *Bioresour. Technol.*, *298*, 122446.

Zhou, X., Zhou, X., Liu, G., Xu, Y., & Balan, V. (2018). Integrated production of gluconic acid and xylonic acid using dilute acid pretreated corn stover by two-stage fermentation. *Biochem. Eng. J.*, *137*, 18–22.

https://doi.org/10.1142/9781800613799_0011

Chapter 11

A Comprehensive Evaluation of the Various Routes of Biochemical Conversion of Pelletized Biomass

Anindita Paul*, Ankita Juneja†, Jaya Shankar Tumuluru‡, and Deepak Kumar*,§

**Department of Chemical Engineering State University of New York-Environmental Science and Forestry Syracuse, New York, USA*

†Department of Biology, Syracuse University, Syracuse, New York, USA

‡Southwestern Cotton Ginning Laboratory United States Department of Agriculture Agriculture Research Service, Las Cruces, New Mexico, USA

§dkumar02@esf.edu

Abstract

Lignocellulosic biomass (LCB) represents abundant biomass that is economically feasible and environmentally sustainable to produce biofuels and bioproducts. Despite these lucrative advantages, LCB faces logistical and technological challenges in its utilization. Low bulk density and high moisture content hinder the biomass's efficient transportation and storage. Densification has become an essential step for efficient transport, storage, and utilization of biomass, pelletization being one of the most common densification methods. Pellets (and other densification techniques) have been studied in great detail for the thermochemical conversion of biomass. However, the extent of knowledge of its effect on sugar yields during biochemical conversions is limited. This chapter summarizes the densification methods used for biomass for biochemical conversion and their possible merits and limitations.

Keywords: Lignocellulosic biomass, densification, biochemical conversion, hydrolysis, fermentation

11.1. Introduction

The role of bio-based, renewable resources in the context of bridging the gap between demand and supply of global energy and consumables has been tremendously impactful and steadily increasing. Biomass is one of the most abundant and sustainable bioresources on the planet. Various biomass feedstocks such as agricultural crop residues, algae, forestry residues, and wood processing residues are the major sources for bioenergy production and have successfully contributed 14% to the global primary energy supply (Parikka, 2004). In the last decade, lignocellulosic biomass (LCB) has gained much attention as a feedstock for bioenergy, biomaterials, and biochemicals, with major advantages coming from the availability and low cost of the feedstock. Due to its economic feasibility and environmental friendliness, lignocellulosic biomass represents one of the most economical, sustainable, and largest feedstocks for bio-based energy production.

Despite the lucrative advantages of biomass-based biofuel production, there are still challenges in the utilization of biomass, both logistical and technological. The first step in producing bio-based energy is harvesting and transporting biomass to the biorefinery, followed by storage until utilized. Due to their inherent characteristics of low bulk density and high moisture content, LCB faces challenges of the high cost of transportation and storage (Malladi and Sowlati, 2018) and being prone to microbial attack during storage (Yuan *et al.*, 2022). Microbial attack due to high moisture content, especially in humid environments, causes biomass degradation, leading to decreased feedstock quality for conversion technologies. In addition to increasing the logistical costs, the low bulk density of biomass also leads to low productivity during pretreatment, which is one of the most energy-intensive and high capital cost operations.

To augment biomass utilization efficiency, densification is an essential step before its use. Biomass densification increases the mass and energy density by using a relatively small amount of energy, reducing the cost of transportation and storage, and can also make the biomass more efficient by improving the feedstock properties. Densification is fundamentally done to increase the bulk density of the biomass [from about 40–200 kg/m^3 up to 600–800 kg/m^3 (Mani *et al.*, 2003)] and provide regular shape to the particles. Various approaches to densification include cubing, screw extrusion, pellet milling, roller pressing, briquette pressing, tablet pressing, and tumble agglomeration (Tumuluru *et al.*, 2011). Pelletization is one of the most commonly used densification methods, making uniform-sized pellets that are stable, both mechanically and biologically. Pelletization enhances the thermal efficiency of the biomass by enabling high conversion efficiency greater than 75% (Xiao *et al.*, 2015), reduces bulk transportation, handling, and storage cost, and eases the automatic feeding process in large-scale operations.

Biomass is used via various routes to be converted to bioenergy, majorly categorized as thermochemical and biochemical routes. Thermochemical conversion of biomass includes pyrolysis, torrefaction, and gasification, to produce liquid fuel such as pyrolysis oil, solid byproducts such as charcoal, and gaseous fuel, namely producer gas or yield heat via direct combustion (Prasad *et al.*, 2015). Most of the woody biomass and agricultural residues used for thermochemical conversions are available at a high moisture content of 30–60% (w.b.) and 20 + 35% (w.b.), respectively. Since the high moisture content of raw biomass is a limitation for thermochemical conversion technologies, the biomass needs to be dried to less than 10% moisture content to achieve better conversion efficiencies (Bridgwater and Peacocke, 2000). For this reason, densification has been used for a long time as a preprocessing step for thermochemical conversion technologies.

Although the variability of moisture content in the feedstock is not a concern in biochemical conversion technologies, the particle size of the feedstock is an important parameter affecting energy consumption and conversion efficiency. In a study by Yang *et al.*

(2018), the authors observed an overall biomass particle size of 2 mm for wheat straw and big blue stem biomass conversion throughout the multi-step bioconversion involving pelletization followed by enzymatic hydrolysis yielded the highest sugars (Yang *et al.*, 2018). Other studies with wheat straw, bagasse, corn stover, and poplar wood masses have shown that the size reduction of particles to a range of 3.2–6.5 mm enhanced pellet tensile strength, density, and durability (Theerarattananoon *et al.*, 2011; Pradhan *et al.*, 2018a). However, the low bulk density of the biomass can also impact the overall process due to feed handling and storage issues. Densification, especially using a pellet mill or briquette press, can be performed on biomass to overcome these limitations. However, the studies on the impact of densification on subsequent sugar yields in biochemical conversion are limited. This chapter aims to study the densification processes used for the biomass (Fig. 11.1) and broad scope and the developments in the various methodologies of biochemical conversion of pelletized biomass while also endeavoring to understand the possible merits and the limitations of the processes.

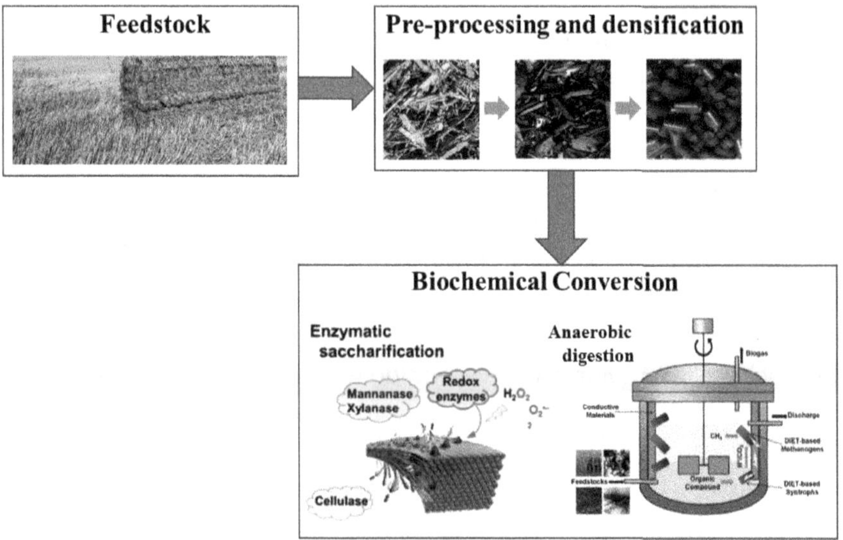

Figure 11.1. Schematic representation of the broad scope of biochemical conversion of pelletized biomass.

11.2. Densification Technologies

Biomass is difficult to use for fuel in large-scale applications in its original form as it is bulky, wet, dispersed, and difficult to transport (Tumuluru *et al.*, 2010a). The low bulk density of biomass is typically 150–200 kg/m^3 for woody biomass (Sahoo *et al.*, 2018) and 80–100 kg/m^3 for herbaceous biomass (Tumuluru *et al.*, 2011), limiting its application at the commercial scale. According to Sahoo *et al.* (2018), the low bulk densities make biomass material difficult to store, transport, and interface with biorefinery infeed systems. Densifying biomass using a mechanical system helps overcome this limitation. According to Tumuluru *et al.* (2011), the densification process is critical for producing a feedstock material suitable as a commodity product. Densification helps overcome physical property variability issues, such as moisture, particle size distribution, and density. Densified biomass has improved handling and conveyance efficiencies throughout the supply system, biorefinery infeed, and feedstock uniformity and density. Common biomass densification systems that are widely used in biomass processing are pellet mill, briquette press, and screw extruder (Tumuluru *et al.*, 2011; Pradhan *et al.*, 2018b). The densification technologies are also called binder technologies as they use high compaction pressure, temperature, and moisture to form a densified product (Tumuluru *et al.*, 2011). Many researchers have studied woody, herbaceous biomass and distiller's dried grains with solubles densification characteristics (Tumuluru, 2014, 2019; Tumuluru *et al.*, 2010b; Tabil Jr and Sokhansanj, 1996; Ndiema *et al.*, 2002; Li and Liu, 2000).

11.2.1. *Screw Press*

The densification of biomass using an extruder involves bringing smaller particles closer so that they bind. The biomass is subjected to mixing, shearing, and frictional heat developed in the extruder barrel. After traversing through the extruder barrel, the biomass is exited through the narrow-constricted die hole. The screw extrusion process involves the following mechanism: First, the biomass is partially compressed in the compression zone, forms a loosely packed biomass, and requires maximum energy. After the initial compression, the material becomes soft due to shearing mixing and frictional heat developed in the die. This changes the elastic properties of the biomass and increases the surface area of the particles and interparticle contact, which helps form the local bridges. After this biomass material transformation, the material is extruded from the narrow-constricted die, creating a densified product such as a pellet or a log.

11.2.2. *Briquette Press*

Briquetting using a piston press is a viable solution to densify larger particle-sized materials. The various briquetting systems used to produce densified products are hydraulic, mechanical, and roller presses. In the briquetting process, the biomass is subjected to high pressure. Due to high pressure and temperature during the briquetting process, the biomass particles bind to form a briquette due to the thermoplastic flow of the lignin. Among the various biomass components, lignin is considered a natural binder that gets activated at pressures, temperatures, and moistures resulting in the formation of high-density briquettes

Compared to pellet mills, the briquette press can process larger particles and moistures and form a densified biomass (Tumuluru and Fillerup, 2020). Briquetting increases the calorific value of the biomass which improves combustion characteristics, reduces particulate emissions, and has a uniform size and shape. In addition, the increase in the bulk density of herbaceous biomass by about four times compared to raw biomass makes briquettes easy to transport. The briquette's bulk density ranges from 350 kg/m^3 to 450 kg/m^3 (Tumuluru *et al.*, 2010). The briquettes produced are typically used as clean and green fuels and are used in furnaces and boilers. There are two types of briquette presses (a) hydraulic briquette press and (b) mechanical briquette press (Tumuluru *et al.*, 2010). Mechanical briquettes presses are typically used for commercial-scale production (Tumuluru *et al.*, 2010). Briquettes are also produced using a roller press. The material is compressed between the two counter-rotating rolls in a roller press. The briquette quality depends on the diameter of the rollers, the gap between the rollers, the roller's force, and the die's shape (Tumuluru *et al.*, 2010).

11.2.3. *Pellet Mill*

Pelleting process is similar to the briquetting and extrusion process, where the raw material is compressed and extruded through a narrow die. The major difference between briquettes and pellets is the size and density. Pellets are smaller than briquettes and have a higher bulk density (typically between 700 kg/m^3 and 750 kg/m^3). There are two types of pellet mills: (a) flat die pellet mill and (b) ring die pellet mill. Before feeding the biomass into the die, it is subjected to steam conditioning. The steam conditioning of the biomass helps activate the biomass components such as lignin, protein, and other water-soluble starches and improves the pelleting characteristics of the biomass. In general, pellet mill consists of a die and two rollers mounted on it. In the pelleting process, either the die or the rollers are rotated. As a result, the biomass is further forced through the die holes, resulting in pellet formation. Pilot-scale pellet mills are available in the range of 0.5–1 t/h, and commercial-scale ones are available in 5–10 t/h throughput.

Processing the material using a briquette press is more efficient than using a pellet mill. The material does not need to be ground to smaller particle size and does not need to be dried to lower moisture content by 10% (w.b.). Also, the briquette presses are easy to be deployed on the field. The briquettes press is less capital intensive and requires a lower operating cost. Briquettes are used for specific applications, whereas pellets are used at a commercial scale for power generation. In terms of transportation and handling, pellets are more efficient and easier to transport. Tumuluru (2014, 2015, 2016) pelleted biomass at higher moisture contents, eliminating the biomass's initial drying from 30 to 10% (w.b.) using a high-temperature rotary dryer (Tumuluru, 2014, 2015, 2016), which makes pelleting deployable in the field. This process is also called a high moisture pelleting process. As shown in Fig. 11.1, the high moisture pelleting process can accommodate a bigger hammer mill grind size

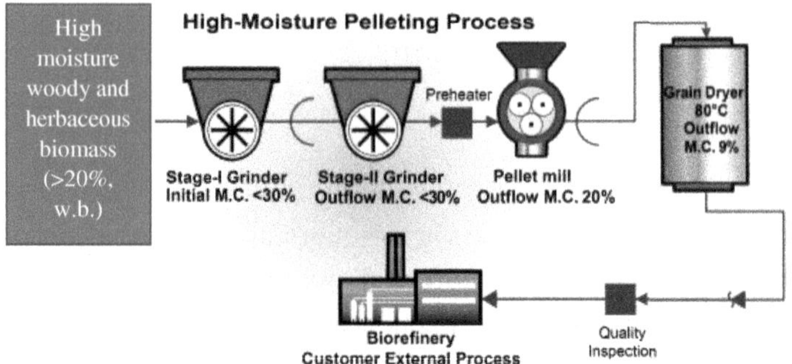

Figure 11.2. High moisture pelleting process (Lamers *et al.*, 2015).

of about 7/16-in. screen size. This process has helped produce pellets with a density of about 400–600 kg/m^3 and at a lower pelleting cost (50% lower than the conventional method followed by the industry) (Lamers *et al.*, 2015; Tumuluru, 2016, 2018). Figure 11.2 gives the high moisture pelleting process developed at the Idaho National Laboratory.

11.3. Impact of Pelletization on Biochemical Conversion Yields

Biomass is exposed to increased temperatures (~100°C) and pressures (~4000 psi) during the densification process, leading to structural changes that can positively impact the conversion performance. Pelleting is one of the most common techniques for biomass densification aimed at biochemical conversion. Some of the densification techniques also act as pretreatment before the biochemical conversion of biomass. Densifying the biomass may include various steps acting as preprocessing, such as washing, leaching, torrefaction, and co-processing, followed by size reduction using chopping, grinding, shredding, or steam explosion (Bajwa *et al.*, 2018).

11.3.1. *Preprocessing of Biomass*

One of the preprocessing methods before pelletization is torrefaction. Torrefaction is a slow pretreatment process where biomass is heated to 200–300°C in an inert environment (Tumuluru *et al.*,

2011). The major decomposition reactions during torrefaction impact hemicelluloses, whereas lignin and cellulose are affected to a lesser degree. The structure of the lignocellulosic biomass is modified as the available lignin sites increase, the hemicellulose matrix is broken down, and the unsaturated fatty structures are formed. The resulting biomass has increased bulk density (750–850 kg/m^3) and energy density (20 GJ/m^3) (Lipinsky *et al.*, 2002).

Another preprocessing technique used before pelleting the biomass is ammonia fiber explosion (AFEX). The process uses aqueous ammonia at elevated temperatures and pressures that recrystallizes cellulose, removes some hemicellulose, and reduces lignin (Teymouri *et al.*, 2005). Since more lignin sites are available for binding, densification characteristics are improved after the AFEX process. In addition, biomass degradation leads to higher hydrolysis yields during processing.

Various biological conversion methods have also been explored through which the feedstock can be made more tenable for or well suited to the pelletization process. For instance, pre-conditioning of the biomass by the deconstruction of lignocellulose recalcitrance through the application of microorganisms producing various lignocellulosic enzymes greatly improves the physical quality of the single fibers, which can then be densified through pelletization. As the lignin fragments become accessible through the degradation process, the chances of interlocking of fibers through the lignin matrix and the improvement of subsequent flowability of the biomass are greatly enhanced (Canam *et al.*, 2011). The said objective is achieved by utilizing fungal biomasses that produce lignin-altering enzymes and subjecting the biomass to treatments such as solid-state fermentation (SSF). Solid-state fermentation is advantageous in various ways. As the process is carried out by growing filamentous fungi over moderately moist biomass mats or sheets, it needs sparse liquid media for fermentation and doesn't complicate the moisture profile of the would-be pellets. Furthermore, most of the lignocellulosic enzyme-yielding fungi are filamentous, which would prefer biomass in solid forms for anchorage. Unlike the liquid fermentation process, SSF is advantageous in the ways of less energy usage, moderate capital investment, less moisture intensive, less aseptic environment maintenance, higher solid loading, more volumetric productivity, etc. The study led by Gao *et al.* (2015) aimed to optimize biological pretreatment conditions that positively affect the quality of wheat straw pellets (Gao *et al.*, 2017). Conducting a multi-lateral study of solid-state fermentation of wheat straw by *Trametes versicolor 52J (TV52J)*, *T. versicolor m4D (TVm4D)*, and *Phanerochaete chrysosporium (PC)*, they were able to show that fermentation incubation time, moisture content (%), and temperature greatly impacted and enhanced the tensile strength, dimensional stability, and pellet density of wheat straw, the results of which have been listed in Table 11.1. Another study led by Guo *et al.* (2015) by applying solid-state fermentation on sawdust of *Castanopsis fissa* mulberry wood chips before briquette formation showed positive amendment in feedstock properties, such as density, water resistance strength, and Meyer's strength (Guo *et al.*, 2015). Fungal species, *P. chrysosporium*, readily metabolized hemicellulose and degraded lignin with their enzymatic complexes like manganese peroxidase, lignin peroxidase, laccase, and versatile peroxidase.

11.3.2. *Enzymatic Hydrolysis and Fermentation*

Enzymatic hydrolysis and subsequent fermentation stand to be one the most widely adopted techniques of biochemical conversion of biomass into several bio-based products such as platform chemicals like bioethanol, xylitol, sorbitol, esters such as poly-hydroxy alkanoates (PHAs) and poly-hydroxybutyrate (PHBs) or widely known as commercial bioplastics, polylactide nanocomposites, or PLA (Hou *et al.*, 2021; Awasthi *et al.*, 2022), nutritionally important compounds like sugars and pigments. Enzymatic hydrolysis involves the application of hydrolytic enzyme complexes such as cellulase and hemicellulase to break down the polymeric chains of the structural polymers such as cellulose and hemicellulose into simple monomeric units such as glucose and xylose, respectively. The mechanism is extremely meticulous and targeted. Cellobiohydrolase I and II complex or exoglucanase, endoglucanase, and β-glucosidase

Table 11.1. Preprocessing technology involving biological treatment precluding pelletization.

Biological preprocessing technology	Feedstock	Process conditions	Observations	References
Solid-state fermentation	Wheat straw	Fungi such as *Trametes versicolor 52J (TV52J), T. versicolor m4D (TVm4D),* and *Phanerochaete chrysosporium (PC)* Temp.: 22–22.1°C Time: 21–35 days Moisture content: 70%	Highest density achieved with *TV52J* treatment: 1047 kg/m^3 Dimensional stability achieved with *PC* treatment: 1.060% Highest tensile strength with *TVm4d* treatment: 0.7054%	Gao *et al.* (2017)
Solid-state fermentation	Sawdust of *Castanopsis fissa* mulberry wood chips	Fungi such as *Phanerochaete chrysosporium (PC)* Fermentation time: 48 days	Water adsorption: 13.25% Density: 1070.88 kg/cm^3 Meyer hardness: 5.23 N/m^2	Guo *et al.* (2015)

enzymes are a class of hydrolytic enzymes that exhibit synergism or tandem-synchronized action and result in effective cellulose degradation. Endocellulases randomly bind to the glucose chain and hydrolyze single or multiple bonds. Accompanying that, exoglucanase (cellobiohydrolase or CBH) with its two subtypes, Type I and Type II, attacks from the reducing end of the chain and the non-reducing end of the glucose chain, respectively, thereby producing a cellobiose molecule (a dimer of glucose). Exoglucanase stays bound to the glucose chain and cleaves a cellobiose molecule until a minimum chain length is achieved. Most endoglucanases, however, are non-processive, with certain exceptions of types newly discovered in some bacteria.

Along with these complexes, certain micro-organisms also produce β-glucosidases which act on cellobiose dimers to yield glucose. Finally, in addition to these cellulolytic enzymes, glycoside hydrolase family 61 (GH61) will assist the enzymes in their actions and improve efficiency (Kumar and Murthy, 2013). Hemicellulose is a group of pentoses forming an intercalated matrix between the cellulose and lignin layers of biomass; it isn't a conventionally fermentable sugar moiety. However, for the benefit of the processing industry in terms of profitable product manufacturing like xylitol and sorbitol, efforts have been made to also hydrolyze hemicellulose sugars by application of hemicellulase enzymes for further downstreaming through fermentation by various engineered microbes for product yields. Hemicelluloses mainly comprise a diverse group of compounds such as D-xylose sugars (~25–50% of dry weight), uronic acids, and some hexoses with a backbone of 1,4-β-linkage. The chains can be linear or branched in fashion, with acetyl and methyl groups attached to the branches, thereby forming a very heterogenous complex structure. The hemicellulose group of holoenzymes is thus either glycoside hydrolases or carbohydrate esterases which break down the glycosidic backbone or acetate/methyl side chain groups. Thus, endo-β-1,4-xylanases can break down β-1,4 glycosidic bonds in the xylan backbone producing xylo-oligomers, while exo-xylanases further degrade the β-1,4 linkages in these xylo-oligomers to form xylo-oligosaccharides or xylobiose. Finally, β-xylosidases work on these exo-xylanase derivatives to produce xylose units. These C6-C5 sugar-producing enzyme cocktails form the basis of the enzymatic hydrolysis and fermentation operation (Álvarez *et al.*, 2016).

Enzymatic hydrolysis far supersedes other forms of conversion of pelletized biomass in the respect that it is a greener, cleaner, and safe mode of operation, as has been discussed previously. However, certain considerations come into play while designing the optimum enzymatic hydrolysis and fermentation process. To begin with, the type of pretreatment of lignocellulosic, pelleted biomass influences the hydrolysis process and the subsequent fermentation. This is because the formation of inhibitory substances due to the reduction of lignin, such as furans like hydroxymethylfurfural (HMFs) and levulinic acid, can negatively impact biochemical conversions (Takada *et al.*, 2020). These can form an intense interconnective matrix through biomass densification through various research into enzymatic hydrolysis that supports pelleted biomass utilization. However, enzyme inhibition through the substrate and the cost of the process pose considerable bottlenecks in the process.

Several forms of pretreatment have preceded enzymatic hydrolysis to aid the easy accessibility of structural sugars for saccharification. However, studies involving hydrolytic breakdown of pretreated and pelletized biomass indicate mixed results. For instance, Guragain *et al.* (2013) evaluated ethanol production from a range of lignocellulosic feedstocks such as wheat straw, corn stover, big bluestem, and sorghum stalk in both pelleted and unpelleted forms post alkali pretreatment. They found that while there were no significant statistical differences in the ethanol yield from both applied methods, the volumetric hydrolytic efficiency increased by nearly 12–23% across the various feedstocks in pelletized samples compared to the unpelleted form (Guragain *et al.*, 2013). It is inferred that pelletization can enhance the hydrolysis process while uncompromising the quality of the fermentable sugars. Especially from bluestem and sorghum biomass, pelletization had a positive effect on ethanol product yield. Delignification of each of the biomass, followed by lignin linkage in the pellet matrix, might have attributed to the positive results achieved in the study. Thus, pelletization can also lead to a reduction in enzyme loading. Another study investigated pelleted and ground corn stover for ethanol production with pretreatment followed by simultaneous saccharification and fermentation ray (Ray *et al.*, 2013). It was observed that pellets achieved about 84% of theoretical ethanol yield, 23.6% higher than ground biomass. Monomeric xylose yields for pelleted biomass were 58% higher than the ground biomass. Pelleting did not increase the recalcitrance of the biomass for pretreatment but enhanced the sugars and ethanol yields.

Similar studies led by Li *et al.* (2014) with dilute sodium hydroxide pretreatment and biomass pelletization corroborated the same views regarding hydrolysis, with the sugar conversion rate increasing by 5.5% when compared to its unpelleted counterpart (Li *et al.*, 2014). In the case of advanced biofuel iso-pentanol production from a mixture of biomass such as corn stover, switchgrass, lodgepole pine, and eucalyptus (1:1:1:1 on a dry weight basis) after pelletization and pretreatment with three distinct methods of ionic liquid (IL), dilute sulfuric acid (SA), and aqueous ammonia soaking (SAA) in parallel, results mostly leaned towards IL-treated pellets of mixed feedstock undergoing simultaneous saccharification and fermentation (SSF) by *E. coli*. Sugar recovery in the case of IL pellet pretreatment was 101.1% for glucan and 91.5% for xylan. Hydrolysis efficiency was nearly 95%, and isopentenol fermentation outperformed the SA and SAA treatments. This indicates that pelletized biomass is optimal and compatible with improved fermentation techniques like SSF (Shi *et al.*, 2015). The severity of aqueous ammonia pretreatment was observed to reduce similar glucose yields with pellets of corn stover in comparison to loose biomass (Pandey *et al.*, 2019). It was also found that with similar conditions of pretreatment and hydrolysis, pelleted biomass needed 80% less enzyme and saved 58% of hydrolysis time to obtain 90% glucose yields.

Wu *et al.* (2022) examined the effects of steam pretreatment and pelletization on enzymatic hydrolysis efficiency and came up with a few mechanical refinements and chemical steps to make the pretreated pellets more efficient for the hydrolysis process (Wu *et al.*, 2022). It was discovered that steam pretreatment increased the durability of the pellets by facilitating the relocation of lignin particles on the exposed surfaces of the fiber and establishing a bridging matrix between fibers. However, the excessive drying up of the fibers hindered the hydrolysis process. Incorporating a chemical reaction like alkaline deacetylation or neutral sulfonation step before pre-steaming followed by mechanical refining of the steamed pellets cumulatively improved the hydrolytic efficiency of the process greatly, more than 90% in the case of aspen wood.

Ammonia fiber explosion treatment (AFEX), coupled with ultrasound microwave treatment or without it, mostly gave neutral to positive results for hydrolytic conversion yields of sugar. In the case of corn stover, Hoover *et al.* (2014) observed that sugar yields of AFEX-treated stover improved as particle size decreased during pelletization and increased mixing during high solid loadings with a minor negative effect of hornification (Hoover *et al.*, 2014). Shi *et al.* (2013) opined that the pressure and heat generated from pelleting are a form of thermochemical pretreatment (Shi *et al.*, 2013). Again, thermal softening or plasticization of lignin during pelletization and shear stress processes like grinding and extrusion are great contributory factors.

Future endeavors in product development through enzymatic hydrolysis of pelleted biomass can be envisioned in the production of bio-engineered plastic polymers like PHAs by the application of fermentation of biomass by engineered microbes such as *Pseudomonas citronellolis* on lignocellulosic wastes like apple pomace (Liu *et al.*, 2021). Results of various studies utilizing pellets for enzymatic hydrolysis are summarized in Table 11.2.

11.3.3. *Anaerobic Digestion*

Anaerobic digestion is a very versatile process of utilization of biomass by channelizing the feedstock towards the production of calorific biogas, a mixture of methane and carbon dioxide, by the metabolic action of micro-organisms in the absence of air (oxygen). The main product consists of 50–75% methane gas and 20–25% carbon dioxide (Sánchez-Muñoz *et al.*, 2020), along with the release of other volatile organic matters such as hydrogen sulfide, ammonia, hydrogen, and water vapor. In certain cases, some residual organic acids like butyric acid and acetic acid from the acetogenesis step can also accumulate due to the action of homoacetogens (Pan *et al.*, 2021). The quality of biogas produced from the feedstock by methanogenic microbes, especially in terms of heating value and composition, greatly depends on the nature of the feedstock used and

Table 11.2. Biochemical conversions through enzymatic hydrolysis.

Biochemical conversion process	Feedstock	Process conditions	Highlights	References
Enzymatic hydrolysis	Wheat straw, corn stover, blue stem, sorghum stalk	Alkali pretreatment Pelleted and unpelleted biomasses taken Hydrolysis done with 5.4% Cellic CTec2 and 0.6% Cellic HTec2 enzymes (6% loading) for 48 h at 50°C and 150 rpm shaking Solid loading 5% Fermentation for 24 h	Boost in mass recovery through alkali pretreatment increased by 14%, 11%, 2%, and 5% in unpelleted biomass Volumetric productivity of enzymatic hydrolysis in alkali pretreated pelleted biomass increased by 14%, 11%, 2%, and 5% Ethanol yields were similar	Guragain *et al.* (2013)

Enzymatic hydrolysis	Corn leaves and stalks	NaOH pretreatment Hydrothermal pretreatment; both pelleted and unpelleted biomasses taken Hydrolysis conditions: cellulase (activity 700 FPU/g at 20 FPU/g glucan loading) and cellobiase (activity 240 IU/g at 40 IU/glucan loading), temp: 50°C Solid loading 2% pre-treated biomass, time: 72 h, speed: 200 rpm	Both pretreatment suited pelletized biomass hydrolysis Dil. NaOH treatment increased sugar conversion rate by 5.5% w.r.t. unpelleted biomass Hydrothermal treatment increased rate by 7%	Li *et al.* (2014)

(*Continued*)

Table 11.2. *(Continued)*

Biochemical conversion process	Feedstock	Process conditions	Highlights	References
Enzymatic hydrolysis	Switchgrass	Pelletization employed to form pellets of durability 95% and bulk density 724 kg/m^3 Both dilute acid and alkali pretreatments conducted Hydrolysis at 1% solid loading, temp: 50°C, time: 72 h, speed: 130 rpm, cellulase, cellobiase, and xylanase loading of 25 FPU/g, 50 CBU/g, and 3500 IU/g glucan, respectively.	Pelleting increased glucose yield by 37% in aqueous ammonia treatment Xylose yields higher by 42% in the same pretreatment	Rijal *et al.* (2012)

Enzymatic hydrolysis	Corn stover	Ammonia fiber expansion treatment Hydrolysis conditions: 20% solid loading, Ctec3 and Htec3 enzyme of 10 mg protein per g glucan, temp: 50°C, speed: 200 rpm, time: 72 h, pH: 5	AFEX pellets of bulk density 588–634 kg/m^3, durability >97.5%, and dimensions 4 mm AFEX corn stover at 40 Hz produced best results of 82% glucose yields and 67% xylose yields	Hoover *et al.* (2014)

its physical properties such as the C:N ratio, the complexity of the biomass structure, the kind of microorganism consortia used, and the anaerobic digestion conditions. While pondering upon the physical nature of the feedstock preluding the anaerobic digestion process, it is interesting to think about the feasibility of biochemical conversions of pelletized biomass through the said process. Certainly, a densified biomass of nearly 788–1296 kg/m^3 (Shaw *et al.*, 2009) for feedstocks such as poplar and wheat straw will not only be beneficial for logistics, transportation, and storage purpose in industries involved in biogas manufacturing but also make the anaerobic digestion experimental setup more compact with increased throughput.

Another significant impact of biomass densification is the formation of solid bridges between the biomass fibers leading to a matrix. Chemical reactions develop these solid bridges, for example, the degradation of lignin and hemicellulose with their subsequent hardening via crystallization (Tumuluru *et al.*, 2011). The moisture present in the biomass gets converted into steam under high pressure and temperature, which leads to the breakdown of the hemicellulose and lignin into lower intermediates or monomers, lignin products, etc. (Li *et al.*, 2018). Thus, the results vary from the traditional approach while utilizing pelletized biomass for anaerobic digestion.

Wang *et al.* (2016) utilized densified corn stover, both in the form of relatively smaller dimensioned pellets and larger-sized briquettes, for conducting anaerobic digestion to draw a comparative analysis of biogas yield and economic feasibility of densification with the traditional approach of producing biogas from non-densified corn stover (Wang *et al.*, 2016). The yield results obtained from anaerobic digestion of densified biomass were mostly in favor of pelletization and briquetting. The biogas production from pelletized biomass was 25.8% higher than the traditional method, with methane production exceeding its conventional counterpart by 50.6 mL/g volatile solids (VS). However, the pelletized biomass had a slightly smaller number of structural sugars and lignin. With densification, the organic loading of feedstock could also be increased to 80 g total solids (TS) per kg. Regarding the economic feasibility of the logistics of the densified corn stover, a nearly one-third reduction in costs was estimated. The main reasons behind the successful application of

Table 11.3. Biochemical conversions through anaerobic digestion.

Biochemical conversion process	Feedstock	Process conditions	Observations	References
Anaerobic digestion	Pelletized corn stover	Batch digestion: Volume: 1 L Pelleted biomass moisture: 15–30% Diameter: 10 mm Length: 15–20 mm Unit density: 1176 kg/m^3 Inoculum conditions: 15 g TS sludge seed (incubated for 15 days at mesophilic temp.) A.D. time: 46 days Organic load: 30–80 TS/kg A.D. temp: 35°C ± 1.0°C	Biogas production of pelleted biomass: 349.9 mL/g VS Methane production from pelleted biomass: 185.7 mL/g VS at 30% loading Economic feasibility: Harvest cost: 13.7% increase Transportation cost: 57.9% decrease Logistics cost: 36.1% decrease	Wang *et al.* (2016)
Anaerobic digestion	Pelletized vines and barley straw	Batch study: A.D. residence time: 70 days Pellet diameter: 6 mm	Methane production: 135% increase from traditional treatment VFA production: 64% increase from traditional treatment	Hidalgo *et al.* (2022)
Anaerobic digestate followed by combustion	Co-digestate of pig manure and corn silage	Co-digestate ratio (PM:CS) = 30:70 Temp.: 38°C A.D. residence time: 42 days	Digestate heating value: 13.065 MJ/kg ± 0.52 MJ/kg	Đurđević *et al.* (2018)

pelletized biomass in anaerobic digestion, as elucidated by the paper, were the following: certain degradation of susceptible hemicellulose chains due to high pressure–temperature conditions of pelletization and briquetting and certain lignin deconstruction and accessibility of the sugars to microbial enzymatic action and some free water retention by briquette biomass at higher organic loading.

Another study led by Hidalgo *et al.* (2022) by utilizing pelleted barley straw and vine shoots of 6 mm diameter showed the highest biogas production around 70 days at an estimate of 151 mL/g VS and 100 mL/g VS for vine shoots and straw, respectively (Hidalgo *et al.*, 2022). They were higher than the anaerobic digestion of untreated feedstock. The VFA production in pelletized barley peaked around day 7 of treatment at a value of 7.7 g VFA/L vs. 4.7 g VFA/L of untreated biomass on the same day. All these results greatly support pelletization for anaerobic biodegradability for biomass valorization. In Croatian biogas facilities, pelleted co-digestate, formed by anaerobic digestion of various organic waste like corn silage and pig manure, has shown to generate a heating value comparable to existing biofuels with an average value of 13.065 MJ/kg $\pm$0.52 MJ/kg (Đurđević *et al.*, 2018). Important observations of studies utilizing pellets for anaerobic digestion are summarized in Table 11.3.

11.4. Conclusion

Densification increases the bulk density of biomass, offsetting the high cost of biomass handling, storage, and transportation. Pelleting improves the bulk density of biomass from about 40–200 kg/m^3 to 600–800 kg/m^3 and has shown tremendous improvement in its utilization for large-scale applications. Being an essential preprocessing step, it is important to understand how it affects the conversion processes. This chapter covers the densification processes currently utilized, preprocessing of biomass before pelleting, and the impact of pelletization on biological conversion processes: enzymatic hydrolysis and anaerobic digestion. In contrast to the expectation of carbohydrate loss and increased resistance due to reduced moisture content and pressure exerted on densified biomass, various studies have shown an improvement in the conversion yields and efficiencies with pellets compared to their undensified form. In conclusion, pelleting of biomass offers low transportation and storage costs, minimized microbial contamination during storage, reduced energy consumption, and increased sugar yields during processing.

References

Álvarez, C., Reyes-Sosa, F. M., & Díez, B. (2016). Enzymatic hydrolysis of biomass from wood. *Microbial Biotechnol.*, *9*(2), 149–156.

Awasthi, M. K., *et al.* (2022). Recent trends and developments on integrated biochemical conversion process for valorization of dairy waste to value added bioproducts: A review. *Bioresour. Technol.*, *344*, 126193.

Bajwa, D. S., *et al.* (2018). A review of densified solid biomass for energy production. *Renewable Sustainable Energy Rev.*, *96*, 296–305.

Bridgwater, A. & Peacocke, G. (2000). Fast pyrolysis processes for biomass. *Renewable Sustainable Energy Rev.*, *4*(1), 1–73.

Canam, T., *et al.* (2011). Biological pretreatment with a cellobiose dehydrogenase-deficient strain of Trametes versicolor enhances the biofuel potential of canola straw. *Bioresour. Technol.*, *102*(21), 10020–10027.

Đurđević, D., Blecich, P. & Lenić, K. (2018). Energy potential of digestate produced by anaerobic digestion in biogas power plants: The case study of Croatia. *Environ. Eng. Sci.*, *35*(12), 1286–1293.

Gao, W., *et al.* (2017). Optimization of biological pretreatment to enhance the quality of wheat straw pellets. *Biomass Bioenergy*, *97*, 77–89.

Guo, J., *et al.* (2015). Effect of pretreatment by solid-state fermentation of sawdust on the pelletization and pellet's properties. *Sheng wu Gong Cheng xue bao= Chin. J. Biotechnol.*, *31*(10), 1449–1458.

Guragain, Y. N., *et al.* (2013). Evaluation of pelleting as a preprocessing step for effective biomass deconstruction and fermentation. *Biochem. Eng. J.*, *77*, 198–207.

Hidalgo, D., *et al.* (2022). Influence of cavitation, pelleting, extrusion and torrefaction pretreatments on anaerobic biodegradability of barley straw and vine shoots. *Chemosphere*, *289*, 133165.

Hoover, A. N., *et al.* (2014). Effect of pelleting process variables on physical properties and sugar yields of ammonia fiber expansion pretreated corn stover. *Bioresour. Technol.*, *164*, 128–135.

Hou, L., *et al.* (2021). Enhanced polyhydroxybutyrate production from acid whey through determination of process and metabolic limiting factors. *Bioresour. Technol.*, *342*, 125973.

Kumar, D. & Murthy, G. S. (2013). Stochastic molecular model of enzymatic hydrolysis of cellulose for ethanol production. *Biotechnol. Biofuels*, *6*(1), 1–20.

Lamers, P., *et al.* (2015). Techno-economic analysis of decentralized biomass processing depots. *Bioresour. Technol.*, *194*, 205–213.

Li, Y. & Liu, H. (2000). High-pressure densification of wood residues to form an upgraded fuel. *Biomass Bioenergy*, *19*(3), 177–186.

Li, Y., *et al.* (2014). Responses of biomass briquetting and pelleting to water-involved pretreatments and subsequent enzymatic hydrolysis. *Bioresour. Technol.*, *151*, 54–62.

Li, W., *et al.* (2018). Methane production through anaerobic digestion: Participation and digestion characteristics of cellulose, hemicellulose and lignin. *Appl. Energy*, *226*, 1219–1228.

Lipinsky, E. S., Arcate, J. R., & Reed, T. B. (2002). Enhanced wood fuels via torrefaction. *Fuel Chem. Div.*, *47*(1), 408–410. Preprints.

Liu, H., *et al.* (2021). Biopolymer poly-hydroxyalkanoates (PHA) production from apple industrial waste residues: A review. *Chemosphere*, *284*, 131427.

Malladi, K. T. & Sowlati, T. (2018). Biomass logistics: A review of important features, optimization modeling and the new trends. *Renewable Sustainable Energy Rev.*, *94*, 587–599.

Mani, S., Tabil, L., & Sokhansanj, S. (2003). Compaction of biomass grinds-an overview of compaction of biomass grinds. *Powder Handling Process.*, *15*(3), 160–168.

Ndiema, C., Manga, P., & Ruttoh, C. (2002). Influence of die pressure on relaxation characteristics of briquetted biomass. *Energy Convers. Manage.*, *43*(16), 2157–2161.

Pan, X., *et al.* (2021). Deep insights into the network of acetate metabolism in anaerobic digestion: Focusing on syntrophic acetate oxidation and homoacetogenesis. *Water Res.*, *190*, 116774.

Pandey, R., *et al.* (2019). Quantifying reductions in soaking in aqueous ammonia pretreatment severity and enzymatic hydrolysis conditions for corn stover pellets. *Bioresour. Technol. Rep.*, *7*, 100187.

Parikka, M. (2004). Global biomass fuel resources. *Biomass Bioenergy*, *27*(6), 613–620.

Pradhan, P., Mahajani, S. M., & Arora, A. (2018b). Production and utilization of fuel pellets from biomass: A review. *Fuel Process. Technol.*, *181*, 215–232.

Pradhan, P., Arora, A., & Mahajani, S. M. (2018a). Pilot scale evaluation of fuel pellets production from garden waste biomass. *Energy Sustainable Dev.*, *43*, 1–14.

Prasad, L., Subbarao, P., & Subrahmanyam, J. (2015). Experimental investigation on gasification characteristic of high lignin biomass (Pongamia shells). *Renewable Energy*, *80*, 415–423.

Ray, A. E., *et al.* (2013). Effect of pelleting on the recalcitrance and bioconversion of dilute-acid pretreated corn stover under low-and high-solids conditions. *Biofuels*, *4*(3), 271–284.

Rijal, B., *et al.* (2012). Combined effect of pelleting and pretreatment on enzymatic hydrolysis of switchgrass. *Bioresour. Technol.*, *116*, 36–41.

Sahoo, K., *et al.* (2019). Economic analysis of forest residues supply chain options to produce enhanced-quality feedstocks. *Biofuels, Bioprod. Biorefin.*, *13*(3), 514–534.

Sánchez-Muñoz, S. *et al.* (2020). Technological routes for biogas production: current status and future perspectives. In N. Balagurusamy &

A. K. Chandel (Eds.), *Biogas Production.* Cham: Springer. (https://link.springer.com/chapter/10.1007/978-3-030-58827-4_1).

Shaw, M., Karunakaran, C., & Tabil, L. (2009). Physicochemical characteristics of densified untreated and steam exploded poplar wood and wheat straw grinds. *Biosyst. Eng.*, *103*(2), 198–207.

Shi, J., *et al.* (2013). Impact of mixed feedstocks and feedstock densification on ionic liquid pretreatment efficiency. *Biofuels*, 4(1), 63–72.

Shi, J., *et al.* (2015). Impact of pretreatment technologies on saccharification and isopentenol fermentation of mixed lignocellulosic feedstocks. *Bioenergy Res.*, *8*(3), 1004–1013.

Tabil Jr., L. & Sokhansanj, S. (1996). Compression and compaction behavior of alfalfa grinds Part I: Compression behavior. *Powder Handling Process.*, *8*(1), 17–24.

Takada, M., *et al.* (2020). The influence of lignin on the effectiveness of using a chemithermomechanical pulping based process to pretreat softwood chips and pellets prior to enzymatic hydrolysis. *Bioresour. Technol.*, *302*, 122895.

Teymouri, F., *et al.* (2005). Optimization of the ammonia fiber explosion (AFEX) treatment parameters for enzymatic hydrolysis of corn stover. *Bioresour. Technol.*, *96*(18), 2014–2018.

Theerarattananoon, K., *et al.* (2011). Physical properties of pellets made from sorghum stalk, corn stover, wheat straw, and big bluestem. *Ind. Crops Prod.*, *33*(2), 325–332.

Tumuluru, J. S., *et al.* (2010b). Effect of process variables on the quality characteristics of pelleted wheat distiller's dried grains with solubles. *Biosyst. Eng.*, *105*(4), 466–475.

Tumuluru, J. S., *et al.* (2011). A review of biomass densification systems to develop uniform feedstock commodities for bioenergy application. *Biofuels, Bioprod. Biorefin.*, *5*(6), 683–707.

Tumuluru, J. S. (2014). Effect of process variables on the density and durability of the pellets made from high moisture corn stover. *Biosyst. Eng.*, *119*, 44–57.

Tumuluru, J. S. (2015). High moisture corn stover pelleting in a flat die pellet mill fitted with a 6 mm die: Physical properties and specific energy consumption. *Energy Sci. Eng.*, *3*(4), 327–341.

Tumuluru, J. S. (2016). Specific energy consumption and quality of wood pellets produced using high-moisture lodgepole pine grind in a flat die pellet mill. *Chem. Eng. Res. Design*, *110*, 82–97.

Tumuluru, J. S. (2018). Effect of pellet die diameter on density and durability of pellets made from high moisture woody and herbaceous biomass. *Carbon Resour. Convers.*, *1*(1), 44–54.

Tumuluru, J. S. (2019). Pelleting of pine and switchgrass blends: Effect of process variables and blend ratio on the pellet quality and energy consumption. *Energies*, *12*(7), 1198.

Tumuluru, J. S. & Fillerup, E. (2020). Briquetting characteristics of woody and herbaceous biomass blends: Impact on physical properties, chemical composition, and calorific value. *Biofuels, Bioprod. Biorefin.*, *14*(5), 1105–1124.

Tumuluru, J. S., Wright, C. T., Kenney, K. L., & Hess, J. R. (2010a). A review on biomass densification technologies for energy applications. [Online]. Tech. Report INL/EXT-10-18420, Idaho National Laboratory, Idaho Falls, Idaho, USA. Available at: https://inldigitallibrary.inl.gov/sites/sti/sti/4559449.pdf (Accessed 7 December 2022).

Wang, D., *et al.* (2016). Effects of biomass densification on anaerobic digestion for biogas production. *RSC Adv.*, *6*(94), 91748–91755.

Wu, J., *et al.* (2022). The use of steam pretreatment to enhance pellet durability and the enzyme-mediated hydrolysis of pellets to fermentable sugars. *Bioresour. Technol.*, 126731.

Xiao, Z., *et al.* (2015). Energy recovery and secondary pollutant emission from the combustion of co-pelletized fuel from municipal sewage sludge and wood sawdust. *Energy*, *91*, 441–450.

Yang, Y., Zhang, M., & Wang, D. (2018 April 9). A comprehensive investigation on the effects of biomass particle size in cellulosic biofuel production. *ASME. J. Energy Resour. Technol.*, *140*(4), 041804.

Yuan, X., *et al.* (2022). Densifying lignocellulosic biomass with sulfuric acid provides a durable feedstock with high digestibility and high fermentability for cellulosic ethanol production. *Renewable Energy*, *182*, 377–389.

https://doi.org/10.1142/9781800613799_bmatter

Index

Q

R

S